ENVIRONMENTAL CONTAMINANTS

Impact, Assessment, and Remediation

ENVIRONMENTAL CONTAMINANTS

Impact, Assessment, and Remediation

Edited by
Pritam Ganguly, PhD
Jajati Mandal, PhD
Mariappan Paramasivam, PhD
Sandip Patra, PhD

AAP | APPLE ACADEMIC PRESS

First edition published 2024

Apple Academic Press Inc.
1265 Goldenrod Circle, NE,
Palm Bay, FL 32905 USA

760 Laurentian Drive, Unit 19,
Burlington, ON L7N 0A4, CANADA

CRC Press
2385 NW Executive Center Drive,
Suite 320, Boca Raton FL 33431

4 Park Square, Milton Park,
Abingdon, Oxon, OX14 4RN UK

Library and Archives Canada Cataloguing in Publication

Title: Environmental contaminants : impact, assessment, and remediation / edited by Pritam Ganguly, PhD, Jajati Mandal, MSc, Mariappan Paramsivam, PhD, Sandip Patra, PhD.
Names: Ganguly, Pritam, editor. | Mandal, Jajati, editor. | Paramsivam, Mariappan, editor. | Patra, Sandip, editor.
Description: First edition. | Includes bibliographical references and index.
Identifiers: Canadiana (print) 20230532225 | Canadiana (ebook) 20230532268 | ISBN 9781774913963 (hardcover) | ISBN 9781774913970 (softcover) | ISBN 9781003412236 (ebook)
Subjects: LCSH: Pollutants. | LCSH: Environmental impact analysis. | LCSH: Bioremediation.
Classification: LCC TD174 .E56 2024 | DDC 363.738—dc23

Library of Congress Cataloging-in-Publication Data

CIP data on file with US Library of Congress

ISBN: 978-1-77491-396-3 (hbk)
ISBN: 978-1-77491-397-0 (pbk)
ISBN: 978-1-00341-223-6 (ebk)

About the Editors

Pritam Ganguly, PhD
Assistant Professor-cum-Junior Scientist
Department of Soil Science & Agricultural Chemistry, Bihar Agricultural University, Sabour, Bhagalpur, Bihar, India

Pritam Ganguly, PhD, is working as Assistant Professor-cum-Junior Scientist in the Department of Soil Science & Agricultural Chemistry, Bihar Agricultural University, Sabour, Bhagalpur, Bihar. He has done his PhD in Agricultural Chemicals from Bidhan Chandra Krishi Viswavidyalaya, India. He has received several young scientist awards from reputed societies. He has sound expertise in analyzing trace amounts of pesticides in different components of the environment. He has published several research as well as review articles in reputed journals. Pesticide residue monitoring in food items and prediction of safety risks are some of his key research interests.

Jajati Mandal, PhD
ICAR Netaji Subhas International Fellow,
School of Science Engineering and Environment, University of Salford, Salford, United Kingdom

Jajati Mandal, PhD, completed his PhD at the School of Science Engineering and Environment, University of Salford, Manchester, United Kingdom, after being selected for a Netaji Subhas International Fellowship from the Indian Council of Agricultural Research (ICAR) in 2020. His area of research is heavy metal contamination (arsenic) and micronutrients and modeling with machine learning algorithms. He joined the Department of Soil Science and Agricultural Chemistry, Bihar Agricultural University, Sabour, in 2012. He has been the principal investigator of one nationally funded project from the Science Engineering and Research Board (SERB), Department of Science and Technology, Government of India, and two state-funded projects. He has supervised one MSc and co-supervised seven MSc students. He was a recipient of a Young Scientist Award from the Indian Chemical Society in Environmental Chemistry.

Mariappan Paramasivam, PhD
Assistant Professor
Department of Soil Science & Agricultural Chemistry, Tamil Nadu Agricultural University (TNAU), Coimbatore, Tamil Nadu, India

Mariappan Paramasivam, PhD, is working as an Assistant Professor (Soil Science & Agricultural Chemistry) at Tamil Nadu Agricultural University (TNAU), Coimbatore, Tamil Nadu, India, after the completion of his PhD in Agricultural Chemicals. He is handling Department of Science and Technology, Government of India-funded projects, in addition to two TNAU-funded projects. He has guided eight MSc and three PhD students as chairman and member of the advisory committee. He has published more than 40 research papers, mostly in refereed journals of international repute, three books, and 10 book chapters written at national and international levels. He is listed by TNAU as the top 9th among scientists with a Scopus H-index and has also been awarded the position of Best Participant of the ICAR-sponsored Winter School. He has reviewed research articles for peer-reviewed journals of Elsevier, Springer, American Chemical Society, and Taylor and Francis.

Sandip Patra, PhD
Scientist (Agricultural Entomology)
Division of Crop Science, ICAR Research Complex for North Eastern Hill Region, Meghalaya, India

Sandip Patra (ARS), PhD, is presently working as a Scientist (Agricultural Entomology) in the Division of Crop Science, ICAR Research Complex for North Eastern Hill Region, Meghalaya, India. He received a national scholarship during undergraduate work and a senior merit scholarship during his master's degree program. He received the Smt. Hemangini Dutta Memorial Book Prize and a Certificate of Merit for securing the highest OGPA during his MSc in Agricultural Entomology. He has received Young Scientist Awards and Excellence in Research Awards from reputed societies. He supervised three MSc (Agriculture) students and has published more than 50 research papers in many national and international journals of repute along with two edited books. He has also authored many popular articles, book chapters, and training manuals. He is a member of various professional societies and reputed journals.

Contents

Contributors

Romila Akoijam
ICAR Research Complex for North Eastern Hill Region, Manipur Centre, Lamphelpat, India

Purabi Banerjee
Department of Agronomy, Faculty of Agriculture, Bidhan Chandra Krishi Viswavidyalaya, Mohanpur, Nadia, West Bengal, India

Parijat Bhattacharya
Department of Agricultural Chemistry and Soil Science, Faculty of Agriculture, Bidhan Chandra Krishi Viswavidyalaya, Mohanpur, Nadia, West Bengal, India

Prakash Ranjan Behera
Siksha 'O' Anusandhan Deemed to be University, Bhubaneswar, Odisha, India

Nanthi Bolan
UWA School of Agriculture and Environment, The University of Western Australia, Perth, WA, Australia
The UWA Institute of Agriculture, The University of Western Australia, Perth, WA, Australia

Verônica Casagrande
Environmental Sanitation Institute, University of Caxias do Sul, Caxias do Sul, Rio Grande do Sul, Brazil

Subhash Chander
ICAR-National Bureau of Plant Genetic Resources, New Delhi, Delhi, India

Nilanjan Chattopadhayaya
Department of Soil Science and Agricultural Chemistry, Bihar Agricultural University, Sabour, Bhagalpur, India

Nilanjana Roy Chowdhury
School of Environmental Studies, Jadavpur University, Kolkata, West Bengal, India

Antara Das
School of Environmental Studies, Jadavpur University, Kolkata, West Bengal, India

Kingsuk Das
Department of Agricultural Biotechnology, Assam Agricultural University, Jorhat, Assam, India
Department of Agricultural Biotechnology, Faculty of Agriculture, Bidhan Chandra Krishi Viswavidyalaya, Mohanpur, Nadia, West Bengal, India

Ayan De
School of Environmental Studies, Jadavpur University, Kolkata, West Bengal, India

Eswar Deepthi
Department of Agriculture, Government of Tamil Nadu, Vellore, Tamil Nadu, India

Debomita Dey
Department of Agricultural Chemistry & Soil Science, Faculty of Agriculture, Bidhan Chandra Krishi Viswavidyalaya, Mohanpur, Nadia, West Bengal, India

Saptaparnee Dey
Division of Soil Science and Agricultural Chemistry, ICAR-Indian Agricultural Research Institute, New Delhi, India

Gurusamy Dhandapani
Department of Botany, Kongunadu Arts and Science College (Autonomous), Bharathiar University, Coimbatore, Tamil Nadu, India

Ayesha Fatima
Department of Agronomy, Dr. Kalam Agricultural College, Bihar Agricultural University, Kishanganj, Bihar, India

Pritam Ganguly
Department of Soil Science & Agricultural Chemistry, Bihar Agricultural University, Sabour, Bhagalpur, Bihar, India

Dibakar Ghosh
ICAR-Indian Institute of Water Management, Bhubaneswar, Odisha, India

Mainak Ghosh
Department of Agronomy, Bihar Agricultural University, Sabour, Bhagalpur, Bihar, India

Soham Hazra
Department of Genetics and Plant Breeding, Faculty of Agriculture,
Bidhan Chandra Krishi Viswavidyalaya, Mohanpur, Nadia, West Bengal, India
Department of Agriculture, Brainware University, Barasat, Kolkata, WB, India

Konappan Jayalakshmi
Department of Veterinary medicine, Veterinary College and Research Institute, Orathanadu, Thanjavur, Tamil Nadu, India
Veterinary and Animal Sciences University, Chennai, Tamil Nadu, India

Madhurima Joardar
School of Environmental Studies, Jadavpur University, Kolkata, West Bengal, India

Upendra Kumar
ICAR-National Rice Research Institute (NRRI), Cuttack, India

Sharmistha Majumder
School of Environmental Studies, Jadavpur University, Kolkata, West Bengal, India

Jajati Mandal
School of Science Engineering and Environment, University of Salford, Salford, United Kingdom

Prithusayak Mondal
Regional Research Station (Terai Zone), Uttar Banga Krishi Viswavidyalaya, Pundibari, Cooch Behar, West Bengal, India

Debojyoti Moulick
Plant Stress Biology and Metabolomics Laboratory, Assam University, Silchar, Assam, India

Deepanjan Mridha
School of Environmental Studies, Jadavpur University, Kolkata, West Bengal, India

Rajeev Padbhushan
Department of Soil Science and Agricultural Chemistry, Bihar Agricultural University, Sabour, Bhagalpur, India

Mariappan Paramasivam
Pesticide Toxicology Laboratory, Department of Agricultural Entomology,
Tamil Nadu Agricultural University, Coimbatore, Tamil Nadu, India

Sandip Patra
ICAR Research Complex for North Eastern Hill Region, Umiam, Meghalaya, India

Ravindu Perera
Department of Livestock Production, Faculty of Agricultural Sciences, Sabaragamuwa University, Belihuloya, Sri Lanka, India

Parthendu Poddar
Department of Agronomy, Uttar Banga Krishi Viswavidyalaya, Pundibari, Coochbehar, West Bengal, India

Pooja Rani
Department of Soil Science and Agricultural Chemistry, Bihar Agricultural University, Sabour, Bhagalpur, India

Sakthi Ambothi Rathnasamy
Department of Plant Biotechnology, Tamil Nadu Agricultural University, Coimbatore, Tamil Nadu, India

Dibakar Roy
ICAR- Directorate of Weed Research, Jabalpur, Madhya Pradesh, India

Tarit Roychowdhury
School of Environmental Studies, Jadavpur University, Kolkata, West Bengal, India

Bedabrata Saha
Weed Research Department, Newe Ya'ar Research Centre, Agricultural Research Organization, Ramat Yishai, Israel

Bholanath Saha
Department of Soil Science, Dr. Kalam Agricultural College, Bihar Agricultural University, Kishanganj, Bihar, India

Sushanta Saha
Directorate of Research, Bidhan Chandra Krishi Viswavidyalaya, Kalyani, Nadia, West Bengal, India

Sanjib Kumar Sahoo
Department of Soil Science & Agricultural Chemistry, Odisha University of Agriculture & Technology, Bhubaneswar, India

Ananda Baskaran Sangeetha
Department of Veterinary Public Health and Epidemiology, Veterinary College and Research Institute, Orathanadu, Tamil Nadu, India
Tamil Nadu Veterinary and Animal Sciences University, Chennai, Tamil Nadu, India

Manickam Sasikala
Department of Veterinary Pathology, Veterinary College and Research Institute, Namakkal,
Tamil Nadu Veterinary and Animal Sciences University, Chennai, Tamil Nadu, India

Chellamuthu Selvi
Pesticide Toxicology Laboratory, Department of Agricultural Entomology, Tamil Nadu Agricultural University, Coimbatore, Tamil Nadu, India

Sudip Sengupta
Department of Agricultural Chemistry and Soil Science, Faculty of Agriculture, Bidhan Chandra Krishi Viswavidyalaya, Mohanpur, Nadia, West Bengal, India
Department of Agriculture, School of Agriculture, Swami Vivekananda University, Barrackpore, Kolkata, West Bengal, India

Akoijam Ratankumar Singh
ICAR Research Complex for North Eastern Hill Region, Umiam, Meghalaya, India

Garima Singh
Department of Agronomy, Bihar Agricultural University, Sabour, Bhagalpur, Bihar, India

Bishwajit Sinha
College of Agriculture, Odisha University of Agriculture & Technology, Bhawanipatna, Odisha, India

Prashant Srivastava
The Commonwealth Scientific and Industrial Research Organisation (CSIRO), Environment, Waite Campus, Urrbrae, SA, Australia

Sushmita Th
Department of Entomology, Assam Agricultural University, Jorhat, Assam, India

Hasintha Wijesekara
Department of Natural Resources, Faculty of Applied Sciences, Sabaragamuwa University, Belihuloya, Sri Lanka

Abbreviations

AAS	atomic absorption spectrometry
ABC	ATP-binding cassette
AE_N	agronomic N use efficiency
AF	aqueous flowables
Al(OH)3	aluminum hydroxide
AMF	arbuscular mycorrhizal fungi
AMPA	aminomethylphosphonic acid
APCI	atmospheric pressure chemical ionization
APGC	atmospheric pressure gas chromatography
API	active pharmaceutical ingredients
APS	American Physical Society
APX	ascorbate peroxidase
ARB	antibiotic-resistant bacteria
ARGs	antimicrobial resistance genes
As	arsenic
ASE	accelerated solvent extraction
AT	active tillering
ATRA	atrazine
CA	conservation agriculture
Ca_xO_y	calcium oxides
CDF	cation diffusion facilitator
CFCs	chlorofluorocarbons
CI	chemical ionization
CNF	carbon nanofiber
CNMs	carbon-based nanomaterials
CNTs	carbon nanotubes
CO	carbon monoxide
CO_2	carbon dioxide
CRHA	citric acid pretreated rice husk ash
CRP	coordinated research programme
CS	microcapsule seed-coating agent
CS	microencapsulation or capsule suspension
CU	Consumers Union
DBA	dichlorobenzoic acid

DC	dispersion concentrates
DCP	dichlorophenol
DE	diethyl-terbuthylazine
DDE	dichlorodiphenyldichloroethylene
DEA	deethylatrazine
DFS	dry flowable seed-coating agent
DDT	dichloro-diphenyl-trichloroethane
DM	dynamic membranes
DMA	dimethylarsinic acid
EC	electrocoagulation
EC	emulsifier concentrates
ECD	electron capture detector
EDDS	ethylene diaminedisuccinic acid
EDTA	ethylenediaminetetraacetic acid
EI	electron ionization
ELS	extra-long slender
ENP	engineered nanoparticles
EPA	Environment Protection Agency
EU	European Union
EuF2	europium fluoride
EVA	ethylene vinyl acetate
EW	emulsion concentrate
EWS	water emulsion seed-coating agent
F	fluoride
FeS	iron sulfide
FFP	farmers' usual fertilizer practices
FFP	real-time N management
FS	flowable seed coating agent
FYM	farmyard manure
GC	gas chromatography
GCB	graphitized carbon black
GHG	greenhouse gas
GHGI	greenhouse gas intensity
GMB	Ganga Meghna Brahmaputra
GR	glutathione reductase
GR	granular formulation
GWP	global warming potential
HCB	hexachlorobenzene
HCBD	hexachlorobutadiene
HDL	high-density lipoproteins

HFμe	hollow fiber microextraction
Hg	mercury
HGT	horizontal gene transfer
HMA	heavy-metal ATPase
HPLC	high-performance liquid chromatography
HS	headspace
HTC	hyperthermophilic composting
IAQ	indoor air quality
ICP-MS	inductively coupled plasma-mass spectrometry
ICP-OES	inductively coupled plasma-optical emission spectrometry
IDLs	instrumental detection limits
IFSM	integrated soil fertility management
IOP	intraocular pressure
IPCC	Intergovernmental Panel on Climate Change
ISEs	ion-selective electrodes
ISO	International Organization for Standardization
K-Ti-NT	potassium titanate
LC	liquid chromatography
LCC	leaf color chart
LD	liquid desorption
LDL	low-density lipoproteins
LDPE	low-density polyethylene
Li_4SiO_4	lithium ortho-silicate
LLE	liquid-liquid extraction
LODs	limit of detection
LOQ	limit of quantification
LR	lipid regulators
LS	long-slender
MAC	maximum acceptable concentration
MAE	microwave-assisted extraction
MAS	molecular absorption spectrometry
ME	micro-emulsion
MIP	microwave-induced plasmas
MnO_x	manganese oxide
MP-AES	microwave plasma atomic emission spectrometry
MRLs	maximum residue limits
MS	mass spectrometry
MS	medium-slender

Na-Ti-NT	sodium titanate
NAWQA	National Water Quality Assessment
NCI	negative chemical ionization
NEERI	National Environmental Engineering Research Institute
NNI	National Nanotechnology Initiative
NO_{3-}	nitrate
NUE	nitrogen use efficiency
NXOY	nitrogen oxides
nZVI	nanoscale zero-valent iron
OCPs	organochlorine pesticides
PAC	phenoxyacetic acid
PAHs	polycyclic aromatic hydrocarbons
PAL	phenylalanine ammonia lyase
PAM	polyacrylamide
PAN	polyacrylonitrile
PCA	pentachloroanisole
PCBs	polychlorinated biphenyls
PCN	polychlorinated naphthalenes
PDBE	polybrominated diphenyl ethers
PEN	productivity of applied N
PET	polyethylene terephthalate
PFASs	perfluorinated alkylated substances
PFP_N	partial factor productivity of applied N
PI	panicle initiation
PLE	pressurized liquid extraction
PO	peroxidase
POPs	persistent organic pollutants
PPCPs	pharmaceuticals and personal care products
PPO	polyphenol oxidase
PS	polystyrene
PSA	primary secondary amine
PSE	pressurized solvent extraction
PTDI	provisional tolerable daily intake
REN	recovery efficiency
RE_N	N recovery efficiency
RO	reverse osmosis
ROS	reactive oxygen species
SB	short-bold
SBSE	stir bar sorptive extraction

SC	suspension concentrates
SCCPs	short-chain chlorinated paraffins
SDME	single drop microextraction
SE	suspo-emulsion
SES	suspended emulsion seed coating agent
SFE	supercritical fluid extraction
SL	soluble concentrates
SOD	superoxide dismutase
SOM	soil organic matter
SPAD	Soil Plant Analysis Development
SPAD	chlorophyll meters
SPE	solid-phase extraction
SPME	solid-phase microextraction
SRFs	slow-release fertilizers
SSNM	site-specific nutrient management
TBA	trichlorobenzoic acid
TCP	trichlorophenol
TDS	total dissolved solids
TER	terbuthylazine
TF	translocation factor
TiO_2	titanium dioxide
TNTs	titanate nanotubes
T2P	trichloro-2-pyridinol
UAE	ultrasonic-assisted extraction
UASE	ultrasound-assisted solvent extraction
USDA-ARS	United States Department of Agriculture, Agriculture Research Service
USEPA	United States Environmental protection Agency
USFDA	United States Food and Drug Administration
USGS	US Geological Survey
VOCs	volatile organic compounds
VOC	volatile organic solvents
WET	Water Extraction Test
WG	water dispersible granule
WP	wettable powder
WWTPs	wastewater treatment plants
Zr-MOFs	zirconium-based metal organic framework
ZVI	zero-valent iron

Foreword

Pollution is increasing on an unprecedented scale due to both release of certain chemicals principally from burning fossil fuels and from the use and release of innumerable toxic substances due to industrial processes, mining activities, improper waste management, and agriculture runoffs. These are just a handful of the many human activities that pollute the environment. Pollution can have an impact on the food we eat, the water we drink, and the air we breathe, putting humans and the planet at risk, while posing formidable challenges for societies, both in developed and developing countries. That said, developing countries are the most vulnerable and suffer the worst. In fact, poor countries often lack the technology and resources to fight environmental pollution.

The United Nations' Sustainable Development Goals (SDGs) place a major emphasis on decreasing pollution. Indeed, target 3.9 of the SDG is to "substantially reduce the number of deaths and illnesses from hazardous chemicals and air, water, and soil pollution and contamination by 2030." To address the serious pollution challenges, several large-scale social and technological transitions are needed. Since pollution today is integrally related to economic production, modern technology, lifestyles, and a host of other factors, effective communication of the latest pollution-related research and scientific knowledge to the students, faculties, policymakers, and other stakeholders is critical.

Heavy metals, pesticide residues, microplastics, pharmaceutical medicines, and other new and emerging environmental pollutants have been covered in depth in this book comprising 13 chapters written by a number of authors. The interaction of these pollutants with various critical environmental processes has been addressed in depth. This book also includes information on the pollutants' toxicity and human health risks. Intensive technologies, bringing most sophisticated science to reduce the detrimental impacts of persistent pollutants, have been discussed. Perceiving pollution as transboundary and global, this interdisciplinary approach provides pertinent details about significant pollutants and their environmental consequences, allowing the readers to build a solid foundation of knowledge.

I would like to thank the authors from various disciplines for their participation in compilation of the chapters and for offering their expertise. I hope this book will inspire and widen the perspective of students, researchers, policymakers, and stakeholders.

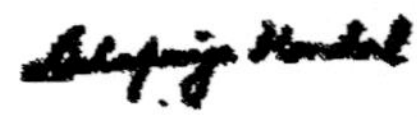

Dr. Debapriya Mondal
Principal Epidemiology Scientist
Environmental Epidemiology Group,
Chemical and Environmental Effects Department,
Radiation, Chemical and Environmental Hazards Directorate,
UK Health Security Agency

Preface

This book, *Environmental Contaminants: Impact, Assessment, and Remediation*, has been compiled through the painstaking effort of the authors as they struggled to gather information on the impact, assessment, and remediation of different environmental contaminants. The number of publications and dissertations on this topic is increasing day by day, and new analytical methods are being developed optimizing the existing techniques. A compilation of these recently designed or customized methods is needed by environmental chemists. An attempt has been made to impart knowledge about the environmental contamination and its remediation which is required to determine the following: what environmental changes may occur due to arsenic, heavy metals, herbicides, fluorides, microplastics, chemical fertilizers contaminants; what may be the remedial measures of these environmental contaminants; and to analyze the trace level concentration of contaminants.

This edition has 13 chapters which includes basics of environment, pollution, and bioremediation techniques, arsenic in water–soil–plant–human continuum, heavy metal (Pb, Cd, and Cr) contamination in soil worldwide: situation impact and assessment, microplastics: an emerging contaminant, herbicide contamination in ground water and their remediation, fluoride contamination in the ecosystem and its remediation strategies, impact of pharmaceutical contaminants in environment and environmental pollution caused by the incorrect application of synthetic fertilizers and their remediation have been covered in a holistic manner. Apart from this, methods of trace analysis for persistent pollutants, phytoremediation: green approach toward removal of contaminants, role of nanotechnology on remediation of environmental contaminants, agronomic intervention for mitigation of environmental pollutants, and advancement in pesticide formulation technology for reducing load of contaminants are also covered.

The readers will find the information provided in these chapters helpful in making them understand the environmental contaminants and its remediation. Data are summarized in charts and tables. A wealth of such information, as well as diagrams and illustrations, enables this book a substantial reference guide. Finally, in the Methods, Techniques, and Instruments section, readers can learn more about the topics with details on the most often used analysis methods: classical and more recently developed ones. Methods for analyzing, quantifying, and detecting are addressed in this section.

This book offers readers detailed descriptions of stepwise processes as well as a significant number of references summed up in tables. Moreover, they are also designed to be used by university graduates and researchers involved in the field of environmental chemistry.

The contributors are strong leaders in fields of specialization at academic and research institutions. We would like to thank them for their exceptional efforts. We are well aware of some of the limitations of the book due to various difficulties and we request the readers to offer their suggestion for further improvement of the book, which will be incorporated in the next edition. We, the authors, feel that this book will be useful to the researchers to expand their knowledge on various aspects.

—Pritam Ganguly
Jajati Mandal
M. Paramasivam
Sandip Patra

Introduction

Life in any form heavily relies on nature and environment. Several environmental processes dictate the fate of living beings and organisms respond to certain changes for their survival on the Earth. It goes without saying that human civilization still largely depends on nature and any development comes with a price that the environment has to pay. Environmental contamination is, certainly, the growing concern on which scientists across the world are focused on. Substances that enter the environment unintentionally or eventually, typically but not every time due to anthropogenic activity, are known as environmental contaminants. Some of these toxic pollutants may have come out as industrial effluents and are difficult to break down due to their high persistence nature. Some examples of harmful contaminants are heavy metals, microplastics, chemical pesticides, fertilizers, synthetic dyes, pharmaceutical wastes, etc. These are now potential threat for lives and can degrade natural resources like soil, water bodies, atmosphere, and other important components.

One of the major reasons of concern about environmental pollution is the entry of contaminants in the food chain. Foods have been assessed to contain a wide range of environmental toxicants. These include metals and "ionic" species such as perchlorate, as well as organic (carbon-based) compounds such as persistent organic pollutants (POPs). These hazardous compounds create several chronic diseases in humans like asthma, respiratory and/or nervous system disorder, cancers, or even birth defects in next generation. Many of these xenobiotic compounds have high-leaching potential, and thereby, contaminate the groundwater reserve. There are certain natural pollutants that are also present in the system and cause detrimental effects in human. Arsenic is the leading element among them. It pollutes the groundwater and thus, restricts its use for both drinking as well as agriculture purpose. The challenges are growing day by day as industrialization/urbanization is expanding at a rapid speed.

As a counter strategy, the first step is to create awareness among the people. Use of synthetic inputs in a judicious way, encouraging bio-based approaches, etc., are some of the remedial measures that can resist pollution. Moreover, developing effective and precise methods for assessing extent of contamination is highly required. Risk assessment based on the available data needs to be carried out to ascertain safety to human beings and other living organisms. Researchers have invented several techniques for mitigation like waste/disposal management, treatment of effluents, use of alternatives/safer products, etc. Among these, bioremediation can be very useful as it involves use of specific microbes and plants to reduce the load of contaminants. A holistic/integrated approach is needed in the society to mitigate the ill effects of contamination.

The present book is edited with the focus to collect recent and relevant information about environmental contaminants and their remedial strategy. It will be helpful for policymakers to make decisions to tackle the growing levels of pollution in the near future.

CHAPTER 1

Basics of Environment, Pollution, and Bioremediation Techniques

PRITAM GANGULY[1], JAJATI MANDAL[2], SUDIP SENGUPTA[3,4], and BISHWAJIT SINHA[5]

[1]*Department of Soil Science & Agricultural Chemistry, Bihar Agricultural University, Sabour, Bhagalpur, Bihar, India*

[2]*School of Science Engineering and Environment, University of Salford, Salford, UK*

[3]*Department of Agricultural Chemistry & Soil Science, Faculty of Agriculture, Bidhan Chandra Krishi Viswavidyalaya, Mohanpur, Nadia, West Bengal, India*

[4]*Department of Agriculture, School of Agriculture, Swami Vivekananda University, Barrackpore, Kolkata, West Bengal, India*

[5]*College of Agriculture, Odisha University of Agriculture & Technology, Bhawanipatna, Odisha, India*

ABSTRACT

Environment is basically the surroundings in which an organism operates its necessary functions. It grows in vigor, multiplies in numbers, and reacts as the conditions of environment get changed or altered. Certain components exist in the environment such as lithosphere, biosphere, atmosphere, and hydrosphere which are inter-related with each other in functions and conditions. Each and every organism is heavily dependent on specific environmental factors, biotic or abiotic, like temperature, humidity, light, gaseous contents, population dynamics (both micro and macro), and so on. The

Environmental Contaminants: Impact, Assessment, and Remediation.
Pritam Ganguly, Jajati Mandal, M. Paramsivam, & Sandip Patra, (Eds.)

circulation of the vital elements through natural processes of living matter is denoted as a biogeochemical cycle. It transports elements in diverse forms from nonliving (abiotic) to living (biotic) components of the biosphere and back. The six most common elements associated with organic molecules are carbon (C), nitrogen (N), hydrogen (H), oxygen (O), phosphorus (P), and sulfur (S), which can be found in various chemical forms in the atmosphere, on land, in water, and beneath the Earth's surface. Pollution is defined as the entry of toxins into the natural environment that have a negative impact. Apart from the chemical substances, noise, heat, or light can also cause pollution to the environment. Both natural and man made (anthropogenic) sources pollute the environment. Among the various control approaches, bioremediation offers a sustainable and rational option to reduce pollution in the environment. The use of biological substances to counteract the pollutants or contaminants is termed bioremediation. Several techniques adopted in this process are broadly divided into two categories, such as in situ and ex situ bioremediation. These are basically bioventing, biosparging, bioaugmentation, landfarming, composting, biopiles, and so on.

1.1 WHAT IS ENVIRONMENT?

The natural world in which people, animals, and plants live is known as the environment. The term "environment" comes from the French words "virer" (to turn) or "in/viron," which means "to encircle," and refers to the interaction of natural surrounds with species, including humans. As a result, environment refers to the settings in which living and nonliving entities coexist, flourish, and die. The conditions that exist in the environment have an impact on the life of the species that live there.

Environmental studies explain the interactions between organisms, the environment, and all of the variables that influence life on Earth, such as atmospheric conditions, food chains, and the water cycle. It is a fundamental science regarding our planet and its daily operations, and as such, it is vital to everyone.

1.2 COMPONENTS OF ENVIRONMENT

The lithosphere, hydrosphere, atmosphere, and biosphere, which correspond to rocks, water, air, and life, are the four major components of the environment.

The outmost layer of the earth's crust, known as the lithosphere, is made up of various minerals. It can be found on both the oceanic crust (at a depth of up to 100 km) and the terrestrial crust. Earth's tectonic plates are the most important part of the lithosphere.

The hydrosphere includes all types of water bodies on the planet, such as seas, oceans, rivers, lakes, ponds, and streams which covers 70% of the planet's surface. The oceans hold 97.5% of the water on the planet in the form of salt water. Freshwater makes up only 2.5% of the total amount of water on the planet. Groundwater accounts for 30.8% of the total, whereas glaciers account for 68.9% of the total. A 0.3% concentration can be found in reservoirs, rivers, and lakes, which is easily accessible to humans.

The gaseous layer that surrounds the Earth is the atmosphere. The presence of oxygen in the atmosphere is distinctive to Earth which is necessary for the existence of life. 78.08% nitrogen, 20.95% oxygen, 0.93% argon, 0.038% carbon dioxide, and traces of hydrogen, helium, and noble gases make up the majority of it. The amount of water vapor in the atmosphere varies.

The term "biosphere" refers to all of the areas on Earth where life can be found. Life-supporting ecosystems can be found in soil, air, water, or land. Geologist Edward Suess invented the word "biosphere" to describe a location on Earth where life can be found. The term "biosphere" refers to the whole of all living beings, also known as "biota." It stretches from the arctic ice caps to the equator, with each zone containing a unique living type adapted to the local environment.

1.3 ENVIRONMENTAL FACTORS

Any abiotic (nonliving) or biotic (living) component that affects living organisms is referred to as an environmental factor, ecological factor, or eco factor. Amount of sunlight, temperature, and pH of the water or soil in which an organism lives are all abiotic parameters. Biotic factors include the occurrence of biological specificity, predators, competitors, and parasites, as well as the accessibility of food for organisms.

A change of weather and also climate represents environmental influences. A change in temperature can have an impact on a variety of businesses, including agriculture. The climate change mainly due to global warming this external aspect is becoming increasingly important for businesses to address.

The global water cycle has an impact on precipitation patterns, water availability in particular location, and the most common water sources. For example, in locations with high rainfall, precipitation may be a main water supply (learn about rainwater collection), whereas in dryer areas, groundwater or invalid link sources may suffice. Water requirement and use for agricultural activities, as well as household consumption and industrial activities, are all influenced by the climate. There may be a greater economic and political motivation to utilize water more efficiently in locations where water shortage is a result of climate change. In arid areas like Jordan and Israel, for example, there are prospects for large-scale wastewater reuse for irrigation (Fardous, 2006).

1.4 BIOGEOCHEMICAL CYCLES

Sunlight, the main source of energy for the ecosystem, enters as wavelengths of various frequencies (for chemoautotrophs as inorganic compounds) transmit through the trophic levels and exit as heat. C, N, H, O, P, and S are the six most abundant elements allied with organic molecules, and they exist in diverse chemical forms in the atmosphere, on land, in water, and beneath the Earth's surface. Weathering, erosion, water drainage, and continental plate subduction are all geologic processes that contribute to element cycling on Earth. Inorganic components get recycled between living species and their nonliving environment is referred to as biogeochemical cycles because geology and chemistry play significant roles in the study of these processes.

The six elements as indicated above are used by the organisms in a number of ways. Organic substances and water have H and O, both of which are compulsory for life. N is a foremost element of nucleic acids and proteins, while C is present in all the organic compounds. P is present in the nucleic acids and the phospholipids that make up cellular membranes. Finally, S is necessary for proteins to maintain their three-dimensional form.

These elements' cycle is interrelated. For example, mobility of water is vital for S and P leaching into rivers, lakes, and oceans. The movement of minerals is observed between the biotic and abiotic components, as well as from one living organism to the next.

1.4.1 WATER CYCLE

The energy from the Sun drives the water cycle by warming the water bodies across the globe. This results in the evaporation (the transformation of water (liquid state) to vapor state) of liquid surface water and sublimation (the conversion of ice to water vapor) of frozen water, resulting in the release of vast volumes of water as water vapor into the atmosphere. Water vapor condenses into clouds as liquid or frozen droplets over time, eventually resulting in precipitation (rain, snow, and hail), which returns water to the Earth's surface. Rain that reaches the surface of the Earth may evaporate, flow over it, or seep into the earth. Surface runoff, the movement of freshwater over land caused by rain or melting ice, is the most visible. Streams and lakes can carry runoff to the sea.

1.4.2 CARBON CYCLE

The C cycle consists of two subcycles: one that deals with quick exchange of C among living organisms, and the other that deals with long-term C cycling through various geologic processes. Even between ecosystems, organisms are linked in a variety of ways. The exchange of C between heterotrophs and autotrophs via atmospheric CO_2 is a notable illustration of this link. CO_2 is used by the autotrophs as a basic building ingredient glucose (high-energy molecules). The covalent bonds that hold C atoms together are mediated by these organisms using the energy from the Sun. The energy is stored in these chemical bonds that are used in the respiration process. CO_2 is obtained directly from the atmosphere by most terrestrial autotrophs, whereas the marine autotrophs obtained it in dissolved form (bicarbonate, HCO_3^-). Through consumption, C is transferred from producers to higher trophic levels.

Carbon movement across land, water, and air is complex, and it occurs far more sluggishly than movement between species in many circumstances. C is stored in C reservoirs such as the atmosphere, bodies of liquid water (primarily oceans), ocean sediment, soil, rocks (including fossil fuels), and the Earth's interior for extended periods of time. CO_2 from the atmosphere dissolves in water and forms ionic compounds when it combines with water molecules. Some of these ions react with calcium ions in saltwater to generate calcium carbonate ($CaCO_3$), which is a key component of marine species' shells. When these animals die, their shells become sediment on the ocean floor. $CaCO_3$ transforms into limestone over time, forming the world's greatest C reserve.

C is stored as organic C in soil on land as a result of organism breakdown or weathering of terrestrial rock and minerals. Fossil fuels, the anaerobically degraded remains of plants and algae that lived millions of years ago, are found deeper beneath. As the formation of fossil fuels significantly outnumbers its usage, it is classified as nonrenewable resources. A nonrenewable resource is either gradually regenerated or certainly not regenerated at all. Volcanic outbreaks and other geothermal systems are alternative way for C to reach the atmosphere from land (even land beneath the surface of the ocean).

1.4.3 NITROGEN CYCLE

N cannot be directly absorbed by the plants and phytoplankton directly from the atmosphere, despite the fact that this molecule makes up roughly 78% of the atmosphere. N enters the living world through free-living and symbiotic bacteria, which use specific biochemical mechanisms to incorporate nitrogen into their organic molecules. N fixation is the process of turning nitrogen gas into ammonia (NH_3), which then spontaneously turns into ammonium (NH_4^+) by some bacteria. Bacteria convert ammonium to nitrites (NO_2^-) and eventually to nitrates (NO_3^-). Plants and other producers use nitrogen-containing molecules to build organic molecules like DNA and proteins at this phase. This nitrogen is now available to the general public.

1.4.4 PHOSPHOROUS CYCLE

P is found in nature as the phosphate (PO43-) ion. In addition to phosphate runoff induced by human activities, natural surface runoff occurs when PO43- is leached from phosphate-containing rock through weathering, sending phosphates into rivers, lakes, and the ocean. Ocean sediments containing phosphate are largely made up of the bodies and excretions of sea animals. On the other hand, volcanic ash, aerosols, and mineral dust could all be significant phosphate suppliers. Over geologic time, the rise of the Earth's surface takes this silt to land. Phosphorus is also exchanged between marine organisms and phosphate dissolved in the water. The transport of phosphate from the ocean to the land and through the soil is extraordinarily sluggish, with an oceanic residence time of 20,000 to 100,000 years for the average phosphate ion.

1.4.5 SULFUR CYCLE

S is an essential element of all living beings' molecules. It is crucial for the synthesis of cysteine which is an important part of the amino acid. Sulfur dioxide (SO_2) is plentiful in the atmosphere which enters first through the decomposition of organic molecules; second, through volcanic activity and geothermal vents, and lastly, through human combustion of fossil fuels. On land S is deposited through four processes: direct atmospheric fallout, precipitation,, rock weathering, and geothermal vents.

1.5 WHAT IS POLLUTION?

The entry of toxins into the natural environment that have a detrimental influence is referred to as pollution. Chemical toxins, sometimes energy like heat, light, or noise, cause pollution. Pollutants are chemicals or energies that are either foreign or naturally occurring that contribute to pollution (Douglas, 1968). Pollution is typically divided into two categories: the point source and the non-point source.

Major types of pollution include air pollution, light pollution, litter, noise pollution, plastic pollution, soil contamination, radioactive contamination, thermal pollution, visual pollution, and water pollution (Spellman, 2009).

1.5.1 POLLUTANTS

A waste material that pollutes the air, water, or land is referred to as a pollutant. A pollutant's severity is determined by three factors: its chemical type, concentration, affected region, and persistence (Tinsley, 2004).

1.5.2 HOW ENVIRONMENT GETS POLLUTED?

Air can get polluted both from natural and anthropogenic (man-made) sources. Pollutants due to anthropogenic activity due to combustion, construction, mining, agriculture (Patra et al., 2018), and warfare, on the other hand, are becoming increasingly important in the global air pollution equation (Surendran et al., 2016).

Exhaust from the vehicles is one of the most significant sources of air pollution. China, the United States, Russia, and India are four of the most powerful countries in the world when it comes to air pollution (Pao and Tsai, 2011). Mexico and Japan are the top leaders in terms of air pollution emissions. Among the most significant stationary pollution sources are coal-fired power plants, chemical plants, oil refineries, nuclear waste disposal activity, petrochemical plants, incinerators, large livestock farms (pigs, dairy cows, poultry, etc.), plastics factories, metals production factories, PVC factories, and other heavy industries (Surendran et al., 2016). Modern agricultural methods such as clear felling and burning of natural flora, as well as pesticide and herbicide application, contribute to agricultural air pollution.

Each year, over 400 million metric tonnes of hazardous garbage are produced. Approximately 250 million metric tonnes are produced in the United States alone. Although Americans make up less than 5% of the global population, they produce nearly 25% of CO_2 and generate roughly 30% of the world's garbage (Kaza et al., 2018). China surpassed the United States as the world's largest CO_2 generator in 2007, yet it remains far behind in terms of per capita pollution (ranking 78th among the world's nations).

Some of the most frequent soil contaminants are pesticides (Ganguly et al., 2017), heavy metals (such as chromium, cadmium—found in rechargeable batteries, and lead—found in lead paint, aviation fuel, and even gasoline in some countries), MTBE, zinc, arsenic, and benzene. In 2001, a series of journalistic exposes, culminating in the release of the book Fateful Harvest, exposed a widespread practice of recycling industrial waste into fertilizer, resulting in metal poisoning of the soil.

In many cases landfills by municipal wastes are a source of many chemical substances entering the soil environment (and often groundwater), resulting from a wide variety of waste (Kaza et al., 2018), particularly illegally discarded substances, or pre-1970 landfills in the United States or the European Union that may have been subject to little control. There have also been some unusual releases of polychlorinated dibenzodioxins (PCDDs), also known as dioxins, such as TCDD.

Natural disasters sometimes also create pollution of a site (Wilbanks and Fernandez, 2014). For example, sewage contamination and petrochemical leaks from burst boats or automobiles are common during hurricanes. Larger-scale and environmental damage is not uncommon when coastal oil rigs or refineries are involved. Some pollution sources, such as nuclear power plants or oil ships, can create enormous and perhaps catastrophic emissions when accidents occur.

The automobile is the most frequent source of noise pollution, accounting for more than 90% of all unwanted noise worldwide.

1.5.3 TYPES OF POLLUTION

The principal types of pollution (Holdgate, 1979; Grossman, 1993; Khopkar, 2007) are described below, along with the specific contaminant that applies to each:

1.5.3.1 AIR POLLUTION

Air pollution is caused by the emission of chemicals and particles into the atmosphere. Common gaseous pollutants produced by industry and motor vehicles include carbon monoxide, chlorofluorocarbons (CFCs), sulfur dioxide, and nitrogen oxides. Photochemical ozone and smog are created when nitrogen oxides and hydrocarbons react with sunlight. Particulate matter, often known as fine dust, is classed as PM10 to PM2.5 and measured in micrometers.

1.5.3.2 ELECTROMAGNETIC POLLUTION

It is an excess of nonionizing electromagnetic radiation, such as radio waves, to which humans are constantly exposed, especially in large cities. However, whether such types of radiation have any effect on human health is still unknown.

1.5.3.3 LIGHT POLLUTION

Light trespass, over-illumination, and astronomical interference are all examples of light pollution.

1.5.3.4 LITTERING

Littering is defined as the intentional dumping of unsuitable man-made things onto public and private property.

1.5.3.5 NOISE POLLUTION

Roadway noise, aviation noise, industrial noise, and high-intensity sonar are all examples of noise pollution.

1.5.3.6 PLASTIC POLLUTION

Plastic pollution occurs when plastic items and microplastics accumulate in the environment, causing harm to wildlife, wildlife habitat, and humans.

1.5.3.7 SOIL POLLUTION

Chemicals are released into the soil as a result of a spill or an underground leak. Hydrocarbons, heavy metals (Raj et al., 2021), MTBE, herbicides

(Bhattacharyya, et al., 2011), pesticides (Ganguly et al., 2020), and chlorinated hydrocarbons are among the most significant soil pollutants.

1.5.3.8 RADIOACTIVE POLLUTION

Radioactive pollution occurs as a result of efforts in atomic physics in the twentieth century, such as nuclear power generation and nuclear weapons research, development, and deployment.

1.5.3.9 THERMAL POLLUTION

A temperature shift in natural water bodies caused by human activity, such as the use of water as a coolant in a power plant, is known as thermal pollution.

1.5.3.10 VISUAL POLLUTION

Overhead power lines, highway billboards, scarred landscapes (as from strip mining), open trash storage, municipal solid waste, or space debris are examples of visual pollution.

1.5.3.11 WATER POLLUTION

Discharges of untreated sewage and chemical contaminants (Goon et al., 2021), such as chlorine, from treated sewage; and releases of trash and contaminants into surface runoff flowing to surface waterways (intentionally or by spills) are the most common sources of water pollution (including urban runoff and agricultural runoff, which may contain chemical contaminants).

1.6 WHAT IS BIOREMEDIATION?

Bioremediation is a method for destroying or rendering harmless different pollutants through natural biological activity. As a result, it involves low-cost, low-tech procedures that are commonly accepted by the general populace and may often be performed on-site. However, it will not always be suitable because the range of contaminants on which it is effective is limited, the time periods required are rather long, and the residual contamination levels reached may not always be sufficient (Vidali, 2001). Despite the fact that the methodologies used are not technically complex, designing and implementing a successful bioremediation programme may necessitate a significant amount of experience and expertise due to the need to thoroughly assess a site for suitability and optimize conditions to achieve a satisfactory result.

As bioremediation appears to be a viable alternative to traditional clean-up technologies, research in this sector is fast expanding, particularly in the United States. Bioremediation has been employed with varied degrees of effectiveness in a variety of locations around the world, including Europe. As more knowledge and experience is gathered, techniques are improving, and there is little doubt that bioremediation offers a lot of potential for dealing with specific forms of site pollution (Crawford and Crawford, 2005). Unfortunately, the ideas, procedures, benefits, and downsides of bioremediation are poorly understood, especially by those who will have to deal directly with bioremediation proposals, such as site owners and regulators. We aimed to assist by providing a clear, practical explanation of bioremediation processes, benefits, and downsides of the technique, and concerns to consider when dealing with a bioremediation proposal.

It is a technique for treating contaminated media such as water, soil, and underground material by altering environmental conditions to induce microorganism growth and breakdown of the target contaminants. Bioremediation is used to clean up oil spills, soils contaminated by acidic mine drainage, underground pipe leaks, and crime scene cleanups. These dangerous compounds are detoxified by enzymes found in microorganisms. The majority of bioremediation processes (Li and Yu, 2015) involve oxidation-reduction reactions, in which an electron acceptor (typically oxygen) is added to stimulate the oxidation of a reduced pollutant (e.g., hydrocarbons) or an electron donor (typically an organic substrate) is added to reduce oxidized pollutants (nitrate, perchlorate, oxidized metals, chlorinated solvents, explosives, and propellants).

Bioremediation is a strategy for decreasing the environmental impact of anthropogenic by products, such as those produced by industrial and agricultural operations. Bioremediation is frequently less expensive and lasts longer than conventional cleanup methods. Other restorative treatments include thermal desorption, vitrification, air stripping, bioleaching, rhizofiltration, and soil cleaning. Bioremediation, often known as biological treatment, is a technique for dealing with pollutants such as wastewater, industrial waste, and solid waste. The ultimate goal of bioremediation is to eliminate or reduce dangerous chemicals to improve soil and water quality (Sasikumar and Papinazath, 2003).

A number of in situ and ex situ bioremediation technologies can be used to eliminate or reduce contaminants. The classification of bioremediation approaches is determined on the treatment location. Contaminated settings are treated with nondestructive and cost-effective in situ solutions. On the other hand, ex situ remedies typically entail the excavation of the polluted site, which increases costs (Mohanty and Rath, 2020). To improve the micro-organisms' situation, additional nutrients, vitamins, minerals, and pH buffers can be introduced to both of these methods. Sometimes, biostimulation (the addition of specialized microbe cultures to speed up biodegradation) is used. Bioremediation-related methods include phytoremediation, biosparging, bioventing, bioattenuation, composting (biopiles and windrows), and land farming.

1.6.1 BIOREMEDIATION STRATEGIES

Depending on the moisture content and aeration of an area, several strategies are used. The term "in situ" refers to treatments that are applied to soil and groundwater on-site with little disturbance. Ex situ procedures are those that are used on soil and groundwater that has been removed from a place through excavation (soil) or pumping (groundwater) (water). Microorganisms with the potential to break down contaminants are added to bioaugmentation processes.

1.6.1.1 IN SITU BIOREMEDIATION

These approaches are the most popular since they are less expensive and cause less disruption because they treat toxins in site rather than excavating and transporting them. The depth of soil that can be adequately treated limits in situ treatment (Kuppusamy et al., 2016). Effective oxygen diffusion for optimum bioremediation rates in many soils extends from a few centimeters to about 30 cm into the soil; however, depths of 60 cm and greater have been successfully treated in other circumstances.

The most important land treatments are as follows.

1.6.1.1.1 Bioventing

Bioventing is one of the most common in situ therapies, which involves supply of nutrient elements and air to polluted soils via wells to foster the growth of indigenous microorganisms (Kumar et al., 2018). Bioventing lessens volatilization and pollutant release into the environment by using low air flow rates and only supplying the amount of O required for biodegradation. It works effectively for simple hydrocarbons and can also be used where pollution are detected deep underground.

1.6.1.1.2 In Situ Biodegradation

The circulation of aqueous solutions through contaminated soils to provide O and nutrients to naturally occurring microorganisms breaks down the organic contaminants. It is really suitable for both soil and groundwater (Kuppusamy et al., 2016). This technique encompasses circumstances such as the penetration of water-carrying nutrients, oxygen, or other electron acceptors for groundwater treatment.

1.6.1.1.3 Biosparging

It is the process of injecting pressurized air below the water table to raise groundwater O levels and to pace up the breakdown of pollutants (biologically) by naturally occurring microorganisms. Biosparging increases the interaction between soil and groundwater by increasing mixing in the saturated zone (Kumar et al., 2018). Installing small-diameter air injection sites is simple and inexpensive, allowing for a lot of flexibility in the system's design and construction.

1.6.1.1.4 Bioaugmentation

The addition of microorganisms, either indigenous or exogenous, to contaminated locations is widely used in bioaugmentation. The utilization of additional microbial cultures in a land treatment unit is limited by two factors: (1) Nonindigenous cultures seldom compete effectively enough with indigenous cultures to develop and sustain viable population levels, and (2) most soils with long-term exposure to biodegradable waste have indigenous microbes that are good degraders if the land treatment unit is adequately managed (Kumar et al., 2018).

1.6.1.2 EX SITU BIOREMEDIATION

These methods entail excavating or removing contaminated soil from the ground (Tomei and Daugulis, 2013).

1.6.1.2.1 Landfarming

Excavating filthy soil, distributing it over a prepared bed, then tilling it periodically until toxins have deteriorated is known as landfarming (Nikolopoulou and Kalogerakis, 2016). The goal is to stimulate native biodegradative microorganisms and make it easier for them to aerobically digest contaminants. Typically, the process is confined to treating the top 10–35 cm of soil. Landfarming has sparked a lot of attention as a disposal alternative since it has the ability to lower monitoring and maintenance costs while also reducing clean-up duties.

1.6.1.2.2 Composting

It involves mixing soil with nonhazardous organic additions such as manure or agricultural waste. The presence of these organic materials encourages the growth of a diverse microbial population and the high temperatures associated with composting (Jørgensen et al., 2000).

1.6.1.2.3 Biopiles

Landfarming and composting are combined in biopiles. Engineered cells are essentially aerated composted piles with aeration. They are a refined version of landfarming that tends to control physical losses of contaminants via leaching and volatilization (Jørgensen et al., 2000). They are typically employed for the treatment of surface pollution with petroleum hydrocarbons. Biopiles provide an ideal environment for aerobic and anaerobic bacteria to thrive.

1.6.1.2.4 Slurry Reactors or Aqueous Reactors

These are used to treat polluted soil and water that has been pumped up from a contaminated plume ex situ. In reactor bioremediation, polluted solid material (soil, sediment, sludge) or water is processed through a designed containment system. A slurry bioreactor is a containment vessel and apparatus used to create a three-phase (solid, liquid, and gas) mixing condition to increase the bioremediation rate of soil bound and water-soluble pollutants as a water slurry of contaminated soil and biomass (usually indigenous microorganisms) capable of degrading target contaminants (Jørgensen et al., 2000) as a water slurry of the contaminated soil and biomass (usually indigenous microorganisms) capable of degrading target contaminants.

KEYWORDS

- **environment**
- **pollution**
- **biogeochemical cycles**
- **pollutants**
- **bioremediation**

REFERENCES

Bhattacharyya, A.; Ganguly, P.; Barik, S. R.; Kundu, C. Studies on the Persistence of Diclosulam in Soybean Crop. In *23rd Asian-Pacific Weed Science Society Conference*, 2011; p 34.

Crawford, R. L.; Crawford, D. L. *Bioremediation: Principles and Applications* (No. 6); Cambridge University Press, 2005.

Douglas, M. Pollution. *Int. Encyclopedia Soc. Sci.* **1968**, *12*, 336–342.

Fardous. *Irrigation Water Strategy For Sustainable Agricultural Development in Jordan*; Symposium on Irrigation Modernization Constraints and Solutions; Damascus: FAO, 2006.

Ganguly, P.; Barik, S. R.; Patra, S.; Roy, S.; Bhattacharyya, A. Persistence of Chlorfluazuron in Cabbage Under Different Agro-Climatic Conditions of India and Its Risk Assessment. *Environ. Toxicol. Chem.* **2017**, *36* (11), 3028–3033.

Ganguly, P.; Mandal, J.; Mandal, N.; Rakshit, R.; Patra, S. Benzophenyl Urea Insecticides–Useful and Ecofriendly Options for Insect Pest Control. *J. Environ. Biol.* **2020**, *41* (3), 527–538.

Goon, A.; Bhattacharyya, A.; Ghosh, B.; Rakshit, R.; Das, A.; Choudury, S. R.; Kundu, C.; Ganguly, P.; Hossain, A. Photodegradation of Flucetosulfuron, a Sulfonylurea-Based Herbicide in the Aqueous Media Is Influenced by Ultraviolet Irradiation. *J. Xenobiot.* **2021**, *11* (4), 142–154.

Grossman, G. Pollution and Growth: What Do We Know? *CEPR Discussion Papers*, 1993, *848*.

Holdgate, M. W. *A Perspective of Environmental Pollution*; Cambridge University Press, 1979.

Jørgensen, K. S.; Puustinen, J.; Suortti, A. M. Bioremediation of Petroleum Hydrocarbon-Contaminated Soil by Composting in Biopiles. *Environ. Pollut.* **2000**, *107* (2), 245–254.

Kaza, S.; Yao, L.; Bhada-Tata, P.; Van Woerden, F. *What a Waste 2.0: A Global Snapshot of Solid Waste Management to 2050*; World Bank Publications, 2018.

Khopkar, S. M. *Environmental Pollution Monitoring and Control*; New Age International, 2007.

Kumar, V.; Shahi, S. K.; Singh, S. Bioremediation: An Eco-Sustainable Approach for Restoration of Contaminated Sites. In *Microbial Bioprospecting for Sustainable Development*; Springer: Singapore, 2018; pp 115–136.

Kuppusamy, S.; Palanisami, T.; Megharaj, M.; Venkateswarlu, K.; Naidu, R. In Situ Remediation Approaches for the Management of Contaminated Sites: A Comprehensive Overview. *Rev. Environ. Contam. Toxicol.* **2016**, *236*, 1–115.

Li, W. W.; Yu, H. Q. Electro-Assisted Groundwater Bioremediation: Fundamentals, Challenges and Future Perspectives. *Bioresour. Technol.* **2015**, *196*, 677–684.

Mohanty, I.; Rath, A. Bioremediation: Current Research Trends and Applications. *Waste to Energy: Prospects and Applications*, 2020; 99–120.

Nikolopoulou, M.; Kalogerakis, N. Ex Situ Bioremediation Treatment (Landfarming). In *Hydrocarbon and Lipid Microbiology Protocols*; Springer: Berlin, Heidelberg, 2016, pp 195–220.

Pao, H. T.; Tsai, C. M. Multivariate Granger Causality Between CO_2 Emissions, Energy Consumption, FDI (Foreign Direct Investment) and GDP (Gross Domestic Product): Evidence from a Panel of BRIC (Brazil, Russian Federation, India, and China) Countries. *Energy*, **2011**, *36* (1), 685–693.

Patra, S.; Ganguly, P.; Barik, S. R.; Samanta, A. Dissipation Kinetics and Risk Assessment of Chlorfenapyr on Tomato and Cabbage. *Environ. Monit. Assess.* **2018**, *190* (2), 1–10.

Raj, A.; Mandal, J.; Golui, D.; Sihi, D.; Dari, B.; Kumari, P. B.; Ghosh, M.; Ganguly, P. Determination of Suitable Extractant for Estimating Plant Available Arsenic in Relation to Soil Properties and Predictability by Solubility-FIAM. *Water Air Soil Pollut.* **2021**, *232* (6), 1–11.

Sasikumar, C. S.; Papinazath, T. Environmental Management: Bioremediation of Polluted Environment. In *Proceedings of the Third International Conference on Environment and Health*; Department of Geography, University of Madras, Chennai and Faculty of Environmental Studies, York University, New York, 2003; pp 465–469.

Spellman, F. R. *The Science of Environmental Pollution*; CRC Press, 2009.

Surendran, A.; Rajagopal, S. S.; Kandasamy, K.; Muthusamy, S, R. S. Is Air Pollution a Leading Risk Factor for Stroke: A Review. *Int. J. ChemTech Res.* **2016**, *9* (9), 381–389.

Tinsley, I. J. *Chemical Concepts in Pollutant Behavior*; John Wiley & Sons, 2004.

Tomei, M. C.; Daugulis, A. J. Ex Situ Bioremediation of Contaminated Soils: An Overview of Conventional and Innovative Technologies. *Crit. Rev. Environ. Sci. Technol.* **2013**, *43* (20), 2107–2139.

Vidali, M. Bioremediation: An Overview. *Pure Appl. Chem.* **2001**, *73* (7), 1163–1172.

Wilbanks, T. J.; Fernandez, S. Climate Change and Infrastructure, Urban Systems, and Vulnerabilities*: Technical Report for the US Department of Energy in Support of the National Climate Assessment*; Island Press, 2014.

CHAPTER 2

Food Chain Arsenic Contamination in West Bengal, India: Accumulation, Distribution, and Human Health Risk

ANTARA DAS, NILANJANA ROY CHOWDHURY, MADHURIMA JOARDAR, DEEPANJAN MRIDHA, AYAN DE, SHARMISTHA MAJUMDER, and TARIT ROYCHOWDHURY

School of Environmental Studies, Jadavpur University, Kolkata, West Bengal, India

ABSTRACT

Groundwater arsenic contamination in West Bengal, India is a prolonged existing environmental hazard where millions of people are exposed to arsenic toxicity. At present, 132 blocks covering 12 districts in the state have groundwater arsenic concentration above its recommended limit in drinking water. Number of arsenic-affected habitations has amplified by 145% in India from the year 2015 to 2020 (https://www.downtoearth.org.in/news/governance/arsenic-affected-habitations-in-india-up-145-in-last-5-yrs-73450). The districts like Maldah, Murshidabad, Nadia, North 24 Parganas, and South 24 Parganas are most adversely affected where numerous people are suffering from arsenic-related diseases including children. Being an agriculture-dependent state, every year a huge amount of groundwater is withdrawn to reach the irrigational demand. Therefore, arsenic contamination in food chain especially through rice grain, pulses, and vegetables has become an additional health burden in recent years which is spread in apparently unexposed areas too due to the transportation of contaminated foodstuffs. However, rainwater plays an important role in scaling down of arsenic concentration in cultivable foodstuffs. According to our study, monsoonal rice grain is safer to consume than premonsoon grains as the

Environmental Contaminants: Impact, Assessment, and Remediation.
Pritam Ganguly, Jajati Mandal, M. Paramsivam, & Sandip Patra, (Eds.)

former one accumulates much lesser arsenic. Post harvesting treatments of paddy grain through two-way boiling process using arsenic-contaminated groundwater increase the arsenic burden in parboiled rice grain in comparison to sunned rice grain. Cooking of raw rice with highly arsenic-contaminated water causes an entry of arsenic from water to rice while cooking with lower arsenic-contaminated water draws out arsenic from rice into gruel. Inorganic arsenic has been observed to be the dominating one among the other species in both raw and cooked rice, while for both rice and vegetables, arsenite, and arsenate are more in amounts than DMA.

2.1 INTRODUCTION

Groundwater arsenic contamination is a yearlong prevailing environmental hazard in West Bengal, India where around 6 million people are exposed to arsenic toxicity. Nine districts including 85 blocks in the Bengal delta were previously reported with massive arsenic-contaminated groundwater above 50 μg/L (Chowdhury et al., 2000, 2001; Rahman et al., 2001, 2005), where the most severely affected districts are Maldah, Murshidabad, North 24 Paragans, South 24 Paragans, and Nadia (Mazumder et al., 2010; Mukherjee et al., 2005; Rahman et al., 2003, 2014). Later on, the magnitude of groundwater arsenic contamination above 10 and 50 μg/L was reported in 148 blocks from 14 districts and 111 blocks from 12 districts, respectively (Chakraborti et al., 2009, 2013; Sengupta et al., 2009). The source of arsenic in groundwater is geogenic and mainly caused due to the reductive dissolution of arsenic-rich iron oxyhydroxide minerals (Nickson et al., 2000). Therefore, this natural occurrence is catalyzed with time due to excessive groundwater withdrawal and sheer negligence of the local government. A considerable amount of population of the rural West Bengal depends on groundwater for drinking, cooking, and other household purposes. In addition to the contaminated drinking water, food chain arsenic contamination has triggered a serious health concern over the last few years (Chowdhury et al., 2018a; Samal et al., 2011; Santra et al., 2013). The huge population suffers from arsenic exposure through contaminated foodstuffs because cultivation with arsenic-contaminated irrigational groundwater compels the toxicant entering into the grains and vegetables where the effect of contaminated soil is also significant (Roychowdhury et al., 2002a, b, 2005; Sanyal et al., 2002). The prime source of arsenic exposure in humans is rice and rice-based products whereas arsenic concentrations in other foodstuffs like pulses, cereals, and vegetables also contribute to human health risk when consumed at a daily rate (Bhattacharya et al., 2010; Biswas et al., 2012; Roychowdhury et al., 2002b; Upadhyay et al., 2019; Mandal et al., 2019a, b). In West Bengal, paddy is the largest irrigational production with approximately 5,900,000 ha cultivation area (Signes et al., 2008). The population of Bengal is majorly dependent on rice as a staple diet and interestingly, rice grain accumulates more arsenic than other frequently harvested crops (Mondal and Polya, 2008; Kumarathilaka et al., 2019; Williams et al., 2007). Accumulation of arsenic in paddy grains fundamentally differs with season, variety, arsenic concentration in water and soil which is elaborately described in two of our recent research reports along with arsenic accumulation and distribution in the different growth phases of paddy plants (Chowdhury et al., 2018a, 2020a). Paddy cultivation requires huge amount of water, which is fulfilled by arsenic-contaminated groundwater during summer (Chowdhury et al., 2018a) and also in monsoon when rainwater is not sufficient (Chowdhury et al., 2020a). Moreover, the post harvesting mechanism of paddy grain into parboiled one has caused an additional entry of arsenic burden in these areas where the parboiled rice grain and the rice by-products are observed with higher arsenic accumulation compared to sunned rice grain (Chowdhury et al., 2018b, 2019). While comparing the arsenic species in rice varieties, it is well established that inorganic arsenic dominates the methylated arsenic (DMA, MMA etc.) in a high ratio (Islam et al., 2017; Pal et al., 2009; Roychowdhury, 2008). It is pertinent to mention here that inorganic arsenic is way more toxic than any of its methylated species (Jomova et al., 2011; Signes Pastor et al., 2008). This is not an only concern for the local inhabitants because the cultivated crops get transported in the widespread urban areas (Biswas et al., 2019). A household as well as market basket survey of food materials from different locations of apparently unexposed Kolkata city shows that the urban population is also under the same threat of arsenic toxicity though at a lower degree (Biswas et al., 2019). The cooked rice is another point of health concern in Bengal delta where arsenic concentration varies with cooking water, rice variety, cooking method, rice and water ratio etc. (Chowdhury et al., 2020b; Mandal et al., 2019c; Mondal et al., 2010; Pal et al., 2009; Rahman et al., 2006, 2011; Sengupta et al., 2006; Signes et al., 2008). Chowdhury et al., (2020b) have depicted the distribution of arsenic accumulation in different fractions of rice during cooking, that is, uncooked rice, cooking water, cooked rice, and gruel. It has also delineated the arsenic flow between water and rice grain while cooking with a varied range of arsenic concentrated water. While cooking rice with low or moderate arsenic-concentrated water, arsenic draws out from rice into water, whereas cooking with high arsenic-concentrated water enhances the rice arsenic concentration significantly resulting in increased health risk (Chowdhury et al., 2020b; Halder et al., 2014; Mandal et al., 2019a; Sengupta et al., 2006). Major portion of total arsenic content in cooked rice is also contributed by inorganic arsenic species (Chowdhury et al., 2020b; Halder et al., 2013, 2014; Roychowdhury, 2008; Laparra et al., 2005). The suggested intake value of inorganic arsenic is in the lower limit on the benchmark dose for a 0.5% increased incidence of lung cancer (BMDL 0.5) as calculated 3.0 μg/kg bw/day, ranged from 2 to 7 μg/kg bw/day (WHO, 2011). The previous recommended Provisional Tolerable Daily Intake (PTDI) value for consumption of inorganic arsenic, that is, 2.1 μg/kg bw/day was withdrawn at the 72nd meeting of the Joint FAO/WHO Expert Committee on Food Additives (JECFA, 2011; EFSA, 2014). In most of the arsenic-exposed populations, researchers have observed that the average daily dietary intake rate is much higher than the recommended limit (Bhowmick et al., 2018; Halder et al., 2013; Joardar et al., 2021; Roychowdhury, 2008, 2010). Therefore, arsenic in drinking water and foodstuffs, being a potential hazard to the mankind, needs to be immediately taken care of (Das et al., 2009; Mazumder, 2003; Rahaman et al., 2013).

In different districts, arsenic toxicity and the sufferings of people vary depending on the water arsenic concentration range. Analysis of the biological samples of the inhabitants shows the extent of arsenic toxicity in their health (Das et al., 1995, Rahman et al., 2001, Roychowdhury 2010). Being a carcinogen, severe arsenic toxicity results in various cancerous diseases, the symptoms vary in keratosis, melanosis, hyper pigmentation, bowens etc. Cases of numerous patients are reported from West Bengal and Bangladesh since many years (IARC, 2012; Khan et al., 2003; Mazumder et al., 1988, 1998, 2011; Rahman et al., 2005). Chronic arsenic toxicity is not limited into the systemic malignant skin manifestations; it causes several other health disorders such as peripheral neuropathy, dyspepsia anemia, diarrhoea, hepatomegaly, gangrene, and lung diseases (Majumdar et al., 2009; Mukherjee et al., 2003). Not only the adults, children from the arsenic endemic sites too suffer from health risk due to subacute and subclinical toxicity (Joardar et al., 2021). Infants and young adults are reported with respiratory troubles, increased mortality rate, impaired intelligence etc. (Biswas et al., 2018; Rahman et al., 2001; Raqib et al., 2009; Watanabe et al., 2007).

The present article will give an idea of arsenic exposure through food chain in the past 20 years on the arsenic-affected sites in West Bengal in comparison with the less affected or unexposed sites. This study will put emphasis on arsenic accumulation and distribution in irrigated food components which brings additional danger upon drinking water contamination in recent times. Arsenic contamination in the various kinds of cereals, spices, raw rice, cooked rice, and vegetables has been briefly explained in this study. It will also exhibit the contaminated dietary intake extent in the human system resulting in future health risk in adults and children.

2.2 STUDY AREA

The present study reports the status of food chain arsenic contamination in some adversely arsenic exposed sites of the districts like Murshidabad, Nadia, North and South 24 Parganas as well as some apparently unexposed or control sites of West Bengal, India. The present groundwater arsenic-contamination status above 10 μg/L includes 132 blocks covering 12 districts of the state (Fig. 2.1).

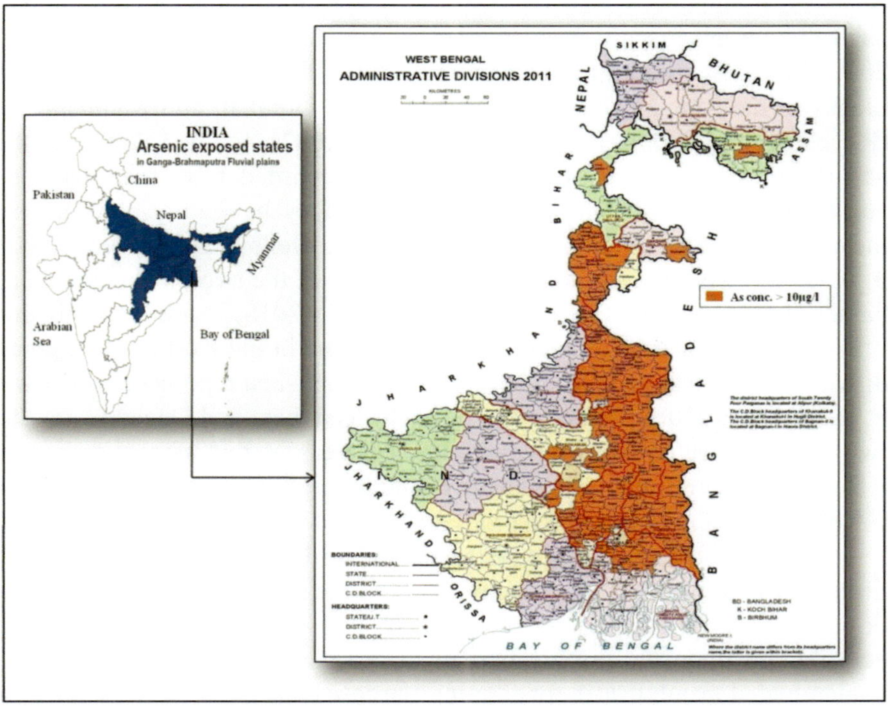

FIGURE 2.1 Present groundwater-arsenic contamination status in West Bengal.

2.3 RESULTS AND DISCUSSION

2.3.1 ARSENIC CONTAMINATION STATUS IN GROUNDWATER AND SOIL

The status of arsenic contamination in groundwater used for irrigational purposes in the studied areas of West Bengal is explained well in Table 2.1. In the rural sites of West Bengal, huge areas of agricultural lands are irrigated every year through groundwater withdrawn from shallow depths. It is a well-known fact that arsenic concentration is generally high in these shallow depth ranges of Ganga Meghna Brahmaputra (GMB) belt. An extensive amount of work has been done in some of the blocks of the three most affected districts (Murshidabad, Nadia and North 24 Parganas), where the groundwater arsenic concentration range varies 20 to 40 times higher than the recommended level (Table 2.1). Nadia is such a district where all the 17 blocks are reported with arsenic contamination in groundwater (Das et al., 2020; Rahman et al., 2014), while 25 out of total 26 blocks in Murshidabad district are arsenic-contaminated above 10 μg/L (Chakraborti et al., 2009, 2013). According to our reported studies, the highly arsenic-affected blocks in the Murshidabad district are Jalangi, Domkal, and Raninagar II (Roychowdhury, 2008; Roychowdhury et al., 2002a, b, 2003, 2005, 2008) and the most adverse condition in Nadia is recorded from the Chakdah, Haringhata, Tehatta-I, and Tehatta-II blocks (Roychowdhury, 2008). The mean arsenic concentration in the shallow tube wells was reported as 410 μg/L (n=24) from the studied agricultural lands of the two blocks in Nadia, namely Chakdah and Haringhata (Biswas et al., 2018). An overall picture of irrigational water arsenic-contamination status in the North 24 Parganas district can be visualized from several of our published reports (Chowdhury et al., 2018a, 2020b; Roychowdhury, 2010). Gaighata and Deganga are the two mostly affected and frequently studied areas of this aforesaid district where the harsh groundwater arsenic contamination (range: 14–493 μg/L) is vividly resulted in the irrigated crops and vegetables.

Arsenic concentration in the surface soil and root soil is needed to be discussed in light of arsenic accumulation in the agricultural crops because besides the groundwater arsenic burden, soil arsenic has a significant contribution upon the arsenic burden in crops. Extended use of arsenic-contaminated groundwater for farming purposes has led to high deposition of arsenic in irrigated soil (Meharg and Rahman, 2003; Roychowdhury et al., 2002a, 2005). Approximately, 6.4 tons of arsenic was withdrawn per

TABLE 2.1 Status of Arsenic Contamination in Irrigational Water and Soil in the Studied Areas of West Bengal, India.

Study area		Groundwater arsenic in shallow tube well (µg/L)	Soil arsenic (µg/kg)	References
District	**Location**			
Exposed sites				
Murshidabad	Block: Domkal	Mean: 78, range: 18–190; n=19	Fallow land: mean: 5310, range: 2680–6790; n=86	Roychowdhury et al. (2002a)
			Agricultural land: mean: 10,700, range: 3340–31600; n=180	
	Block: Domkal and Jalangi	Mean: 85, range: 18–200; n=6	Agricultural land: mean: 11,350, range: 3340–31600; n=36	Roychowdhury et al. (2002b)
	Block: Domkal	Mean: 85, range: 18–200; n=6	–	Roychowdhury et al. (2003)
	Block: Jalangi	Mean: 108, range: 40–182; n=4		
	Block: Hariharpara	Mean: 74.5, range: 29–120; n=2	Agricultural land: mean: 15,515, range: 10,000–19,400; n=57	Roychowdhury et al. (2005)
	Block: Raninagar-II	Mean: 41, range: 19–63; n=2	Agricultural land: mean: 12,814, range: 9500–17900; n=42	
	Block: Jalangi, Domkal, Hariharpara, Raninagar-II	Mean: 75, range: 18–200; n=23	Agricultural land: mean: 14,300, range 3340–31600; n=279	Roychowdhury (2008)
		Mean: 18.7, range: 4–82; n=8	Agricultural land: mean: 15,600, range 5000–95300; n=189	Roychowdhury et al. (2008)
Nadia	Block: Tehatta-I, Tehatta-II, Karimpur-II	Mean: 21, range: 4–82; n=18	Agricultural land: mean: 16,700, range: 5000–95300; n=162	Roychowdhury (2008)
	Block: Chakdah, Haringhata	Mean: 410; n=24	Agricultural land: mean: 7060; n=20	Biswas et al. (2018)
North 24 Parganas	Block: Gaighata	Mean: 95, range: 14–160; n=55	–	Roychowdhury (2010)

TABLE 2.1 *(Continued)*

Study area		Groundwater arsenic in shallow tube well (µg/L)	Soil arsenic (µg/kg)	References
District	**Location**			
	Block: Deganga	Range: 70–300; n=10	Mean: 29440; n=10	Chowdhury et al. (2018a)
	Block: Gaighata	Mean: 332, range: 171–493	Mean: 18,215, range: 4709–40584; n=4	Chowdhury et al. (2020b)
Control sites				
Hooghly	–	Mean: 10; n=24	Mean: 1030; n=20	Biswas et al. (2018)
West Midnapore	Pingla	Mean: 10	Mean: 1527	Chowdhury et al. (2020b)

year and deposited on soil from all the shallow tubewells (n=3072), used for agricultural irrigation at the Deganga block, North 24 Parganas district (Mandal, 1998). The translocation theory of plants, especially paddy plants along with iron plaque formation in the roots, helps to understand the process of binding of arsenic in the rhizosphere followed by its distribution in the upward parts (Chowdhury et al., 2018a, 2020a). However, it is noticed that the accumulation follows a decreasing trend from soil–root juncture to grain (Abedin et al., 2002; Chowdhury et al., 2018a). The soil arsenic is attributed by several mineralogical and geogenic factors in the Bengal delta which was explained in our previous studies (Chakraborti et al., 2001; Das et al., 1996; Roychowdhury et al., 1999, 2002a). Moreover, in premonsoonal season (summer cultivation), when water is logged in the agricultural lands for paddy cultivation, arsenic in soil becomes mobile and bioavailable through anaerobiosis (reduction of arsenate to arsenite) followed by enhanced uptake of arsenic by plants (Chowdhury et al., 2018a, 2020a). The deposition of arsenic in the soil system from the exposed Bengal districts is shown in Table 2.1 in comparison to two control sites.

2.3.2 ARSENIC CONTAMINATION IN FOOD CHAIN

2.3.2.1 RAW RICE GRAIN AND RICE BY-PRODUCTS

Rice grain arsenic accumulation depends on origin of cultivation or geographical locations, cultivars, genotypes, groundwater, and soil arsenic concentration (Chowdhury et al., 2018a; Tuli et al., 2010). Apart from these, different rice types like polished or unpolished, brown, or white vary with different arsenic concentrations. Highest accumulation of arsenic has been reported in brown rice as the thin outer layer of the rice kernels accumulate arsenic which is removed to produce white rice (Meharg et al., 2008). Depending on the average length and length to breadth ratio, the available rice grains in rural Bengal are classified into four categories, like short-bold (SB), medium-slender (MS), long-slender (LS) (all these three categories fall under brown-colored rice), and extra-long slender (ELS) (falls under white-colored Indian Basmati) (Halder et al., 2012). A large part of the population in Bengal delta uses parboiled rice mainly short-bold (SB) brown rice for cooked rice preparation. A maximum level of 0.2 mg/kg has been suggested for inorganic arsenic in white (polished) rice which covers 79% of international market by Codex Alimentarius, coordinated by the Food and Agriculture Organization of the United Nations (FAO) and the World Health Organization (WHO) (Codex Alimentarius Commission, 2014) while 0.12

mg/kg arsenic has been proposed by the Consumers Union (CU) for both white and brown rice and rice products (Consumers Union, 2014). West Bengal has 5.8 million ha under rice cultivation, which includes both irrigated and rainfed areas, with an average productivity of 2.6 tons/ha (https://icar.org.in/files/state-specific/chapter/125.htm). Approximately, 78% of total area under rice in West Bengal is concentrated under high and medium rice productivity groups, which accounts for 84% of total production of rice in the state. Our study areas like Murshidabad and North 24 Parganas districts fall under "medium productivity" groups, while Nadia falls under "high rice productivity" groups (http://drdpat.bih.nic.in/PA-Table-25-West%20Bengal.htm). The accumulation of arsenic in rice grain, rice by-products, and cooked rice in these arsenic-exposed districts (Table 2.2) is a serious concern because most of the accumulation is above the recommended level of arsenic in rice, that is, 100 μg/kg for contaminated sites (Meharg et al., 2006). Moreover, this huge rice production gets commercialized or transported in other districts or states also which is why the rice samples collected from apparently control or unexposed sites showed substantial amount of arsenic concentration (Biswas et al., 2019; Joardar et al., 2021). Table 2.2 also shows the collective data of rice arsenic in some of the studied control sites. However, the monsoonal cultivation has a lesser arsenic accumulation pattern. Chowdhury et al. (2020a) have shown that the monsoonal rice grains accumulated only one-third amounts of premonsoonal rice grains irrespective of the variety of cultivar and the geographical area based on arsenic exposure which are way safer to consume in view of nutritional status. Exposed and control area monsoonal rice grain contained an average of 350 μg/kg and 224 μg/kg arsenic, respectively, while the exposed premonsoonal rice grain contained 1120 μg/kg arsenic.

The mean total arsenic level in 46 rice samples was reported as 358 μg/kg (range: 46–1110 μg/kg dry weight) in Bangladesh (Smith et al., 2006). Another study in Bangladesh reported arsenic concentration range in rice grains as 70–740 μg/kg (n=76) (Norton et al., 2009). A study shows that only rice contributes about 94.1 and 92.4%, and 93.5 and 91.7% of arsenic with respect to the total intake of arsenic from all foodstuffs (rice, vegetables, and spices) by adults and children in the Jalangi and Domkal blocks, respectively (Roychowdhury et al., 2002b).

After harvesting, the paddy grain is processed in a two-way boiling step for making parboiled rice grain and it is observed that when arsenic-contaminated water is used for parboiling, the parboiled rice grain has an increased arsenic concentration compared to the sunned rice grain (Bae et al., 2002;

Chowdhury et al., 2018b; Mondal et al., 2010). An order of increasing arsenic concentrations is observed in water samples collected at raw, half-boiled, and full-boiled stages of parboiling irrespective of their initial concentration due to the reduction in final volume during the parboiling process. Chowdhury et al. (2018b, 2019) revealed that rice by-products obtained from parboiled rice grains contained more arsenic concentrations than those made from sunned rice grain (Table 2.2).

2.3.2.2 COOKED RICE

While looking into the cooked rice, higher accumulation of arsenic was observed compared to raw rice grains (Table 2.2) (Roychowdhury et al., 2002a, 2008). This observation led to further research which found that the cooking water arsenic concentration has a major role behind the cooked rice arsenic concentration. When rice was cooked with a varied range of arsenic concentrated water, from low to high, a two-directional flow of arsenic between rice and water was observed. Positive regression correlation value (R^2= 0.392) between "% of arsenic increase or decrease" in wet cooked rice and "water arsenic concentration" helps to understand the significance of cooking water (Chowdhury et al., 2020b). In most of the rice grains, irrespective of variety and cultivars, arsenic moves from rice to water when low (<3 μg/L) or moderate arsenic-contaminated water (37–58 μg/L) is used for cooking and a significant decrease of arsenic (23–89%) is observed in cooked rice. However, the flow reverses when high arsenic concentrated cooking water (84–105 μg/L) is used which results in a significant increase of arsenic (24–337%) in cooked rice. Hence, it can be said that there exists a threshold value of water arsenic concentration above which the toxicant enters into cooked rice. In another study from West Bengal, cooked rice (n=100) contributed 31.5–80 μg/kg and 219–664 μg/kg of arsenic compared to raw rice grain arsenic concentration of 105–510 μg/kg when cooked with arsenic safe water (<0.5 μg/L) and arsenic-rich groundwater samples (90–230 μg/L), respectively (Mandal et al., 2019a). Arsenic accumulation and dispersion in the different fractions of rice cooking is tabulated in Table 2.3. A study from severely arsenic-affected areas of Bangladesh showed that cooking rice with limited water where the gruel was completely absorbed in rice, resulted in 44.9% increase of arsenic than rice cooked with excess water (Rahman et al., 2006). The consumption of gruel which is typically used as a daily diet in rural Bengal (Signes-Pastor et al., 2012) contributes a considerable amount of arsenic (16–108.5 μg/L) (Chowdhury et al., 2020b).

TABLE 2.2 Status of Arsenic Contamination in Rice Grain, Rice By-Products, and Cooked Rice from the Studied Areas in West Bengal, India.

Study area		Foodstuffs		Arsenic: Mean and range (μg/kg)	References
District	**Location**				
Exposed sites					
Murshidabad	**Block:** Jalangi	Rice grain	Raw (n= 11)	226, 43–443	Roychowdhury et al. (2002b)
			Cooked (n=6)	374, 290–490	
	Block: Domkal	Rice grain	Raw (n= 23)	245, 61–662	
			Cooked (n=12)	667, 198–1930	
	Block: Jalangi	Rice grain (n=11)		232, 40.8–605	Roychowdhury et al. (2003)
	Block: Domkal	Rice grain (n=23)		233, 78.8–546	
	Block: Jalangi, Domkal, Hariharpara, Raninagar-II	Rice grain	Raw rice grain (n= 34)	239, 43–662	Roychowdhury (2008)
			Cooked rice grain(n=18)	569, 198–1030	
Nadia	**Block:** Tehatta-I, Tehatta-II, Karimpur-II	Rice grain	Raw (n= 18)	190, 45–386	Roychowdhury (2008)
			Cooked (n=4)	290, 105–410	
	Block: Tehatta-I, Karimpur-II	Rice grain (n= 18)		190, 45–386	Roychowdhury et al. (2008)
	Block: Chakdah, Haringhata	Rice grain (n= 60)		550	Biswas et al. (2018)
North 24 Parganas	–	Rice grain (n= 12)		328, 119–540	Sengupta et al. (2006)
	Block: Gaighata	Rice grain (n= 16)		195, 85–330	Roychowdhury (2010)
	Block: Deganga	Premonsoonal Rice grain (n= 10)		1120, 360–1560	Chowdhury et al. (2018a)
	Block: Deganga	Raw rice grain	Sunned (n=8)	78.4, 52–127	Chowdhury et al. (2018b)
			Parboiled (n= 11)	166, 58–259	
		Rice flour (n= 16)		566, 119–1145	

TABLE 2.2 *(Continued)*

Study area		Foodstuffs	Arsenic: Mean and range (µg/kg)	References
District	**Location**			
		Puffed rice (n= 13)	226, 103–354	
		Parched rice (n= 6)	351, 128–743	
		Beaten rice (n= 13)	289, 110–516	
	Block: Gaighata	Monsoonal Rice grain (n= 12)	342, 278–389	Chowdhury et al. (2020b)
Control sites				
Hooghly	–	Rice grain (n= 60)	20	Biswas et al. (2018)
Kolkata	–	Rice grain Market survey (n=27)	120, 80–197	Biswas et al. (2019)
		Family survey (n=30)	105, 46–199	
West Midnapore	Block: Pingla	Monsoonal Rice grain (n= 3)	224, 185–265	Chowdhury et al. (2020b)

TABLE 2.3 Arsenic Fractionation in Cooked Rice.

Rice grain type	Case types	Arsenic concentration						
		Raw Water (µg/L)		Raw rice grain (µg/kg)	Wet cooked rice (µg/kg)	Dry cooked rice (µg/kg)	Gruel (µg/L)	Supernatant liquid (µg/L)
Sunned rice	Uncontaminated site rice + uncontaminated water (lab based)	Low	<3	88	33	45	16	10.5
	Contaminated site rice + uncontaminated water (lab based)		<3	182	44	109	43	34
Parboiled rice	Uncontaminated site rice + uncontaminated water		<3	208	94	187	37	36
	Contaminated site rice + uncontaminated water (Lab based)		<3	312	67	171	64	50
	Contaminated site rice + uncontaminated water		<3	173	130	331	47	20
Sunned rice	Contaminated site rice + contaminated water (spiked 38 µg/L) (lab based)	Moderate	38	182	137	237	108.5	76
Parboiled rice	Contaminated site rice + contaminated water		38	158	134	341	79.5	55
	Contaminated site rice + contaminated water		37	162	97	249	49	40
	Contaminated site rice + contaminated water (spiked 58 µg/L) (lab based)		58	312	140	357	94	68
Sunned rice	Contaminated site rice + contaminated water (spiked 84 µg/L) (lab based)	High	84	182	242	320	126	111
Parboiled rice	Contaminated site rice + contaminated water (spiked 105 µg/L) (lab based)		105	137	246	403	123	115

2.3.2.3 VEGETABLES, SPICES, CEREALS, AND OTHER FOODSTUFFS

In addition to the staple food rice, vegetables and green leafy materials attribute a major portion of the daily diet of Bengal population. Human populace are strikingly exposed to arsenic toxicity through arsenic-contaminated cereals and vegetables (Roychowdhury, 2008; Roychowdhury et al., 2002a, 2003, 2008; Biswas et al., 2019). Although rice accumulates larger amount of arsenic, it is observed that the arsenic concentration in wheat grain, different spices, and vegetables contributes significantly in the arsenic burden of human health which is shown in Table 2.4. In comparison to other foodstuffs, lentils accumulate lesser amount of arsenic (3.9–43 μg/kg). The outer skin of the vegetables accumulates more arsenic than the inner flesh part (Biswas et al., 2019; Roychowdhury et al., 2008). A similar kind of observations was found in other studies from Bangladesh (Alam et al., 2003; Das et al., 2004). The mean total arsenic was reported as 333 μg/kg (range: 19–2334 μg/kg dry weight) in 39 vegetable samples from Bangladesh (Smith et al., 2006). Islam et al. (2014) found arsenic concentrations as 254 μg/kg in cereals and 250 μg/kg in vegetables.

2.3.2.4 ARSENIC SPECIES DISTRIBUTION

Humans are exposed to arsenic in both forms, organic and inorganic, although inorganic arsenic is more toxic in nature than the former. Inorganic arsenic contributes 90% of the total content of arsenic in cereals and vegetables grown in West Bengal (Chowdhury et al., 2020b; Joardar et al., 2021; Roychowdhury, 2008) and Bangladesh (Ahmed et al., 2016), whereas arsenic speciation in rice grain is also dominated by inorganic one and dimethylarsinic acid (DMA) (Williams et al., 2005). Inorganic arsenic in rice and vegetables calculated as arsenite and arsenate made up to 87% and 96%, respectively, with respect to total arsenic in Bangladesh (Smith et al., 2006). Concentration of different arsenic species in various foodstuffs from the studied areas is shown in Table 2.5. It is visibly stated that in both rice and vegetables, the species like arsenite and arsenate are more in amount than DMA. Similarly, inorganic arsenic has been observed to be dominating among other species in both raw (93.8%) and cooked rice (88.1%) from rural Bengal (Chowdhury et al., 2020b; Halder et al., 2014).

TABLE 2.4 Status of Arsenic Contamination in Foodstuffs from the Studied Areas of West Bengal

Study area		Foodstuffs	Arsenic: Mean and range (µg/kg)	References
District	**Location**			
Exposed sites				
Murshidabad	**Block:** Jalangi	Lentils (n=8)	3.9, <0.04–16	Roychowdhury et al. (2002b)
		Wheat grain (n= 2)	7, 4–10	
		Spices (n= 41)	92.2, <0.04–479	
		Vegetables (n= 65)	91.7, <0.04–690	
	Block: Domkal	Lentils (n=3)	43, 2–117	
		Wheat grain (n= 1)	362	
		Spices (n= 25)	207.6, 13–481	
		Vegetables (n= 77)	123, <0.04–690	
	Block: Jalangi	Wheat grain (n= 2)	5.22, 0.95–9.5	Roychowdhury et al. (2003)
		Lentils (n= 5)	4.39, <0.04–15.2	
		Spices (n= 78)	60.3, <0.04–605	
		Vegetables (n= 50)	20.9, <0.04–138	
	Block: Domkal	Lentils (n= 3)	41.1, 1.9–111	
		Spices (n= 120)	102, <0.04–546	
		Vegetables (n= 62)	21.2, <0.04–212	
	Block: Jalangi, Domkal, Hariharpara, Raninagar-II	Vegetables (n= 126)	85, 43–690	Roychowdhury (2008)
		Spices (n= 66)	136, 74–481	

TABLE 2.4 *(Continued)*

Study area		Foodstuffs		Arsenic: Mean and range (µg/kg)	References
District	**Location**				
Nadia	**Block:** Tehatta-I, Tehatta-II, Karimpur-II	Vegetables (n= 48)		61, 21–108	Roychowdhury (2008)
		Spices (n= 16)		90, 40–321	
	Block: Tehatta-I, Karimpur-II	Wheat grain (n= 6)		29.2, 21–38	Roychowdhury et al. (2008)
		Wheat flour (n= 14)		61.9, 25–119	
		Potato	Skin (n= 17)	382, 181–792	
			Flesh (n=16)	61.1, 21–108	
Control sites					
Kolkata	–	Vegetables	Market survey (n=98)	104, 10–227	Biswas et al. (2019)
			Household survey (n=112)	149, 11–727	

TABLE 2.5 Status of Different Arsenic Species in Foodstuffs from the Studied Areas.

Type of sample	Concentration of different arsenic species (µg/kg)				References	Study conducted in
	As (III+V)	DMA	MMA	Sum of as species		
Paddy grain	380 ± 21	96 ± 6	21 ± 1	497 ± 22	Chakraborti et al. (2016)	Deganga; North 24 Parganas
Parboiled rice grain	141.7 (89.5% of total As)	16.7	–	–	Joardar et al. (2021)	Gaighata; North 24 Parganas
Raw rice grain	286	0	0	286	Chowdhury et al. (2020b)	
Cooked rice	114.3	7.5	0	121.8		
Raw rice grain	95.5 (58–140.5, n= 6)	11.95 (2.68–23.9, n=6)	–	154.2 (85–244, n=6)	Roychowdhury (2008)	Jalangi, Domkal, Hariharpara, Raninagar-II; Murshidabad
Vegetable leaves	114 (83.9–159.7, n=4)	15.6 (5.96–35.5, n=4)	–	187 (126–290, n= 4)		Tehatta-I, Tehatta-II, Karimpur-II; Nadia

2.3.2.5 DIETARY INTAKE OF ARSENIC AND HEALTH RISK

Dietary intake of arsenic depends on the daily consumption rate of drinking water and foodstuffs and respective arsenic concentrations estimated. In the two blocks of Murshidabad district, Jalangi and Domkal, dietary intakes of arsenic (μg/day) from the foodstuffs for adults and children were 171.2, 189.1 and 91.89, 101.6, respectively (Roychowdhury et al., 2002a). The calculated average daily dietary intakes of arsenic from all the sources from the same two blocks of the district were reported as 801 and 658 μg (adult male), 718 and 588 μg (adult females), and 432 and 351 μg (children), respectively (Roychowdhury et al., 2003). Maximum amount of arsenic was attributed by drinking water among all the dietary intakes. The daily dietary intakes of arsenic from water and food composites of another studied population from the same district living in mild, moderate, and high arsenic-affected zones were 137, 228, and 568 μg, respectively (Uchino et al., 2006). The cumulative average daily dietary intakes of arsenic by an individual from water, rice, and vegetables in the Murshidabad and Nadia districts were 729 and 401 μg, respectively (Roychowdhury, 2008). The dietary intakes of inorganic arsenic (μg/kg body wt./day) by an adult from water, rice, and vegetables in the two districts were 11.8 and 6.43 (male) and 15.8 and 8.58 (female), respectively (Roychowdhury, 2008), which are higher than the WHO recommended daily intake value of inorganic arsenic. In another study from Nadia, the reported estimated average daily dietary intakes of arsenic from all sources were 359 μg (adult male), 357 μg (adult females), and 210 μg (children), respectively (Roychowdhury et al., 2008). The daily dietary intake rates of inorganic arsenic in a survey from Gaighata population, North 24 Parganas, were reported as 11.7 and 13.1 μg/kg body wt./day for adult male and female, respectively (Roychowdhury, 2010). The observed dietary intake rates of arsenic for school children from an arsenic-exposed area of Gaighata block, North 24 Parganas district through drinking water and rice were 5.26 and 4.7 μg/kg bw/day, which are higher compared to the WHO recommended value of 3.0 μg/kg bw/day. Even the total dietary intake rates of arsenic for the apparently control school children from the Kolkata district were 0.22 and 3.96 μg/kg bw/day from the same two sources (Joardar et al., 2021). This observation is clear inference that the control population is exposed to arsenic-contaminated rice grain which is cultivated and transported from contaminated sites, apart from drinking arsenic safe drinking water (Biswas et al., 2019).

Being a class 1 carcinogen, prolonged consumption of a substantial amount of arsenic can cause malignancy in many forms. In many reports, it is established that intake of arsenic-contaminated water and food brings lifetime health risk in the exposed population when the daily intake of arsenic is calculated. Joardar et al. (2021) showed that even children do have cancer risk in the future through the consumption of contaminated rice only. Exposure to lower levels of arsenic over longer periods of time can also cause skin changes, liver and kidney impairment, and a deficiency of red and white blood corpuscles, which can further lead to exhaustion and an amplified risk of infections.

2.4 CONCLUSIONS

Arsenic calamity in Bengal delta is an unavoidable threat to a large population every day. Since years, many awareness programs have been campaigned in the locality. Boring of deep tube wells and installation of arsenic removal plants by the local governments in the rural areas has ensured that a large part of the population gets arsenic safe drinking water. However, the state like West Bengal is increasingly facing problems of using arsenic-contaminated groundwater for drinking (through domestic hand tube wells) and irrigational purposes. With time, the contamination has been spread out in deeper depths due to excessive withdrawal of groundwater. Due to lack of awareness, and lack of regularity in water quality monitoring, people often consume arsenic-contaminated water for drinking and cooking purposes. Consumption of contaminated vegetables and cooked rice with contaminated water visibly deteriorates their health quality which is understood when their biological samples are analyzed. Chronic arsenic toxicity is expressed through the huge arsenic accumulation in their hair and nail samples. There are numerous patients suffering from cancer in these three severely arsenic-affected districts and the disease is now genetically expanding in some families. The poor socioeconomic status and malnutrition of the rural inhabitants are additional factors for not coping up with this natural hazard. Years after years, they are trapped in the same poisonous history and losing their physical and mental health which is needed to be immediately attended. The state has huge surface water bodies and on an average 2000 mm annual rainfall. A proper watershed management through the treatment of surface water, implementation of rainwater harvesting system in community level, and awareness campaigns among these endangered populations is badly needed to combat this severe arsenic crisis situation. The state needs to find out a policy of alternative agricultural management system to restrict the entry of arsenic in food chain using vast surface water bodies and less amount of groundwater under aerobic condition, if required.

ACKNOWLEDGMENTS

Financial supports from "Department of Science & Technology," Government of West Bengal for providing research project grant (Memo No. 262(Sanc.)/ST/P/S&T/1G-64/2017, dated 25/3/2018) and Inter University Research Project, RUSA (R-11/1092/19, dated 06/08/2019) are highly acknowledged.

KEYWORDS

- **food chain arsenic**
- **paddy cultivation**
- **cooked rice**
- **vegetables**
- **health risk**

REFERENCES

Abedin, M. J.; Feldmann, J.; Meharg, A. A. Uptake Kinetics of Arsenic Species in Rice Plants. *Pl.Physio.* **2002,** *128* (3), 1120–1128.

Ahmed, M. K.; Shaheen, N.; Islam, M. S.; Habibullah-Al-Mamun, M.; Islam, S.; Islam, M. M.; Kundu, G. K.; Bhattacharjee, L. A Comprehensive Assessment of Arsenic in Commonly Consumed Foodstuffs to Evaluate the Potential Health Risk in Bangladesh. *Sci. Total Environ.* **2016,** *544*, 125–133.

Alam, M. G. M.; Snow, E. T.; Tanaka, A. Arsenic and Heavy Metal Contamination of Vegetables Grown in Samta village, Bangladesh. *Sci. Total Environ.* **2003,** *308* (1–3), 83–96.

Bhattacharya, P.; Samal, A. C.; Majumdar, J.; Santra, S. C. Arsenic Contamination in Rice, Wheat, Pulses, and Vegetables: A Study in an Arsenic Affected Area of West Bengal, India. *Water Air Soil Pollut.* **2010,** *213* (1–4), 3–13.

Bae, M.; Watanabe, C.; Inaoka, T.; Sekiyama, M.; Sudo, N.; Bokul, M. H.; Ohtsuka, R. Arsenic in Cooked Rice in Bangladesh. *Lancet* **2002,** *360* (9348), 1839–1840.

Bhowmick, S.; Pramanik, S.; Singh, P.; Mondal, P.; Chatterjee, D.; Nriagu, J. Arsenic in Groundwater of West Bengal, India: A Review of Human Health Risks and Assessment of Possible Intervention Options. *Sci. Total Environ.* **2018,** *612*, 148–169.

Biswas, A.; Biswas, S.; Santra, S. C. Risk from Winter Vegetables and Pulses Produced in Arsenic Endemic Areas of Nadia District: Field Study Comparison with Market Basket Survey. *Bull. Environ. Contam. Toxicol.* **2012,** *88* (6), 909–914.

Biswas, A.; Das, A.; Roychowdhury, T.; Guha Mazumder, D. N. Low Arsenic Exposure Risk in Endemic Population, Cohort Study for Consecutive Years. *Expos. Health.* **2018,** *10*, 273–286.

Biswas, A.; Swain, S.; Chowdhury, N. R.; Joardar, M.; Das, A.; Mukherjee, M.; Roychowdhury, T. Arsenic Contamination in Kolkata Metropolitan City: Perspective of Transportation of Agricultural Products from Arsenic-Endemic Areas. *Environ. Sci. Pollut. Res.* **2019,** *26* (22), 22929–22944.

Chakraborti, D.; Das, B.; Rahman, M. M.; Chowdhury, U. K.; Biswas, B.; Goswami, A. B.; Nayak, B.; Pal, A.; Sengupta, M. K.; Ahamed, S.; Hossain, A. Status of Groundwater Arsenic Contamination in the State of West Bengal, India: A 20-Year Study Report. *Mol. Nutr. Food Res.* **2009,** *53* (5), 542–551.

Chakraborti, D.; Rahman, M. M.; Mitra, S.; Chatterjee, A.; Das, D.; Das, B.; Nayak, B.; Pal, A.; Chowdhury, U. K.; Roychowdhury, T.; Ahmed, S.; Biswas, B. K.; Sengupta, M.; Lodh, D.; Das, A.; Chakraborty, S.; Chakraborty, R.; Dutta, R. N.; Saha, K. C.; Mukherjee, S. C.; Pati, S.; Kar, P. B. Groundwater Arsenic Contamination in India: A Review of Its Magnitude, Health, Social, Socioeconomic Effects and Approaches for Arsenic Mitigation. *J. Ind. Soc. Agric.* **2013**.

Chakraborti, D.; Rahman, M. M.; Chatterjee, A.; Das, D.; Das, B.; Nayak, B.; Pal, A.; Ahmed, S.; Biswas, B. K.; Sengupta, M. K.; Lodh, D.; Samanta, G.; Chakraborty, S.; Roy, M. M.; Dutta, R. N.; Saha, K. C.; Mukherjee, S. C.; Pati, S.; Kar, P. B. Fate of Over 480 Million Inhabitants Living in Arsenic and Fluoride Endemic Indian Districts: Magnitude, Health, Socio-Economic Effects and Mitigation Approaches. *J. Trace Elem. Med. Biol.* **2016,** *38*, 33–45.

Chowdhury, N. R.; Das, R.; Joardar, M.; Ghosh, S.; Bhowmick, S.; Roychowdhury, T. Arsenic Accumulation in Paddy Plants at Different Phases of Pre-Monsoon Cultivation. *Chemosphere* **2018a**, *210*, 987–997.

Chowdhury, N. R.; Ghosh, S.; Joardar, M.; Kar, D.; Roychowdhury, T. Impact of Arsenic Contaminated Groundwater Used During Domestic Scale Post Harvesting of Paddy Crop in West Bengal: Arsenic Partitioning in Raw and Parboiled Whole Grain. *Chemosphere* **2018b**, *211*, 173–184.

Chowdhury, N. R.; Joardar, M.; Ghosh, S.; Bhowmick, S.; Roychowdhury, T. Variation of Arsenic Accumulation in Paddy Plant with Special Reference to Rice Grain and Its Additional Entry During Post-Harvesting Technology. In *Ground Water Development-Issues and Sustainable Solutions*; Springer: Singapore, 2019; pp 289–303.

Chowdhury, N. R.; Das, A.; Mukherjee, M.; Swain, S.; Joardar, M.; De, A.; Mridha, D.; Roychowdhury, T. Monsoonal Paddy Cultivation with Phase-Wise Arsenic Distribution in Exposed and Control Sites of West Bengal, Alongside Its Assimilation in Rice Grain. *J. Hazard. Mater.* **2020a**, *400*, 123206.

Chowdhury, N. R.; Das, A.; Joardar, M.; De, A.; Mridha, D.; Das, R.; Rahman, M. M.; Roychowdhury, T. Flow of Arsenic Between Rice Grain and Water: Its Interaction,

Accumulation and Distribution in Different Fractions of Cooked Rice. *Sci. Total Environ.* **2020b**, *731*, 138937.

Chowdhury, U. K.; Biswas, B. K.; Chowdhury, T. R.; Samanta, G.; Mandal, B. K.; Basu, G. C.; Chanda, C. R.; Lodh, D.; Saha, K. C.; Mukherjee, S. K.; Roy, S.; Kabir, S.; Quamruzzaman, Q.; Chakraborti, D. Groundwater Arsenic Contamination in Bangladesh and West Bengal, India. *Environ. Health Perspect.* **2000,** *108* (5), 393–397.

Chowdhury, U. K.; Rahman, M. M.; Mandal, B. K.; Paul, K.; Lodh, D.; Biswas, B. K.; Basu, G. K.; Chanda, C. R.; Saha, K. C.; Mukherjee, S. C.; Roy, S. Groundwater Arsenic Contamination and Human Suffering in West Bengal, India and Bangladesh. *Environ. Sci.* **2001,** *8* (5), 393–415.

Codex Alimentarius Commission, Report of the Eighth Session of the Codex Committee on Contaminants in Food; The Hague, 31 Marche4 April 2014. Rep14/CF. Rome, Italy: *Food and Agriculture Organization of the United Nations; Geneva, Switzerland: World Health Organization*, **2014,** The Netherlands. http://bit.ly/1vprMSO. (Accessed 19 Dec 2014).

Consumers Union, **2014.** CU Letter to the FDA Regarding Arsenic in Rice [website]. Consumers Union, Yonkers, NY (20 September 2012). http://consumersunion.org/news/cu-letter-to-the-fda-regarding-arsenic-in-rice/. (Accessed 19 Dec 2014).

Das, D.; Chatterjee, A.; Mandal, B. K.; Samanta, G.; Chakraborti, D.; Chanda, B. Arsenic in Ground Water in Six Districts of West Bengal, India: The Biggest Arsenic Calamity in the World. Part 2. Arsenic Concentration in Drinking Water, Hair, Nails, Urine, Skin-Scale and Liver Tissue (Biopsy) of the Affected People. *Analyst* **1995,** *120* (3), 917–924.

Das, D.; Samanta, G.; Mandal, B. K.; Chowdhury, T. R.; Chanda, C. R.; Chowdhury, P. P.; Basu, G. K.; Chakraborti, D. Arsenic in groundwater in Six Districts of West Bengal, India. *Environ. Geochem. Health.* **1996,** *18* (1), 5–15.

Das, H. K.; Mitra, A. K.; Sengupta, P. K.; Hossain, A.; Islam, F.; Rabbani, G. H. Arsenic Concentrations in Rice, Vegetables, and Fish in Bangladesh: A Preliminary Study. *Environ. Int.* **2004,** *30* (3), 383–387.

Das, B.; Rahman, M. M.; Nayak, B.; Pal, A.; Chowdhury, U. K.; Mukherjee, S. C.; Saha, K. C.; Pati, S.; Quamruzzaman, Q.; Chakraborti, D. Groundwater Arsenic Contamination, Its Health Effects and Approach for Mitigation in West Bengal, India and Bangladesh. *Water Qual. Expo. Health* **2009,** *1* (1), 5–21.

Das, A.; Das, B.; Mukherjee, S. C.; Pati, S.; Dutta, R. N.; Saha, K. C.; Quamruzzaman, Q.; Rahman, M. M.; Chowdhury, T. R.; Chakraborti, D. Groundwater: Arsenic Contamination. *Encyclopedia Environ. Manage.* **2013,** 1262–1280.

Das, A.; Das, S. S.; Chowdhury, N. R.; Joardar, M.; Ghosh, B.; Roychowdhury, T. Quality and Health Risk Evaluation for Groundwater in Nadia District, West Bengal: An approach on Its Suitability for Drinking and Domestic Purpose. *Groundw. Sustain. Dev.* **2020,** *10*, 100351.

EFSA (European Food Safety Authority), Dietary Exposure to Inorganic Arsenic in the European Population. European Food Safety Authority (EFSA), Parma, Italy. *EFSA J.* **2014,** *12* (3), 3597.

Halder, D.; Bhowmick, S.; Biswas, A.; Mandal, U.; Nriagu, J.; Guha Mazumdar, D. N.; Chatterjee, D.; Bhattacharya, P. Consumption of Brown Rice: A Potential Pathway for Arsenic Exposure in Rural Bengal. *Environ. Sci. Technol.* **2012,** *46* (7), 4142–4148.

IARC. A Review of Human Carcinogens: Arsenic, Metals, Fibres, and Dusts. IARC Monographs on the Evaluation of Carcinogenic Risks to Humans International Agency for Research on Cancer, World Health Organization, Lyon, France, 2012.

JECFA, **2011,** Safety Evaluation of Certain Contaminants in Food. Seventy-Second Meeting of the Joint FAO/WHO Expert Committee on Food Additives. WHO Food Additives Series: 63; FAO JECFA Monographs: 8. FAO, Rome.

Joardar, M.; Das, A.; Mridha, D.; De, A.; Chowdhury, N. R.; Roychowdhury, T. Evaluation of Acute and Chronic Arsenic Exposure on School Children from Exposed and Apparently Control Areas of West Bengal, India. *Expos. Health.* **2021,** *13,* 33–50.

Kumarathilaka, P.; Seneweera, S.; Ok, Y. S.; Meharg, A.; Bundschuh, J. Arsenic in Cooked Rice Foods: Assessing Health Risks and Mitigation Options. *Environ. Int.* **2019,** *127,* 584–591.

Laparra, J. M.; Vélez, D.; Barberá, R.; Farré, R.; Montoro, R. Bioavailability of Inorganic Arsenic in Cooked Rice: Practical Aspects for Human Health Risk Assessments. *J. Agric. Food Chem.* **2005,** *53* (22), 8829–8833.

Mazumder, D. N. Chronic Arsenic Toxicity: Clinical Features, Epidemiology, and Treatment: Experience in West Bengal. *J. Environ. Sci. Health. Part A* **2003,** *38* (1), 141–163.

Halder, D.; Bhowmick, S.; Biswas, A.; Chatterjee, D.; Nriagu, J.; Guha Mazumder, D. N.; Slejkovech, Z.; Jacks, G.; Bhattacharya, P. Risk of Arsenic Exposure from Drinking Water and Dietary Components: Implications for Risk Management in rural Bengal. *Environ. Sci. Technol.* **2013,** *47* (2), 1120–1127.

Halder, D.; Biswas, A.; Šlejkovec, Z.; Chatterjee, D.; Nriagu, J.; Jacks, G.; Bhattacharya, P. Arsenic Species in Raw and Cooked Rice: Implications for Human Health in Rural Bengal. *Sci. Total Environ.* **2014,** *497*, 200–208.

Islam, M. S.; Ahmed, M. K.; Habibullah-Al-Mamun, M.; Islam, K. N.; Ibrahim, M.; Masunaga, S. Arsenic and Lead in Foods: A Potential Threat to Human Health in Bangladesh. *Food Addit. Contam. Part A*, **2014,** *31* (12), 1982–1992.

Islam, S.; Rahman, M. M.; Rahman, M. A.; Naidu, R. Inorganic Arsenic in Rice and Rice-Based Diets: Health Risk Assessment. *Food Control* **2017,** *82*, 196–202.

Jomova, K.; Jenisova, Z.; Feszterova, M.; Baros, S.; Liska, J.; Hudecova, D.; Rhodes, C. J. & Valko, M. Arsenic: Toxicity, Oxidative Stress and Human Disease. *J. Appl. Toxicol.* **2011,** *31* (2), 95–107.

Khan, M. M. H.; Sakauchi, F.; Sonoda, T.; Washio, M.; Mori, M. Magnitude of Arsenic Toxicity in Tube-Well Drinking Water in Bangladesh and Its Adverse Effects on Human Health Including Cancer: Evidence from a Review of the Literature. *Asian Pac. J. Cancer Prev.* **2003,** *4* (1), 7–14.

Majumdar, K. K.; Mazumder, D. G.; Ghose, N.; Ghose, A.; Lahiri, S. Systemic Manifestations in Chronic Arsenic Toxicity in Absence of Skin Lesions in West Bengal. *Indian J. Med. Res.* **2009,** *129* (1), 75.

Mandal, B. K. Status of Arsenic Problem in Two Blocks Out of Sixty in Eight Groundwater Arsenic-Affected Districts of West Bengal, India. Ph.D Dissertation, Jadavpur University, India, 1998.

Mandal, J.; Golui, D.; Datta S. P. Assessing Equilibria of Organo-Arsenic Complexes and Predicting Uptake of Arsenic by Wheat Grain from Organic Matter Amended Soils. *Chemosphere* **2019a,** *234*, 419–426.

Mandal, J.; Golui, D.; Raj, A.; Ganguly, P. Risk Assessment of Arsenic in Wheat and Maize Grown in Organic Matter Amended Soils of Indo-Gangetic Plain of Bihar, India. *Soil Sediment Contam.* **2019b**, *28* (8), 757–772.

Mandal, U.; Singh, P.; Kundu, A. K.; Chatterjee, D.; Nriagu, J.; Bhowmick, S. Arsenic Retention in Cooked Rice: Effects of Rice Type, Cooking Water, and Indigenous Cooking Methods in West Bengal, India. *Sci. Total Environ.* **2019c,** *648*, 720–727.

Mazumder, D. G.; Chakraborty, A. K.; Ghose, A.; Gupta, J. D.; Chakraborty, D. P.; Dey, S. B.; Chattopadhyay, N. Chronic Arsenic Toxicity from Drinking Tubewell Water in Rural West Bengal. *Bull. World Health Organ.* **1988,** *66* (4), 499.

Mazumder, D. N. G.; Haque, R.; Ghosh, N.; De, B. K.; Santra, A.; Chakraborty, D.; Smith, A. H. Arsenic Levels in Drinking Water and the Prevalence of Skin Lesions in West Bengal, India. *Int. J. Epidemiol.* **1998,** *27* (5), 871–877.

Mazumder, D. N. G.; Ghosh, A.; Majumdar, K. K.; Ghosh, N.; Saha, C.; Mazumder, R. N. G. Arsenic Contamination of Ground Water and Its Health Impact on Population of District of Nadia, West Bengal, India. *Indian J. Community Med.* **2010,** *35* (2), 331.

Mazumder, D. G.; Dasgupta, U. B. Chronic Arsenic Toxicity: Studies in West Bengal, India. *Kaohsiung J. Med. Sci.* **2011,** *27* (9), 360–370.

Meharg, A. A.; Adomaco, E.; Lawgali, Y.; Deacon, C.; Williams, P. Food Standards Agency Contract C101045: Levels of Arsenic in Rice: Literature Review, 2006.

Meharg, A. A.; Sun, G.; Williams, P. N.; Adomako, E.; Deacon, C.; Zhu, Y. G.; Raab, A. Inorganic Arsenic Levels in Baby Rice Are of Concern. *Environ. Pollut.* **2008,** *152* (3), 746–749.

Mondal, D.; Polya, D. A. Rice Is a Major Exposure Route for Arsenic in Chakdaha Block, Nadia District, West Bengal, India: A Probabilistic Risk Assessment. *J. Appl. Geochem.* **2008,** *23* (11), 2987–2998.

Mondal, D.; Banerjee, M.; Kundu, M.; Banerjee, N.; Bhattacharya, U.; Giri, A. K.; Ganguli, B.; Roy, S. S.; Polya, D. A. Comparison of Drinking Water, Raw Rice and Cooking of Rice as Arsenic Exposure Routes in Three Contrasting Areas of West Bengal, India. *Environ. Geochem. Health.* **2010,** *32* (6), 463–477.

Mukherjee, S. C.; Rahman, M. M.; Chowdhury, U. K.; Sengupta, M. K.; Lodh, D.; Chanda, C. R.; Saha, K. C.; Chakraborti, D. Neuropathy in Arsenic Toxicity from Groundwater Arsenic Contamination in West Bengal, India. *J. Environ. Sci. Health A.* **2003,** *38* (1), 165–183.

Mukherjee, S. C.; Saha, K. C.; Pati, S.; Dutta, R. N.; Rahman, M. M.; Sengupta, M. K.; Ahamed, S.; Lodh, D.; Das, B.; Hossain, M. A.; Nayak, B.; Mukherjee, A.; Chakraborti, D.; Dutta, S. K.; Palit, S. K.; Kaies, I.; Barua, A. K.; Asad, K. A. Murshidabad—One of the Nine Groundwater Arsenic-Affected Districts of West Bengal, India. Part II: Dermatological, Neurological, and Obstetric Findings. *Clin. Toxicol.* **2005,** *43* (7), 835–848.

Nickson, R. T.; McArthur, J. M.; Ravenscroft, P.; Burgess, W. G.; Ahmed, K. M. Mechanism of Arsenic Release to Groundwater, Bangladesh and West Bengal. *J. Appl. Geochem.* **2000,** *15* (4), 403–413.

Norton, G. J.; Islam, M. R.; Deacon, C. M.; Zhao, F. J.; Stroud, J. L.; McGrath, S. P.; Isalm, S.; Jahiruddin, M.; Feldmann, J.; Price, A. H.; Meharg, A. A. Identification of Low Inorganic and Total Grain Arsenic Rice Cultivars from Bangladesh. *Environ. Sci. Technol.* **2009,** *43* (15), 6070–6075.

Pal, A.; Chowdhury, U. K.; Mondal, D.; Das, B.; Nayak, B.; Ghosh, A.; Maity, S.; Chakraborti, D. Arsenic Burden from Cooked Rice in the Populations of Arsenic Affected and Nonaffected Areas and Kolkata City in West-Bengal, India. *Environ. Sci. Technol.* **2009,** *43* (9), 3349–3355.

Rahman, M. M.; Chowdhury, U. K.; Mukherjee, S. C.; Mondal, B. K.; Paul, K.; Lodh, D.; Biswas, B. K.; Chanda, C. R.; Basu, G. K.; Saha, K. C.; Roy, S.; Das, R.; Palit, S. K.;

Quamruzzaman, Q.; Chakraborti, D. Chronic Arsenic Toxicity in Bangladesh and West Bengal, India—A Review and Commentary. *J. Toxicol. Clin. Toxicol.* **2001,** *39* (7), 683–700.

Rahman, M. M.; Mandal, B. K.; Chowdhury, T. R.; Sengupta, M. K.; Chowdhury, U. K.; Lodh, D.; Chanda, C. R.; Basu, G. K.; Mukherjee, S. C.; Saha, K. C.; Chakraborti, D. Arsenic Groundwater Contamination and Sufferings of People in North 24-Parganas, One of the Nine Arsenic Affected Districts of West Bengal, India. *J. Environ. Sci. Health A.* **2003,** *38* (1), 25–59.

Rahman, M. M.; Sengupta, M. K.; Ahamed, S.; Chowdhury, U. K.; Hossain, M. A.; Das, B.; Lodh, D.; Saha, K. C.; Pati, S.; Kaies, I.; Barua, A. K.; Chakraborti, D. The Magnitude of Arsenic Contamination in Groundwater and Its Health Effects to the Inhabitants of the Jalangi—One of the 85 Arsenic-Affected Blocks in West Bengal, India. *Sci. Total Environ.* **2005,** *338* (3), 189–200.

Rahman, M. A.; Hasegawa, H.; Rahman, M. A.; Rahman, M. M.; Miah, M. M. Influence of Cooking Method on Arsenic Retention in Cooked Rice Related to Dietary Exposure. *Sci. Total Environ.* **2006,** *370* (1), 51–60.

Rahman, M. A.; Hasegawa, H. High Levels of Inorganic Arsenic in Rice in Areas Where Arsenic-Contaminated Water Is Used for Irrigation and Cooking. *Sci. Total Environ.* **2011,** *409* (22), 4645–4655.

Rahaman, S.; Sinha, A. C.; Pati, R.; Mukhopadhyay, D. Arsenic Contamination: A Potential Hazard to the Affected Areas of West Bengal, India. *Environ. Geochem. Health.* **2013,** *35* (1), 119–132.

Rahman, M. M.; Mondal, D.; Das, B.; Sengupta, M. K.; Ahamed, S.; Hossain, M. A.; Samal, A. C.; Saha, K. C.; Mukherjee, S. C.; Dutta, R. N.; Chakraborti, D. Status of Groundwater Arsenic Contamination in all 17 Blocks of Nadia District in the State of West Bengal, India: A 23-Year Study Report. *J. Hydrol.* **2014,** *518*, 363–372.

Raqib, R.; Ahmed, S.; Sultana, R.; Wagatsuma, Y.; Mondal, D.; Hoque, A. W.; Nermell, B.; Yunus, M.; Roy, S.; Persson, L. A.; Arifeen, S. E.; Moore, S.; Vahter, M. Effects of in Utero Arsenic Exposure on Child Immunity and Morbidity in Rural Bangladesh. *Toxicol. Lett.* **2009,** *185* (3), 197–202.

Roychowdhury, T.; Uchino, T.; Tokunaga, H.; Ando, M. Arsenic and Other Heavy Metals in Soils from an Arsenic-Affected Area of West Bengal, India. *Chemosphere* **2002a**, *49* (6), 605–618.

Roychowdhury, T.; Uchino, T.; Tokunaga, H.; Ando, M. Survey of Arsenic in Food Composites from an Arsenic-Affected Area of West Bengal, India. *Food Chem. Toxicol.* **2002b**, *40* (11), 1611–1621.

Roychowdhury, T.; Tokunaga, H.; Ando, M. Survey of Arsenic and Other Heavy Metals in Food Composites and Drinking Water and Estimation of Dietary Intake by the Villagers from an Arsenic-Affected Area of West Bengal, India. *Sci. Total Environ.* **2003,** *308* (1–3), 15–35.

Roychowdhury, T.; Tokunaga, H.; Uchino, T.; Ando, M. Effect of Arsenic-Contaminated Irrigation Water on Agricultural Land Soil and Plants in West Bengal, India. *Chemosphere* **2005,** *58* (6), 799–810.

Roychowdhury, T. Impact of Sedimentary Arsenic Through Irrigated Groundwater on Soil, Plant, Crops and Human Continuum from Bengal Delta: Special Reference to Raw and Cooked Rice. *Food Chem. Toxicol.* **2008,** *46* (8), 2856–2864.

Roychowdhury, T.; Uchino, T.; Tokunaga, H. Effect of Arsenic on Soil, Plant and Foodstuffs by Using Irrigated Groundwater and Pond Water from Nadia District, West Bengal. *Int. J. Environ. Pollut.* **2008,** *33* (2–3), 218–234.

Roychowdhury, T. Groundwater Arsenic Contamination in One of the 107 Arsenic-Affected Blocks in West Bengal, India: Status, Distribution, Health Effects and Factors Responsible for Arsenic Poisoning. *Int. J. Hyg. Environ. Health.* **2010,** *213* (6), 414–427.

Samal, A. C.; Kar, S.; Bhattacharya, P.; Santra, S. C. Human Exposure to Arsenic Through Foodstuffs Cultivated Using Arsenic Contaminated Groundwater in Areas of West Bengal, India. *J. Environ. Sci. Health A.* **2011,** *46* (11), 1259–1265.

Santra, S. C.; Samal, A. C.; Bhattacharya, P.; Banerjee, S.; Biswas, A.; Majumdar, J. Arsenic in Food Chain and Community Health Risk: A Study in Gangetic West Bengal. *Procedia Environ. Sci.* **2013,** *18*, 2–13.

Sanyal, S. K.; Nasar, S. K. T. Arsenic Contamination of Groundwater in West Bengal (India): Build-Up in Soil-Crop Systems. In *International Conference on Water Related Disasters*; Kolkata, **2002,** 5th–6th Dec.

Sengupta, M. K.; Hossain, M. A.; Mukherjee, A.; Ahamed, S.; Das, B.; Nayak, B.; Pal, A.; Chakraborti, D. Arsenic Burden of Cooked Rice: Traditional and Modern Methods. *Food Chem. Toxicol.* **2006,** *44* (11), 1823–1829.

Sengupta, M. K.; Hossain, A.; Ahamed, S.; Das, B.; Nayak, B.; Pal, A.; Mukherjee, A.; Rahman, M. M.; Chowdhury, U. K.; Biswas, B. K.; Roychowdhury, T.; Mondal, B. K.; Samanta, G.; Chatterjee, A.; Das, D.; Lodh, D.; Chakraborti, D. Groundwater Arsenic Contamination Situation in West-Bengal, India: A Nineteen Year Study. *Bhujal News* **2009,** *24* (2 and 3).

Signes, A.; Mitra, K.; Burlo, F.; Carbonell-Barrachina, A. A. Contribution of Water and Cooked Rice to an Estimation of the Dietary Intake of Inorganic Arsenic in a Rural Village of West Bengal, India. *Food Addit. Contam.* **2008,** *25* (1), 41–50.

Signes-Pastor, A. J.; Mitra, K.; Sarkhel, S.; Hobbes, M.; Burló, F.; De Groot, W. T.; Carbonell-Barrachina, A. A. Arsenic Speciation in Food and Estimation of the Dietary Intake of Inorganic Arsenic in a Rural Village of West Bengal, India. *J. Agric. Food Chem.* **2008,** *56* (20), 9469–9474.

Signes Pastor, A. J.; Al Rmalli, S. W.; Jenkins, R. O.; Carbonell Barrachina, Á. A.; Haris, P. I. Arsenic Bioaccessibility in Cooked Rice as Affected by Arsenic in Cooking Water. *J. Food Sci.* **2012,** *77* (11), T201–T206.

Smith, N. M.; Lee, R.; Heitkemper, D. T.; Cafferky, K. D.; Haque, A.; Henderson, A. K. Inorganic Arsenic in Cooked Rice and Vegetables from Bangladeshi Households. *Sci. Total Environ.* **2006,** *370* (2–3), 294–301.

Tuli, R.; Chakrabarty, D.; Trivedi, P. K.; Tripathi, R. D. Recent Advances in Arsenic Accumulation and Metabolism in Rice. *Mol. Breed.* **2010,** *26* (2), 307–323.

Uchino, T.; Roychowdhury, T.; Ando, M.; Tokunaga, H. Intake of Arsenic from Water, Food Composites and Excretion Through Urine, Hair from a Studied Population in West Bengal, India. *Food Chem. Toxicol.* **2006,** *44* (4), 455–461.

Upadhyay, M. K.; Shukla, A.; Yadav, P.; Srivastava, S. A Review of Arsenic in Crops, Vegetables, Animals and Food Products. *Food Chem.* **2019,** *276*, 608–618.

Watanabe, C.; Matsui, T.; Inaoka, T.; Kadono, T.; Miyazaki, K.; Bae, M. J.; Ono, T.; Ohtsuka, R.; Mozammel Haque Bokul, A. T. M. Dermatological and Nutritional/Growth Effects Among Children Living in Arsenic-Contaminated Communities in Rural Bangladesh. *J. Environ. Sci. Health A.* **2007,** *42* (12), 1835–1841

WHO. Evaluation of Certain Contaminants in Food. Seventy Second Report of the Joint FOA/WHO Expert Committee on Food Additives, WHO Technical Report Series No. 959. World Health Organization (WHO): Geneva, 2011.

Williams, P. N.; Price, A. H.; Raab, A.; Hossain, S. A.; Feldmann, J.; Meharg, A. A. Variation in Arsenic Speciation and Concentration in Paddy Rice Related to Dietary Exposure. *Environ. Sci. Technol.* **2005,** *39* (15), 5531–5540.

Williams, P. N.; Villada, A.; Deacon, C.; Raab, A.; Figuerola, J.; Green, A. J.; Feldmann, J.; Meharg, A. A. Greatly Enhanced Arsenic Shoot Assimilation in Rice Leads to Elevated Grain Levels Compared to Wheat and Barley. *Environ. Sci. Technol.* **2007,** *41* (19), 6854–6859.

CHAPTER 3

Lead, Cadmium, and Chromium Contamination in Soil–Plant System: Global Situation, Impact, and Dietary Risk Delineation

PARIJAT BHATTACHARYA[1], PURABI BANERJEE[2], SOHAM HAZRA[3,5], SUDIP SENGUPTA[1,6], and SAPTAPARNEE DEY[4]

[1]*Department of Agricultural Chemistry and Soil Science, Faculty of Agriculture, Bidhan Chandra Krishi Viswavidyalaya, Mohanpur, Nadia, West Bengal, India*

[2]*Department of Agronomy, Faculty of Agriculture, Bidhan Chandra Krishi Viswavidyalaya, Mohanpur, Nadia, West Bengal, India*

[3]*Department of Genetics and Plant Breeding, Faculty of Agriculture, Bidhan Chandra Krishi Viswavidyalaya, Mohanpur, Nadia, West Bengal, India*

[4]*Division of Soil Science and Agricultural Chemistry, ICAR-Indian Agricultural Research Institute, New Delhi, India*

[5]*Department of Agriculture, Brainware University, Barasat, Kolkata, West Bengal, India*

[6]*Department of Agriculture, School of Agriculture, Swami Vivekananda University, Barrackpore, Kolkata, West Bengal, India*

Environmental Contaminants: Impact, Assessment, and Remediation.
Pritam Ganguly, Jajati Mandal, M. Paramsivam, & Sandip Patra, (Eds.)

ABSTRACT

Heavy metal contamination has emerged as a prime global problem having adverse consequences on the aquatic, terrestrial, and atmospheric environment. These metals may be natural or anthropogenic in origin and do not degrade easily. Anthropogenic activities tend to increase such concentrations by several times through wastewater irrigation and industrial effluents dumped on soil or discharged in water. The concentration of lead, cadmium, and chromium is significantly higher thereby necessitating further emphasizing. Crops tend to accumulate these metals in high amount which severely impacts their growth and results in accumulation in edibles and the unwanted transfer of these heavy metals in our food chain and food web. Several acute and chronic diseases occur in humans due to exposure to these heavy metals through ingestion, inhalation, or dermal contact. Being potentially carcinogenic, cancer occurrences of severe magnitude are also not uncommon through exposure. A significant emphasis is now provided to advances in instrumental elemental characterization with the involvement of atomic absorption spectrometry (AAS), microwave plasma-atomic emission spectrometry (MP-AES), inductively coupled plasma-optical emission spectrometry (ICP-OES), inductively coupled plasma-mass spectrometry (ICP-MS), and so on for precise detection of the load of the contaminant. Taking into consideration all these aspects, the current chapter tries to undermine the global contamination status of lead, cadmium, and chromium, their sources, dynamics in soil, plant uptake and associated dietary risk, and the advanced protocols for precise and easy determination of the contaminants in different spheres of the environment.

3.1 INTRODUCTION

The global population has been subjected to severe environment-related problems in the form of heavy metal pollution (Hembrom et al., 2020). This has become a serious threat to societal well-being and designated as a frontline problem in the aftermath of rapid industrialization and urbanization (Yang and Sun, 2009). Economic prosperity in terms of increased industries has catapulted the problem by several orders and the primary effect is envisaged in the environment (Raju et al., 2013; Zojaji et al., 2014). Heavy metals have derived their name from the properties, such as higher density, atomic weights, or atomic numbers. Based on their occurrence in the periodic table, they are often termed as transition/post-transition metals (Bhattacharyya et al., 2021).

The common toxic metals having serious health consequences for human beings include cadmium, mercury, arsenic, lead, and chromium. These metals having their colorless and odorless properties can be easily absorbed in the body through ingestion, inhalation, or dermal means. Upon entry at a more than tolerable level, they are ionized, which is bioavailable, active, and severely detrimental to ecological systems. Even lower concentrations can result in acute and chronic diseases and these incidences are on the rise over the years. Heavy metals may be released into the atmosphere by various means, through gases, dust, and smoke from various industries, energy production, metallurgy, transportation, and even construction materials. The soil may be a source of heterogeneous contamination of various heavy metals of varying intensity of lethality, period of degradation, and biological half-life (Abii and Okorie, 2011). Apart from the natural sources, human activity can also build up such contaminants in the environment. Anthropogenic sources of heavy metals may arise from pesticide and fertilizer application in the crop field, untreated effluent discharge from industrial activities, mining activities, and sewage irrigation (Zhang et al., 2011).

Lead is most commonly accumulated as a result of urban industrial effluents. It may be transported from solid dumping grounds to the surrounding areas or even through improper disposal of industrial wastewater effluents. This problem is more severe in areas where there is a dearth of freshwater as this effluent-laden polluted water is not only used for irrigating the crops in the field but also for domestic purposes including drinking (Singh et al., 2010). Wastewater irrigation has been reported to be associated with high accumulation of heavy metals in edible parts of the crop especially cereals and vegetables (Marshall et al., 2007). WHO and FAO (2007) have reported elevated levels of heavy metals in milk and drinking water as well, possibly suggesting bioaccumulation in the food web and ramifications in the food chain. The heavy metals may also get accumulated in the aquatic ecosystem and thereby have tremendous implications on the fish and other aquatic organisms. Some of the heavy metals often are found to be beneficial at lower concentrations to support vital physiological function but may inhibit enzymatic activities and poses toxic effects at higher concentrations (Koropatnick and Leibbrandt, 1995). Among the heavy metals, cadmium, chromium, and lead have easy access to ramify in the food chain and mostly affect children through accumulation in tissues that retard their normal growth and adversely affect the kidneys, cardiovascular system, and auditory system (Rahimi, 2013).

Taking into consideration of all such detrimental consequences of cadmium, chromium, and lead in the ecological niche, the current chapter aims to give a robust outlook of the extent of contamination by these heavy metals across the globe, their source, impact on plants and animal life, dynamics in soil and improved assessment of its pathway through plant and food chain, and through improved instrumental protocols. Early and precise detection can always be the cue for management decision-making.

3.2 GLOBAL SCENARIO OF LEAD, CADMIUM, AND CHROMIUM CONTAMINATION

Generally, heavy metals occur in soil environments naturally by undergoing pedogenetic processes, that is, weathering of parent materials at tracer levels (normally <1000 mg kg^{-1}) and pose to be rarely toxic (Kabata-Pendias and Pendias, 2001). In spite of this, most of the soils of both rural and urban areas may accumulate any one or more of those heavy metals high enough to cause risks to human health, plants, animals, or ecosystems owing to continuous interference of humankind which gradually leads to acceleration of the slow-going natural geochemical cycles of metals (D'Amore et al., 2005).

On average, the concentrations of Pb, Cd, and Cr in the earth's crust are 20 mg kg^{-1} (Hu and Gao, 2008), 0.1–0.5 mg kg^{-1} (http://cadmium.org), and 100 mg kg^{-1} (http://wikipedia.org) respectively. To date, so many countries have been exposed to the deposition of these three heavy metals in the soil which are listed in Table 3.1.

TABLE 3.1 Pb, Cd, and Cr Affected Countries.

Heavy metals	Affected country	References
Pb	Mexico, China, France, Australia, United States, Iran, Nigeria, Kenya, Senegal	Obeng-Gyasi (2019)
Cd	Australia, Britain, Canada, Denmark, Finland, Germany, New Zealand, Norway, Sweden, United States, Poland, Austria, Denmark, Greece, Ireland, United Kingdom	Kubier et al. (2019)
Cr	China, India, Russia, Australia, Finland, France, Italy, Norway, Sweden, Taiwan, Canada, Brazil, USA, Greece	Chrysochoou et al. (2016)

3.3 SOURCE OF LEAD, CADMIUM, AND CHROMIUM IN SOIL

There may be different sources of lead, cadmium, and chromium in soil that may be summed up as follows (Table 3.2).

TABLE 3.2 Sources of Pb, Cd, and Cr Accumulations in Soil.

Heavy metals	Sources
Pb	Volcanic eruptions, burning of gasoline containing lead, lead battery manufacturing, automobile exhaust, mining and smelting of metallic ores, municipal sewage, industrial wastes, paints
Cd	Different geogenic sources, metal smelting, refining, burning of fossil fuel, phosphatic fertilizers from agricultural fields, sewage sludge
Cr	Electroplating-related industries, wastes from tanneries, sewage sludge, industrial solid wastes rich in Cr

Source: Adapted from Lone et al. (2008).

3.3.1 NATURAL SOURCE

Numerous research efforts have indicated that natural sources of heavy metals in the environment are usually of little importance in comparison with anthropogenic activities (Dixit et al., 2015). However, the primary source of heavy metals in soils is the indigenous parent material from which they originated. In general, basaltic ingenious rocks are rich in heavy metals with special reference to Cd, whereas shales are consisting of higher amounts of Pb and Cd (Li et al., 2019). In connection, Właśniewski and Hajduk (2012) pointed out that higher cadmium content is primarily associated with the granulometric composition of the local soil. Volcanic eruptions along with windblown dust particles and harmful and toxic gases are the high-level emitters of Pb (Nagajyoti et al., 2010). Available literature also established that the concentrations of Pb and Cd are much higher in wetlands in a recently formed region compared with wet ditches and reclaimed wetlands because of their richness in mangrove soils (Bai et al., 2011).

3.3.2 ANTHROPOGENIC SOURCE

The worldwide recent advancements in the agricultural sector, urbanization, and industrialization have substantially accelerated heavy metal contamination in soil (He et al., 2015). In accordance with the major sources of Pb, Cd, and Cr, multivariate analysis reveals that these three metals are of anthropogenic origin (Zwolak et al., 2019). A soil may be predominantly contaminated by varying concentrations of these heavy metal accumulations through different sources of emissions from growing industrial areas, lead-based gasoline and paints, transferred from mines, excess use of animal manures, fertilizers, pesticides, and wastewater irrigation in agriculture, improper disposal of sewage sludge, spillage and burning of petrochemicals, accumulation of coal combustion residues, and atmospheric deposition (Khan et al., 2008; Zhang et al., 2010; Muradoglu et al., 2015; Khan et al., 2016).

3.3.2.1 BIOSOLIDS

Excess use of several biosolids including animal dung and municipal sewage sludge as compost materials in agricultural lands inadvertently leads to the accumulation of Cd, Cr, and Pb like many other heavy metals (e.g., As, Cu, Hg, Ni, Se, Mo, Zn, Tl, and Sb) in soil (Basta et al., 2005; Sun et al., 2013). In fact, one of the principal reasons behind cadmium enrichment is the application of organic manure, sewage sludge, and limes beyond the limit (Niassy and Diarra, 2012).

3.3.2.2 FERTILIZERS

These days, larger amounts of inorganic fertilizers especially N, P, and K are frequently added to soil in the case of intensive farming systems. Wuana and Okieimen (2011) and Tóth et al. (2016) reported that the fertilizer compounds which are used to supply these nutrient elements contain tracer amounts of heavy metals including Pb, Cr, and Cd as impurities. Extensive fertilizer application triggers significant increments in concentrations of these two heavy metals in soil (Atafar et al., 2010). Especially, Cd concentration has been found to be escalated in soils with long-term higher doses of phosphatic fertilizers application (Su et al., 2014). Also, some reports reveal that Cr in larger amounts accumulates in urban soils, where the industrial wastewater runs into rivers contaminating the alongside alluvial soils (Vodyanitskii et al., 2008).

3.3.2.3 PESTICIDES

In the recent past, several common pesticides consisting of substantial concentrations of heavy metals are used in larger quantities (McLaughlin et al., 2000; Ogunlade and Agbeniyi, 2011). For instance, Lead arsenate was applied to control some parasitic insects in fruit orchards for a long time. Formulations containing Cr have been used to preserve timbers in many countries.

3.3.2.4 *METAL MINING AND INDUSTRIAL WASTES*

Extensive Pb ore mining and smelting have been reported to contaminate the soil. Different materials generated by various industries (tanning, textile, petrochemicals from oil spills, and pharmaceuticals) serve as the potential source of Cr and Pb (Sumner, 2000; Chen et al., 2015). Similarly, the chief source of Cd is found to be coal mines polluting the adjacent soils (Srivastava et al., 2017).

3.3.2.5 *AIR-BORNE SOURCES*

Metals such as Cd and Pb have been found in very higher concentrations in soils and plants adjacent to smelting works. Another major source of Pb contamination in soil is its aerial emission from the combustion of tetraethyl lead-containing petrol, which substantially contributes to enhanced Pb content in soils in urban areas those are adjacent to main roads (Nieć et al. 2013; Zhou et al. 2016). Likewise, Cd from tires and lubricant oils of road traffic may also accumulate in soils adjacent to roads (Gope et al., 2017).

3.4 IMPACT OF HEAVY METALS (LEAD, CADMIUM, AND CHROMIUM) ON PLANTS AND HUMANS

Heavy metals, such as lead, cadmium, and chromium can severely impact the normal functioning of plant and human activities.

3.4.1 *EFFECTS ON PLANT*

3.4.1.1 *LEAD*

Lead is an extremely toxic heavy metal that disturbs various plant physiological processes although has a very insignificant biological function. A plant with a high level of Pb concentration results in a surge of production of reactive oxygen species (ROS) which damages the lipid membrane and ultimately results in the damage of chlorophyll and photosynthetic processes. As a result, plant growth and development are affected to a great extent (Najeeb et al., 2014). Several scientists suggested that lead reduced the biomass and yield of tea and also reduced its quality (Yongsheng et al., 2011). Further, it also affects the ion uptake and results in a decline in metabolic function and affects plant growth (Jaishankar et al., 2014). It was apparent from different studies that with an increase in the level of Pb treatment, antioxidant enzymes, such as guaiacol peroxidase, superoxide dismutase (SOD), ascorbate peroxidase (APX), and glutathione reductase (GR) activities increased compared with control in rice plants (Bishnoi et al., 1993). Pb-mediated oxidative stress arises primarily due to an imbalance in free radical production and the antioxidant mechanism of the plants that detoxify the reactive intermediates like glutathione which protects the cell from free radicals like H_2O_2. Elevated Pb severely impairs the normal functioning of the reduced form of glutathione in order to provide its reducing equivalents ($H^+ + e^-$) from its thiol groups of cysteine to ROS. Under oxidative stress, the concentration of oxidized glutathione (GSSG) exceeds the concentration of its reduced form (GSH). Further, free radicals collect electrons from the lipid molecules in the cell membrane and result in lipid peroxidation (Flora et al., 2008). The unique ability of ionic Pb to replace bivalent cations, such as Ca^{2+}, Mg^{2+}, Fe^{2+}, and monovalent cations like Na^+, can destabilize the entire biological metabolism and affect plant growth and behavior tremendously.

3.4.1.2 CADMIUM

Cadmium is mainly found in fruits and vegetables (Satarug et al., 2011). Cadmium drastically affects enzymatic systems; also causes oxidative stress and ultimately induces several nutrient deficiencies in plants (Irfan et al., 2013). Cd affects the photosynthetic functions of higher plants (Baszynski, 1986) either directly or indirectly. Such phenomenon has been reported earlier by scientists that Cd affects the enzymes of the photosynthetic carbon reduction cycle in *Cajanus cajan* (Sheoran et al., 1990). The reports from other crops also had the same pattern as in the case of *Beta vulgaris* (Greger and Orgen, 1991), radish (Krupa et al., 1987), and Silene and Thlaspi (Ouzounidou, 1993). Cd stress reduces the total chlorophyll content in pigeon peas (Sheoran et al., 1990). From earlier studies, it was evident that the Chlorophyll a/b ratio was altered under Cd stress as it decreased in *Cajanus cajan* (Sheoran et al., 1990) while increased in *Brassica juncea* (Singh and Tewari, 2003). Cd has a serious effect on the inhibition of photosynthetic CO_2 fixation, Hill reaction activity, and electron transport of chloroplasts (Barua and Jana, 1986; Tripathi and Mohanty, 1980). Studies have also put forward that Cd inhibits net photosynthesis by increasing both stomatal and mesophyll resistance to CO_2 uptake (Abo-Kassem et al., 1995). In other cases, studies reported that heavy metal toxicity adversely affects growth, biomass, and yield in several crops through binding with cysteine-rich proteins such as metallothionein. Cd has this unique ability of impairing several functioning resulting in deficiencies of Fe and Zn by scavenging with cellular constituents of cysteine, glutamate, histidine, and aspartate ligands (Castagnetto et al., 2002).

3.4.1.3 CHROMIUM

Higher level of chromium has a negative impact on plants. Cr phyto-toxicity reduces root growth, causes leaf chlorosis, severely affects seed germination, and ultimately reduces biomass yield. Earlier studies revealed that chromium toxicity greatly affects the biological processes in various plants, such as maize, wheat, barley, cauliflower, *Citrullus*, and vegetables. It results in chlorosis and necrosis in plants (Ghani, 2011). Chromium toxicity significantly affects iron-containing enzymes, such as catalase, peroxidase, and cytochrome oxidase, and thus has natural impedance on photosynthesis, nitrate reductase activity, protein content in algae, and photosynthetic pigments (Nath et al., 2008). Several studies established that chromium toxicity adversely affects seed germination, seedling growth, senescence, pigment status, and nutrient contents of crop plants (Bishnoi et al., 1993). Chromium (III) enters the cell through a simple diffusion process and has weak membrane permeability. On the contrary, Cr(VI) is more active and thereby can easily pass through the cell membrane (Chandra and Kulshreshtha, 2004) through passages for isoelectric and isostructural anions such as SO_4^{2-} and HPO_4^{2-} channels and even through phagocytosis. Cr(VI) also results in the formation of ephemeral species of pentavalent and tetravalent chromium and is a natural mechanism of detoxification, carried out by glutathione. However, if the intracellular target site has biological reductants such as thiols and ascorbate, then there is an easy binding that results in the production of ROS, such as superoxide ion, hydrogen peroxide, and hydroxyl radical, culminating in oxidative stress resulting in damage to DNA and proteins (Stohs and Bagchi, 1995).

3.4.2 EFFECTS ON HUMAN HEALTH

3.4.2.1 LEAD

Exposure of human beings to Pb occurs predominantly through air and food and enters through ingestion in the gastrointestinal tract and also through inhalation in the respiratory tract. Children are the most vulnerable to such contamination (Jarup, 2003). The toxicity of Pb is exemplified by the displacement of certain essential elements from several metalloenzymes and the impedance to their vital physiological activities. Several studies showed that the kidney and liver are considered potential targets of Pb toxicity before storage in bones (Athar and Vohora, 1995). Pb toxicity also leads to decreased hemoglobin synthesis, impairment of neurobehavioral and psychological functions, peripheral neuropathy, indirect effect on the heart, renal tubular damage, and reproductive problems (Brown and Kodama, 1986).

3.4.2.2 CADMIUM

Food and cigarette smoking are the most important sources of Cd in addition to its exposure to water. The dietary intake of Cd ranges from 40 to 50 g per day (WHO, 1987). Cd accumulates within the kidney and liver over a long time (McLaughlin et al., 1999). Long-term low-level exposure results in cardiovascular disease and cancer. It also affects the renal tubular function of reabsorbing protein, sugar, and amino acids (McLaughlin et al., 1999). Several studies have found that Cd exposure in conjunction with Ca, Fe, Zn, protein, fat, and vitamin D deficiencies, results in osteomalacia and bone fractures in postmenopausal women. This problem was prevalent in polluted Jintsu Valley, Japan commonly referred to as Itai-Itai disease (Noda and Kitagawa, 1990). It was apparent from several studies that cadmium can affect calcium, phosphorus, and bone metabolisms in both industrial workers and people exposed to Cd in the general environment (Jarup et al., 1998).

3.4.2.3 CHROMIUM

Human intake of heavy metal chromium in adult humans usually ranges from 100 and 300 g/day (WHO, 1984). The hexavalent form of Cr mainly shows toxicity to humans. Chromium toxicity causes liver necrosis, nephrites, and gastrointestinal irritation (Athar and Vohora, 1995). Hexavalent Cr is responsible for the occurrence of cutaneous, nasal, and mucous membrane ulcers and also dermatitis by skin contact. Cr through oral exposure impairs the activity of gastric juice. Chromium has potent carcinogenic effects on human beings and other animals (O'Brien et al., 2003).

3.5 DYNAMICS OF LEAD, CADMIUM, AND CHROMIUM IN SOIL

Lead, cadmium, and chromium show huge similarities in their distribution among soil constituents. These metals are present in about seven fractions such as (a) soluble plus exchangeable, (b) organically bound, (c) carbonate-bound, (d) amorphous iron oxide bound, (e) crystalline iron oxide bound, (f) easily reducible oxide bound, and (g) residual fraction. These fractions (Fig. 3.1) directly and indirectly govern the phytoavailability (Uduma and Jimoh, 2013).

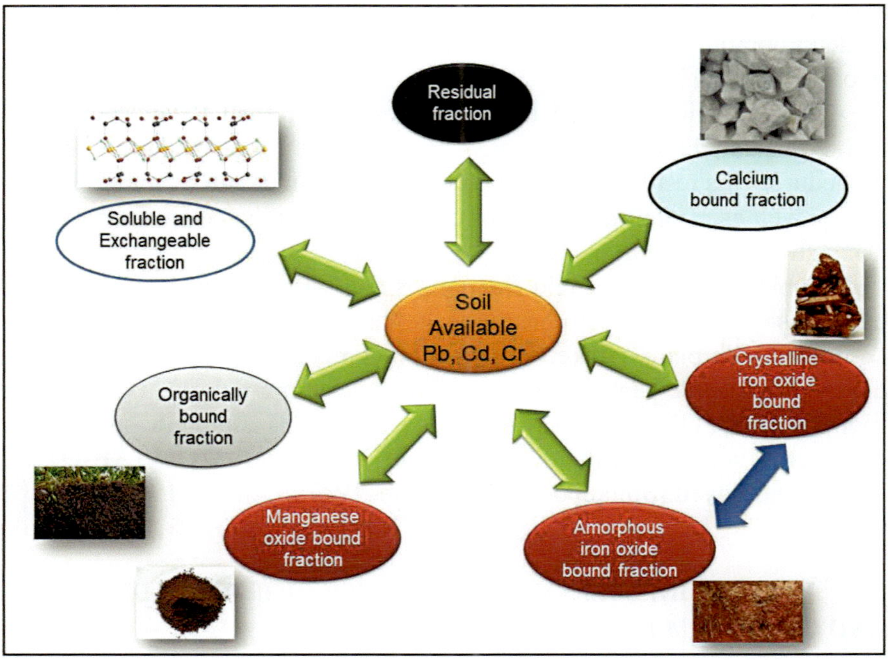

FIGURE 3.1 Dynamic equilibrium between different pools of lead, cadmium, and chromium in soil.

Pb, Cd, and Cr are widely distributed on minerals and clay surfaces which constitute the soluble and exchangeable fraction. This fraction is generally extracted using 0.01 M $CaCl_2$ or 1 M NH_4NO_3 solution (pH 7). Various forms of SOM act as an important sink for these elements. The organically bound pools are generally assessed using strong oxidizing agents, that is, hydrogen peroxide, NaOCl, $Na_4P_2O_7$, and $K_4P_2O_7$ are applied for its extraction (Tessier et al., 1979; Osakwe and Okolie, 2015).

Carbonate-bound forms constitute a larger portion of the solid-phase species and is generally extracted using NaOAc adjusted to pH 5 (Osakwe and Okolie, 2015). Calcium carbonate plays a vital role in the adsorption–desorption mechanism of these metals, more specifically for Cd (Zhao et al., 2014). Sesquioxides hold a significantly higher proportion of these trace metals *via* strong surface adsorption and co-precipitation (Frohne et al., 2014). Easily reducible fractions, that is, manganese oxide bound; moderately reducible fractions, that is, amorphous Fe oxides bound, and poorly reducible fractions (crystalline Fe oxides) constitute the oxide bound fractions (Du Laing, 2010), and these are extracted sequentially by hydroxylamine hydrochloride, oxalic acid/ammonium oxalate, and ascorbic acid–ammonium oxalate mixture (Golui et al., 2017).

After extraction of these fractions, there still remains some portion of heavy metals which are potentially inaccessible by the standard analytical procedure. This residual pool comprises of metals in complex mineral structure as well as recalcitrant fractions of SOM and can be determined by only using strong acid solutions (Iwegbue, 2013).

Soluble and exchangeable fractions along with organically bound and oxide-associated fractions constitute the "available" pool which is highly correlated with plant uptake. This fraction is most widely determined using diethylenetriamine pentaacetic acid (DTPA), calcium chloride ($CaCl_2$), and triethylamine (TEA) (Lindsay and Norvell, 1978) although some other extractants, such as Mehlich-3, ammonium bicarbonate DTPA (AB-DTPA) are also employed (Soltanpour and Schwab, 1977; Mehlich, 1984).

3.6 FACTORS INFLUENCING PREVALENCE OF LEAD, CADMIUM, AND CHROMIUM IN SOIL

Different factors influence the occurrence of lead, cadmium, and chromium in soil. These factors can be summed up as followed:

3.6.1 LEAD

3.6.1.1 SOIL pH

Soil pH or reaction plays a pivotal role in regulating the bioavailability of lead (Pb). At pH values lower than 5.0, preferential binding of the proton on the clay surface enhances the solubility of Pb. When the soil reaction shifts toward the neutral and alkaline range, Pb is held more tightly, and its bioavailability gets substantially reduced.

3.6.1.2 SOIL TEXTURE

Pb has a significant positive correlation with the granulometric fraction, its concentration is higher in heavy loamy or clayey soil as compared with sandy soil. Due to its resemblance with the alkaline-earth metals, Pb can replace Ba, Sr, or Ca from the minerals and adsorption sites.

3.6.1.3 SOIL ORGANIC MATTER (SOM)

The concentration of lead is higher in the topsoil as compared with deeper layers possibly due to its sorption by SOM. The adsorption of Pb on SOM sharply increases with increasing soil pH.

3.6.1.4 SESQUI OXIDES (FE–MN HYDROXIDES)

An excessively higher amount of Pb (up to 20,000 mg/kg) is reported in Fe–Mn nodules. Apart from that $CaCO_3$, phosphate particles, and feldspar surface act as sinks for soil Pb, especially higher pH values.

3.6.1.5 OTHERS

A highly reduced environment leads to the formation of lead sulfides (PbS) and reduces the availability of Pb. In alkaline soil, the formation of soluble Pb-organic and hydroxy complexes further increases the availability.

3.6.2 CADMIUM

3.6.2.1 SOIL pH

Due to an increase in the pH-dependent negative charges in the pH interval of 4.0–7.7, Cd sorption capacity of soil increases approximately threefold for one unit increase in pH. At pH > 7.5, Cd is strongly adsorbed by soil constituents, and its availability is sharply reduced. Cd shows its maximum mobility in the pH values 4.5–5.5. The formation of soluble organic chelates may sometimes enhance the mobility of Cd in the alkaline range.

3.6.2.2 $CACO_3$ CONTENT

$CaCO_3$ concentration is one of the most important factors determining the bioavailability of Cd (Zhao et al., 2014). This reaction mechanism mostly depends on Eh–pH system of the soil system. Higher concentration of $CaCO_3$ concentration in soil solution leads to the precipitation of cadmium in the form of insoluble carbonate ($CdCO_3$) in pH >7.8.

3.6.2.3 COMPETITIVE IONS

In extremely higher pH range, Cd adsorption decreases, possibly due to competition from Ca^{2+} and Mg^{2+} ions. Ionic competition with Zn, Ca, Co, Cu, Ni, and Pb affects the adsorption of Cd.

3.6.2.4 ORGANIC LIGANDS

Low molecular weight organic matter fractions bound more Cd as compared with high molecular weight fractions. As because, free hydrated Cd ions pose higher toxicity to the plants than its organic complexes, organic ligands substantially reduce the toxic effects of Cd on plants though adsorption and chelation.

3.6.2.5 SOIL TEXTURE

Cd fixation in soil is directly controlled by its adsorption onto clay fractions. Cd content in heavy clayey soils is higher as compared with lighter soils.

3.6.2.6 *CHLORIDE ION (CL⁻) CONCENTRATION*

When saline soils have higher concentrations of Cl^- ions, Cd forms very stable soluble complexes with chloride ligands ($CdCl^+$, $CdCl^0$, $CdCl_3^-$ and $CdCl_4^{2-}$) and thereby enhances its mobilization and subsequently its phytoavailability.

3.6.2.7 *REDOX STATUS*

In soils in highly oxidized conditions, Cd occurs in the forms of CdO and $CdCO_3$ and is most likely fixed by phosphates while in a reduced environment, its solubility vis-a-vis availability decreases due to precipitation of Cd sulfide from the corresponding sulfate. Higher proportion of the available fraction of Cd is present under aerobic or oxidizing conditions of the soil having low moisture.

3.6.3 *CHROMIUM*

3.6.3.1 *SOIL pH AND Eh*

The reduction of Cr^{+6} to Cr^{+3} is much more frequent than the reverse process making trivalent Cr the more stable form in the natural system. Cr^{3+} remains stable in pH 0–8 and at an Eh from approximately −0.4 to +1.2 V, and generally bound to organic substances. On the other hand, Cr^{6+} forms soluble inorganic complexes and thereby enhances its mobility as well as availability to plants (Ertani et al., 2017). At pH> 12.0 and negative Eh values soluble chromium hydroxide anions [$Cr(OH)_4^-$] form. The important reactions are as follows:

$$Cr^{3+} + H_2O \text{ ------ } CrOH^{2+} + H^+$$
$$CrOH^{2+} + H_2O \text{ ------ } Cr(OH)_2^+ + H^+$$
$$Cr(OH)_2^+ + H_2O \text{ ------ } Cr(OH)_3 + H^+$$
$$Cr(OH)_3 + H_2O \text{ ------} Cr(OH)_4^- + H^*$$

3.6.3.2 *SESQUIOXIDES*

Iron and managenese oxides are the most important factors for the rapid conversion of the redox status of Cr. Mn^{4+} act as an oxidization agent for the conversion of Cr^{+3} to Cr^{6+} wherein Mn^{4+} is reduced to Mn^{2+}.The governing reaction is as follows:

$$2Cr^{3+} + 3\,MnO_2 + 2\,H_2O \text{ -----------} 2CrO_4^{2-} + 3Mn^{2+} + 4H^{+}$$

In highly reduced environments, anaerobic bacteria-mediated bioreduction of Cr^{6+} is also evident.

3.6.3.3 SOIL ORGANIC MOLECULES

Reduction by organic molecules has the potential to form soluble Cr^{3+}-organic complexes by acting as appropriate electron donors. The reaction is as follows:

$$HCrO_4^{-} + 7H^{+} + 3e^{-} \text{ ---------} Cr^{3+} + 4H_2O$$

Chromium has a huge capability to form various organic complexes. Amino acids, sugars, and fulvic acid fractions may produce some intermediate Cr(V) species in due course of reduction to Cr^{3+}.

3.7 UPTAKE MECHANISM AND ACCUMULATION OF LEAD, CADMIUM, AND CHROMIUM IN PLANT EDIBLES

Heavy metal uptake by plants is a complex process and involves several specialized mechanisms (Fig. 3.2). Plants mainly uptake heavy metals by roots, however, foliar uptake of heavy metals are also reported (Shahid et al., 2017). The plant root system is highly evolved and efficient in accumulating micronutrients from the soil. The plant produces chelating agents, inducing redox reactions and pH changes in the soil which aids the plant roots to take up the micronutrient even when present in trace amounts in the soil (Sinha et al., 2004). This same mechanism is utilized by plants to uptake heavy metals such as Pb, Cr, and Cd (Tangahu et al., 2011). The translocation and storage pathways of heavy metals by plants are similar to those of the micronutrients since heavy metals simulate the properties of the micronutrients and essential elements.

The specialized proteins embedded in the plasma membrane of the plant cell involve three basic transport mechanisms which include proton pumps, co- and anti-transporters, and channels for ion uptake and translocation. Proton pumps are the ATPases that generate electrochemical gradients after consuming energy. The electrochemical gradient generated by the ATPases is utilized by the co- and anti-transporter proteins to drive the uptake of ions, while the channel proteins facilitate the transport of the ions into the cell. Heavy metal present in the soil sap enters the apoplast by crossing the plasma membrane, the apoplastic solution later on enters the symplast and crosses the cells via plasmodesmata (Saifullah et al., 2009). Once the metal ions enter the roots, they are either stored or transferred to the shoot through the xylem vessels (Jan et al., 2016). To understand the translocation mechanism of metal ions from root to the shoot, several specific and nonspecific metal transporter gene families have been identified which include cation diffusion facilitator (CDF), heavy metal ATPase (HMA), ATP-binding cassette (ABC) superfamily, natural resistance-associated macrophage protein (NRAMP), and ZRT, IRT-like protein (ZIP) (Shahid et al 2016).

Lead (Pb) present in the soil sap gets adsorbed to the root system and binds to the carboxylic (-OOH) groups of uronic acid-composed root mucilage or gets directly attached to the outermost cell layer of the root known as rhizodermis (Porrunt et al., 2013). After adsorption in the rhizoderm layer, lead may enter the root by passive movement. The highest concentration of lead can be observed in root apices where the presence of young cells with thin cell walls facilitates Pb uptake (Seregin et al., 2004). Lead solubility and availability to plants is very little in a pH range of 5.5–7.5 (Mitra et al., 2020). Since the soil pH is low in the rhizodermic region, it favors metal solubility and hence corresponds to high lead concentration in this region. The majority of the amount of lead taken up by the plants remains within the roots, while a very small portion of the absorbed lead is translocated to the shoots. The translocation factor (TF), which is the ratio between amounts of lead in the aboveground part and the roots is used to describe the degree of translocation of lead from root to shoot (Bhatti et al., 2018). The TF value for lead has been reported to be less than one by several researchers since higher percentage of lead is accumulated in the roots (around 95%) than in the shoots (around 5%) (Kiran and Prasad, 2017; Chandra et al., 2018). However, the molecular mechanism behind lead uptake and its translocation is still not well explained.

No specific mechanism for Cr uptake by plants is reported and the mechanism of chromium uptake is still unclear (Oliveira, 2012). Cr uptake ability by plants depends upon the plant type and Cr species (Shukla et al., 2007). Reports suggest that plants can take up Cr(III) and Cr(VI) but the uptake mechanisms for both the Cr species are completely different. Cr(III) is taken up by the plants via a passive process requiring no expenditure of energy, while the uptake of Cr(VI) is an active process involving phosphate or sulfate transporters (de Oliveira et al., 2016). The use of metabolic inhibitors

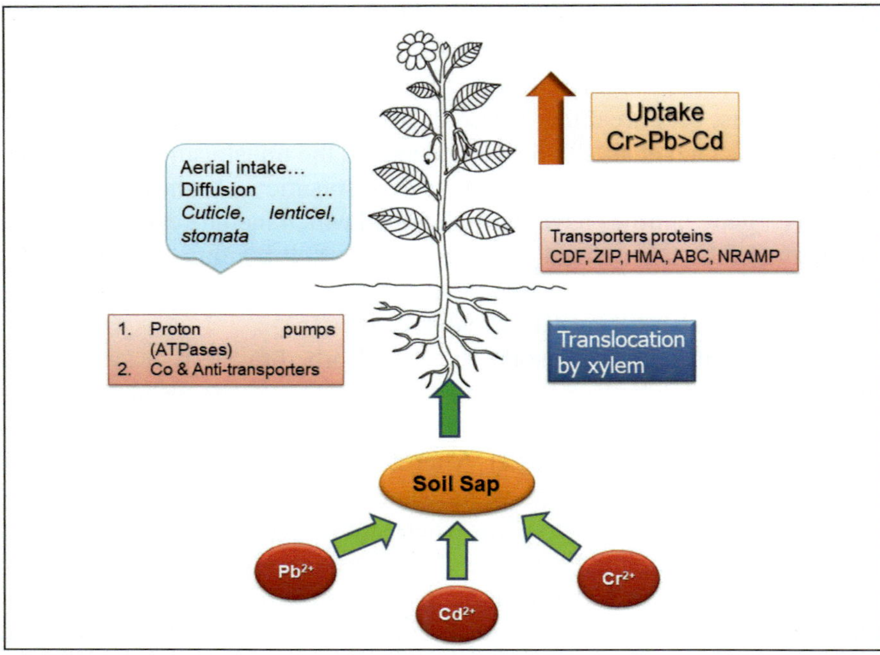

FIGURE 3.2 Uptake mechanism of cadmium, chromium, and lead in plant.

hinders Cr(VI) uptake but has no effect on Cr(III) uptake since it is a passive uptake process. Cr being the least mobile among all the heavy metals, its concentration in the root can be 100 times higher than in the shoots (Shankar et al., 2005). The concentration of Cr has been reported to be highest in the cell walls of the root and cytoplasm and intracellular spaces of the rhizome (Shahid et al., 2017). Sulfur and iron accumulators like Brassicaceae family have been reported to uptake and translocate high levels of Cr to aerial parts (Singh et al., 2013).

Unlike lead and chromium, cadmium is readily taken up by the root system and translocated to the aboveground parts. Cd availability to plants is largely dependent on soil pH, soil organic matter (SOM) content, cation exchange capacity, and concentration of other cations (Benavides et al., 2005). Cadmium uptake is facilitated by cation transport which includes ZIP, HMA, and NRAMP families (Qin et al., 2020). Ogo et al., (2014) reported ZIP8 and AtIRT1 to be responsible for Cd uptake by roots. Nramp1 protein has been suggested to be involved in the cellular uptake of Cd and its tanslocation through the plant (Qin et al., 2020). Cd uptake can also occur through YSL (yellow stripe 1) like proteins belonging to the oligopeptide transporters in the form of cd-chelates.

Elaborate studies regarding foliar uptake of heavy metals have not been reported unlike that of root uptake. Heavy metals emitted from the smelters can travel up to several kilometers and get accumulated in the foliar regions of the plant. Reports suggest that plants grown near the smelter belts show high levels of heavy metals in the foliar region (Shahid et al., 2016). Foliar uptake of heavy metals occurs through stomata, cuticles, lenticels, aqueous pores, and ectodesmata (Fernandes et al., 2013). Foliar uptake is considered mainly as a surface phenomenon and the metals absorbed by the foliar parts of the plant are mainly retained by the trichomes and waxy layer present in those areas. Foliar uptake can be broadly categorized into two broad pathways depending on the phobicity, (a) diffusion through cuticle for lipophilic elements and (b) through the aqueous pores of cuticle and stomata for hydrophilic elements. However, a detailed pathway of foliar uptake of heavy metals is still not clear and requires more investigation.

3.8 HUMAN HEALTH RISK ASSESSMENT THROUGH DIETARY INTAKE OF LEAD, CADMIUM, AND CHROMIUM

The toxicity of heavy metals, such as cadmium, chromium, and lead especially through consumption of contaminated foodstuffs grown in contaminated sites has been witnessed over several years. Several crops have been subjected to such contaminations over the years (Table 3.3).

Exposure over a certain period of time may result in sudden, acute, or even chronic symptoms of exposure that vary depending on the metal type, absorption level, and human age. Particulate or fumes of cadmium can result in acute pulmonary disorders even leading to death (Barbee and Prince, 1999). In addition to this, high exposure levels of cadmium may result in bone and kidney damage. The severity of skeletal damage was first recorded in Japan in the 1950s, called itai-itai disease as a result of consumption of cadmium-contaminated rice grown on a field irrigated with the contaminated water. Subjected to such disease, International Agency for Research on Cancer (IARC) has grouped cadmium as a group I human carcinogen, the most common being prostate cancer. Consumption of high lead-containing diet results in its accumulation in the liver and kidney. Apart from this, a large number of physiological and mental disorders have been associated with lead contamination, such as headache, irritation, abdominal

TABLE 3.3 Concentration of Pb, Cd, and Cr (Mg/Kg) in Edibles of Some Important Crops.

Sl. no	Crop	Area of Study	Pb	Cd	Cr	Reference
1.	Wheat grains	Argentina	0.088	0.017	0.540	Bermudez et al. (2011)
2.	Rice grains	China	0.098	0.064	0.193	Mao et al. (2019)
3.	Maize Grains	China	0.084	0.010	0.60	Ma et al. (2020)
4.	Soybean seeds	Argentina	0.80	<0.05	1.20	Lavado et al. (2001)
5.	Spinach leaves	India, Serbia	3.1	2.9	>0.5	Chary et al. (2008); Pajević et al. (2018)
6.	Pumpkin	Bangladesh	0.25	0.01	1.45	Islam et al. (2014)
7.	Cabbage	China, Ghana	0.016	0.05	0.35	Khan et al. (2008); Ametepey et al. (2018)
8.	Broccoli	Brazil	0.93	0.08	0.48	Guerra et al. (2012)
9.	Cauliflower	China	0.03	0.02	0.014	Song et al. (2009)
10.	Lettuce	China, Spain	0.016	<0.005	0.23	Khan et al. (2008); Montserrat et al. (2018)
11.	Carrot	Bangladesh	0.304	<0.1	<0.1	Tasrina et al. (2015)
12.	Beetroot	Brazil	0.58	0.09	0.21	Guerra et al. (2012)
13.	Red amaranth	Bangladesh	<0.1	<0.1	1.036	Tasrina et al. (2015)
14.	Amaranth	Bangladesh	0.89	0.30	1.71	Naser et al. (2018)
15.	Beetroot	Brazil	0.58	0.09	0.21	Guerra et al. (2012)
16.	Cucumber	Iran	0.04	0.23	0.18	Taghipor and Jalali (2019)
17.	Potato	China	0.067	0.015	0.03	Song et al. (2009)
18.	Sweet potato	Brazil	0.46	0.14	0.04	Guerra et al. (2012)
19.	Garlic	Brazil	2.50	0.09	0.29	Guerra et al. (2012)
20.	Onion	Brazil	0.49	0.02	0.07	Guerra et al. (2012)
21.	Yam	Brazil	0.72	0.06	0.20	Guerra et al. (2012)
22.	Celery	Brazil	0.47	0.05	0.14	Guerra et al. (2012)
23.	Kidney bean	Saudi Arabia	0.3	0.2	2	Eid et al. (2017)
24.	Asparagus bean	China	0.07	0.013	0.16	Zhou et al. (2016); Xu et al. (2015)

pain, encephalopathy, behavioral disorder, confusion, and psychosis. It has also been classified as a "possible human carcinogen" by International Agency for Research on Cancer (IARC), in 1987 with a high prevalence of lung cancer, stomach cancer, and gliomas (Steenland and Boffetta, 2000). Chromium in its occurrence in the human body can result in the formation of ulcers, the majority of which persist for months together and take time to heal, the most common form being ulcers on the nasal septum. Elevated levels of chromium in blood also result in the inhibition of erythrocyte glutathione reductase activity thereby affecting the reduction of methemoglobin to hemoglobin (Jaishankar et al., 2014).

There may be different paths of exposure to heavy metals in human beings, including ingestion, dermal exposure, and inhalation (Asaduzzaman et al., 2017; Li et al., 2017). Ingestion is the primary route as most of the heavy metals are ingested through diet, and it may account for 90% of the daily dietary intake in comparison to inhalation and dermal exposure (Loutfy et al., 2006). All three heavy metals Pb, Cd, and Cr may find their human entry through this process. Environmental occupational health hazards may occur as a result of inhalation of chromium in metal plating and other industries. Upon inhalation, the toxic Cr(III) gets accumulated in human lung tissue due to the reduction of Cr(VI) to Cr(III). Dermal exposure to heavy metals occurs when a significant portion of the skin comes in contact with the metal-laden dust and fine particles. They get deposited in dermal layers and enter into the circulatory system (Li and Zuo, 2013, Keshavarzi et al., 2015).

Thus, the accumulation of these heavy metals through diet has the potential to result in severe carcinogenic and noncarcinogenic human health hazards which necessitates the understanding of its safe limit and risk assessment (Table 3.4).

TABLE 3.4 Recommended Safe Limits of Heavy Metals for Human Beings.

Heavy Metal	Recommended safe limit
Pb	Approximately 9–278 g/day (children) and 20–282 g/day (adults)
Cd	Approximately 2–25 g/day (children) and 15–50 g/day (adults)
Cr	Approximately 50–200 g/day

Source: Adapted from Sharma and Agrawal (2005).

Present research activities have given a major thrust to global food security, nutritional requirement, and dietary risk to augment awareness among the local community regarding environmental toxicology and associated health hazards (Li et al., 2017). A serious impetus in this area has also been given by the health and environmental organizations and regularity authorities at local, regional, and global levels to enable human sustenance under the elevated dietary risk dilemmas. The fact that these heavy metals are potential carcinogens and the major dietary entry comes from water and edible grains tracing out the possible water and plant–human transfer is a must.

Several parameters have been developed to illustrate the dietary risk of heavy metals to human health of which provisional tolerable weekly intake (Guha Mazumder et al., 2013), hazard quotient (Mandal et al. 2019), and target cancer risk (Bhattacharyya et al. 2021) deserve special significance. All of these studies have revealed a close correlation between dietary uptake and health hazard.

3.9 ADVANCED INSTRUMENTATION PROTOCOLS FOR HEAVY METAL (LEAD, CADMIUM, AND CHROMIUM) DETECTION

The alarming extent of heavy metal (especially Pb, Cd, and Cr) contamination has continued to be reported with the advancement of human civilization. Industrial effluents, fertilizer, tanneries, sewage, and sludge act as the major sources of such pollution which ultimately manifests into biomagnification in human food chain (Bhattacharyya et al., 2021). For prompt and efficient monitoring of these trace metals in the environment, an advanced instrumentation technique is necessary. In this regard, (i) atomic absorption spectrometry (AAS), (ii) microwave plasma-atomic emission spectrometry (MP-AES), (iii) inductively coupled plasma—optical emission spectrometry (ICP-OES), and (iv) inductively coupled plasma—mass spectrometry (ICP-MS) are employed for assessing heavy metal contamination as well as human health risk assessment on a global scale (Donati et al., 2017; Fujihara and Nishimoto, 2020).

3.9.1 ATOMIC ABSORPTION SPECTROMETRY

In the latest advancement of optical spectrometry, the AAS was first developed in 1976 followed by further modification (Koirtyohann, 1980). The radiation source, nebulizer, atomizer, monochromatic lens, wavelength selector, detector, photomultiplier, and indicator device make up the modern AAS. The analyte along with the solvent is converted into aerosol/mist wherein the analyte is excited by the flame and the solvent is vaporized. Valence electron from the analyte is excited into higher energetic orbitals absorbing characteristic radiation at element-specific signature (Jeffery et al., 1989). It follows Lambert-Beer's law and the concentration of an unknown sample is determined by known standard concentration. Based on different types of atomizers, AAS is classified as flame-AAS, graphite furnace-AAS, cold vapor-AAS, hydride-AAS, and glow discharge-AAS. Line source-AAS (LS-AAS) and continuum source-AAS (CS-AAS) types based on radiation sources are also in use. The most common type, flame-AAS is operated using air-acetylene (~2300ºC) and nitrous oxide-acetylene (~2700ºC) (Firouzabadi et al., 2017).

3.9.2 INDUCTIVELY COUPLED PLASMA—OPTICAL EMISSION SPECTROMETRY

The determination of the concentration of multiple elements led to the gradual development of ICP-OES during 1970–1980s (Donati et al., 2017). In contrast to AAS, ICP-OES measures the emitted radiation of the target elements, and their concentration is determined by comparing with the standards with high precision levels. Plasma (a cloud of electrons and highly ionized gases) is produced in neutral gas (generally argon) under a strong electrical field using radiofrequency (RF) following Faraday-Lenz's law of induction (Montaser and Golightly, 1992). Following the nebulization and aerosol formation, using the thermal energy from plasma (10000 K), it atomizes and ionizes the analyte sample. Diffraction grating-mediated multiple wavelengths aid in multi-elemental analysis in one go. Cd, Cr, and Pb along with other trace elements can be precisely determined using ICP- OES (Støving et al., 2013).

3.9.3 MICROWAVE PLASMA—ATOMIC EMISSION SPECTROMETRY

Microwave plasma-atomic emission spectrometry (MP-AES) is very much similar to that of ICP-AES. The only difference lies in the production mechanism of plasma. The plasma is generated from the ionization of gaseous medium (Ar, He, N) in the presence of electromagnetic microwave fields (usually 2.45 GHz) (Skogerboe and Coleman, 1976). It can be classified as a capacitively coupled system (CMP) having a coaxial conductive electrode and microwave-induced plasmas (MIP) having no such electrode. MP-AES is of lower cost as compared with the ICP-OES when combined with N as supporting gas. Moreover, MP-AES generally operates at relatively lower temperatures (4800–5000 ºC) in comparison with ICP-OES (6000–8000ºC).

3.9.4 INDUCTIVELY COUPLED PLASMA—MASS SPECTROMETRY

ICP-MS is indeed the latest advancement for rapid, simultaneous quantitative analysis of Pb, Cd, Cr, and other trace metals. ICP-MS is the amalgam of very high-temperature Ar-ICP and quadruple mass spectrometer. Analytes present in the sample are vaporized by Ar-ICP and rapidly converted to positively charged ions. The concentration of these ionic species is determined by a mass spectrometer taking into account the mass/charge ratio (Ammann, 2007). ICP-MS is the most competent analytical instrument for the determination of most elements from the periodic table including their isotopes. Detection limits of lead, cadmium, and chromium among F-AAS, GF-AAS are 5.0, 1.0, and 4.0 μg L^{-1}; 0.2, 0.02, and 0.06 μg L^{-1} (Boumans et al. 1975; Skogerboe and Coleman, 1976; Balaram, 2016 and Ferreira et al. 2018).

3.10 CONCLUSIONS

Heavy metal contamination has emerged as one of the most potent threats to global food security and environmental sustainability. Heavy metals are present on the earth's surface naturally. However, human activities are responsible for augmenting their release into the environment leading to environmental pollution. The occurrence in the soil in different dynamic fractions in nondegradable forms results in their easy entry into the plant system. The consumption of these contaminated foodstuffs accumulates the heavy metals in the human body in amounts often higher than the maximum tolerable daily intake resulting in different carcinogenic and noncarcinogenic risks. The food web-mediated transfer of the contaminant further aggravates the problem. The use of wastewater for irrigating the crops must be restricted and proper treatments must be made prior to its field-based application. The industries must also ensure the safe disposal of solid wastes and liquid effluents to minimize the level of contamination in the environment. Stringent regulation and awareness program among local individuals must also be made as unless a participatory approach of everyone is not guaranteed, a problem of this stature cannot be solved efficiently.

KEYWORDS

- **heavy metals**
- **soil pools**
- **plant uptake**
- **dietary risk**
- **advanced instrumentations**
- **global contamination**

REFERENCES

Abii, T. A.; Okorie, D. O. Assessment of the Level of Heavy Metals [Cu, Pb, Cd and Cr] Contamination in Four Popular Vegetables Sold in Urban and Rural Markets of Abia State Nigeria: Continental. *J. Water. Air Soil Pollut.* **2011,** *2* (1), 42–47.

Abo-Kassem, E.; Sharaf-el-din, A.; Rozema, J.; Foda, E. A. Synergistic Effect of Cadmium, NaCl on the Growth, Photosynthesis and Irons Content on Wheat Plants. *Biol. Plant.* **1995,** *37*, 241–249.

Ametepey, S. T.; Cobbina, S. J.; Akpabey, F. J.; Duwiejuah, A. B.; Abuntori, Z. N. Health Risk Assessment and Heavy Metal Contamination Levels in Vegetables from Tamale Metropolis, Ghana. *Int. J. Food Contam.* **2018,** *5*, 1–8.

Ammann, A. A. Inductively Coupled Plasma Mass Spectrometry (ICP MS): A Versatile Tool. *J. Mass. Spectrom.* **2007,** *42*, 419–427.

Asaduzzaman, K.; Khandaker, M. U.; Baharudin, N. A. B.; Amin, Y. B.M; Farook, M. S.; Bradley, D. A.; Mahmoud, O. Heavy Metals in Human Teeth Dentine: A Bio Indicator of Metals Exposure and Environmental Pollution. *Chemosphere* **2017,** *176*, 221–230.

Atafar, Z.; Mesdaghinia, A.; Nouri, J.; Homaee, M.; Yunesian, M.; Ahmadimoghaddam, M.; Mahvi, A. H. Effect of Fertilizer Application on Soil Heavy Metal Concentration. *Environ. Monit. Assess.* **2010,** *160* (1–4), 83–89. DOI: 10.1007/s10661-008-0659-x.

Athar, M.; Vohora, S. B. Heavy Metals and Environment. In: *Man and Environment Series*; Ray, P. K.; Ed.; Wiley Eastem Ltd.: New Delhi, 1995; pp 1–195.

Balaram, V. Recent Advances in the Determination of Elemental Impurities in Pharmaceuticals– Status, Challenges and Moving Frontiers. *Trac-Trend Anal. Chem.* **2016,** *80*, 83–95.

Barbee, J. Y. Jr.; Prince, T. S. Acute Respiratory Distress Syndrome in a Welder Exposed to Metal Fumes. *South Med. J.* **1999,** *92*, 510–512.

Barua, B.; Jana, S. Effect of Heavy Metals on the Dark Induced Changes in Hill Reaction Activity, Chlorophyll and Protein Contents, Dry Matter and Tissue Permeability in Detached *Spinacia oleracea* L. Leaves. *Photosynthetica* **1986,** *20*, 74–76.

Basta, N. T.; Ryan, J. A.; Chaney, R. L. Trace Element Chemistry in Residual-Treated Soil: Key Concepts and Metal Bioavailability. *J. Environ. Qual.* **2005,** *34* (1), 49–63.

Baszynski, I. Interference of Cd in Functioning of Photosynthetic Apparatus of Higher Plants. *Act. Soc. Bot. Pol.* **1986,** *55*, 219–304.

Benavides, M. P.; Gallego, S. M.; Tomaro, M. L. Cadmium Toxicity in Plants. *Braz. J. Plant Physiol.* **2005,** *17*, 21–34.

Bermudez, G. M.; Jasan, R.; Plá, R.; Pignata, M. L. Heavy Metal and Trace Element Concentrations in Wheat Grains: Assessment of Potential Non-Carcinogenic Health Hazard Through Their Consumption. *J. Hazard. Mater.* **2011,** *193*, 264–271.

Bhattacharyya, K.; Sengupta, S.; Pari, A.; Halder, S.; Bhattacharya, P.; Pandian, B. J.; Chinchmalatpure, A. R. Characterization and Risk Assessment of Arsenic Contamination in Soil–Plant (Vegetable) System and Its Mitigation Through Water Harvesting and Organic Amendment. *Environ. Geochem. Health* **2021,** 1–16. https://doi.org/10.1007/s10653-020-00796-9

Bhatti, S. S.; Kumar, V.; Sambyal, V.; Singh, J.; Nagpal, A. K. Comparative Analysis of Tissue Compartmentalized Heavy Metal Uptake by Common Forage Crop: A Field Experiment. *Catena* **2018,** *160*, 185.

Bishnoi, N. R.; Dua, A.; Gupta, V. K.; Sawhney, S. K. Effects of Chromium on Seed Germination, Seedling Growth and Yield of Peas. *Agric. Eco. Environ.* **1993,** *47*, 47–57.

Boumans, P. W. J. M.; De Boer, F. J.; Dahmen, F. J.; Hoelzel, H.; Meier, A. A Comparative Investigation of Some Analytical Performance Characteristics of an Inductively-Coupled Radio Frequency Plasma and a Capacitively-Coupled Microwave Plasma for Solution Analysis by Emission Spectrometry. *Spectrochim. Acta B* **1975,** *30*, 449–469.

Brown, S. S.; Kodama, Y. *Toxicology of Metals, Clinical and Experimental Research*; Ellis Howard Ltd, Chichester, 1986.

Castagnetto, J. M.; Hennessy, S. W.; Roberts, V. A.; Getzoff, E. D.; Tainer, J. A.; Pique, M. E. MDB: The Metalloprotein Database and Browser at the Scripps Research Institute. *Nucleic Acids Res.* **2002,** *30* (1), 379–382.

Chandra, P.; Kulshreshtha, K. Chromium Accumulation and Toxicity in Aquatic Vascular Plants. *Botanical Rev.* **2004,** *70* (3), 313–327.

Chandra, R.; Kumar, V.; Tripathi, S.; Sharma, P. Heavy Metal Phytoextraction Potential of Native Weeds and Grasses from Endocrine-Disrupting Chemicals Rich Complex Distillery Sludge and Their Histological Observations During In Situ Phytoremediation. *Ecol. Eng.* **2018,** *111*, 143–156.

Chary, N. S.; Kamala, C. T.; Raj, D. S. S. Assessing Risk of Heavy Metals from Consuming Food Grown on Sewage Irrigated Soils and Food Chain Transfer. *Ecotoxicol. Environ. Saf.* **2008,** *69*, 513–524.

Chen, Z. F.; Zhao, Y.; Fan, L. D.; Xing, L. T.; Yang, Y. J. Cadmium (Cd) Localization in Tissues of Cotton (*Gossypium hirsutum* L.) and Its Phytoremediation Potential for Cd-Contaminated Soils. *Bull. Environ. Contam. Toxicol.* **2015,** *95*, 784–789. DOI: 10.1007/s00128-015-1662-x

Chrysochoou, M.; Theologou, E.; Bompoti, N.; Dermatas, D.; Panagiotakis, I. Occurrence, Origin and Transformation Processes of Geogenic Chromium in Soils and Sediments. *Curr. Pollut. Rep.* **2016**, *2*, 224–235. DOI: 10.1007/s40726-016-0044-2

D'Amore, J. J.; Al-Abed, S. R.; Scheckel, K. G.; Ryan, J. A. Methods for Speciation of Metals in Soils: A Review. *J. Environ. Qual.* **2005**, *34* (5), 1707–1745.

de Oliveira, L. M.; Gress, J.; De, J.; Rathinasabapathi, B.; Marchi, G.; Chen, Y.; Ma, L. Q. Sulfate and Chromate Increased Each Other's Uptake and Translocation in As-Hyperaccumulator Pteris vittata. *Chemosphere* **2016,** *147*, 36–43.

Dixit, R.; Wasiullah, M. D.; Pandiyan, K.; Singh, U. B.; Sahu, A.; Shukla, R.; Singh, B. P.; Rai, J. P.; Sharma, P. K.; Lake, H.; Paul, D. Bioremediation of Heavy Metals from Soil and Aquatic Environment: An Overview of Principles and Criteria of Fundamental Processes. *Sustainability* **2015**, *7*, 2189–2212. DOI: 10.3390/su7022189

Donati, G. L.; Amais, R. S.; Williams, C. B. Recent Advances in Inductively Coupled Plasma Optical Emission Spectrometry. *J. Anal. Atom. Spectrom*. **2017,** *32*, 1283–1296.

Du Laing, G. Analysis and Fractionation of Trace Elements in Soils. In: *Trace Elements in Soils*; Hooda, P. S., Ed.; Wiley Online Library, Blackwell Publishing Ltd: New Jersey, 2010; pp 53–80.

Eid, E. M.; Shaltout, K. H.; Alamri, S. A. M.; Sewelam, N. A.; Galal, T. M. Evaluating the Uptake of Ten Heavy Metals by Kidney Bean (*Phaseolus vulgaris* L.) Grown in a Soil-Sludge Mixture Using a Regression Model. *Appl. Ecol. Environ. Res.* **2020,** *18*, 7021–7039.

Ercilla-Montserrat, M.; Muñoz, P.; Montero, J. I.; Gabarrell, X.; Rieradevall, J. A Study on Air Quality and Heavy Metals Content of Urban Food Produced in a Mediterranean City (Barcelona). *J. Clean Prod.* **2018,** *195*, 385–395.

Ertani, A.; Mietto, A.; Borin, M.; Nardi, S. Chromium in Agricultural Soils and Crops: A Review. *Water Air Soil Pollut.* **2017,** *228*, 190–202.

Fernández, V.; Sotiropoulos, T.; Brown, P. Foliar Fertilization: Scientific Principles and Field Practices. Paris, France. *Int. Fert. Ind. Assoc.*, 2013; pp 12–70.

Ferreira, S. L.; Bezerra, M. A.; Santos, A. S.; dos Santos, W. N.; Novaes, C. G.; de Oliveira, O. M.; Oliveira M. L.; Garcia, R. L. Atomic Absorption Spectrometry—A Multi Element Technique. *Trac-Trend Anal. Chem.* **2018,** *100*, 1–6.

Firouzabadi, Z. D.; Shabani, A. M. H.; Dadfarnia, S.; Ehrampoush, M. H. Preconcentration and Speciation of Thallium by Ferrofluid Based Dispersive Solid Phase Extraction and Flame Atomic Absorption Spectrometry. *Microchem. J.* **2017,** *130*, 428–435.

Flora, S. J. S.; Mittal, M.; Mehta, A. Heavy Metal Induced Oxidative Stress & Its Possible Reversal by Chelation Therapy. *Indian J. Med. Res.* **2008,** *128*, 501–523.

Frohne, T.; Rinklebe, J.; Diaz-Bone, R. A. Contamination of Floodplain Soils Along the Wupper River, Germany, with As, Co, Cu, Ni, Sb, and Zn and the Impact of Pre-Definite Redox Variations on the Mobility of These Elements. *Soil Sediment Contamin. Int. J.* **2014,** *23*, 779–799.

Fujihara, J.; Nishimoto, N. Total Antimony Analysis by Hydride Generation-Microwave Plasma-Atomic Emission Spectroscopy with Applications. *Microchem. J.* **2020,** *157*, 104992. https://doi.org/10.1016/j.microc.2020.104992

Ghani, A. Effect of Chromium Toxicity on Growth, Chlorophyll and Some Mineral Nutrients of *Brassica juncea* L. *Egypt. Acad. J. Biol. Sci.* **2011,** *2* (1), 9–15.

Golui, D.; Barman, M.; Datta, S. P.; Shukla, A. K. Fractionation of Micronutrients and Metals in Soil. Manual on Advance Techniques for Analysis of Nutrients and Pollutant Elements. In: *Soil, Plant and Human*; Datta, S. P., Meena, M. C., Dwivedi, B. S., Shukla, A. K., Eds.; Westville Publishing House: New Delhi, 2017.

Gope, M.; Masto, R. E.; George, J.; Hoque, R. R.; Balachandran, S. Bioavailability and Health Risk of Some Potentially Toxic Elements (Cd, Cu, Pb and Zn) in Street Dust of Asansol, India. *Ecotoxicol. Environ. Saf.* **2017**, *138*, 231–241. DOI: 10.1016/j.ecoenv.2017.01.008

Greger, M.; Ogren, E. Direct and Indirect Effects of Cd on Photosynthesis in Sugar Beet (Beta vulgaris). *Physiol. Plant.* **1991**, *83*, 129–135.

Guerra, F.; Trevizam, A. R.; Muraoka, T.; Marcante, N. C.; Canniatti-Brazaca, S. G. Heavy Metals in Vegetables and Potential Risk for Human Health. *Sci. Agric.* **2012,** *69*, 54–60.

Guha Mazumder, D. N.; Deb, D.; Biswas, A.; Saha, C.; Nandy, A.; Ganguly, B.; Ghose, A.; Bhattacharya, K.; Majumdar, K. K. Evaluation of Dietary Arsenic Exposure and Its Biomarkers: A Case Study of West Bengal, India. *J. Environ. Sci. Health, Part A* **2013,** *488*, 896–904.

He, Z.; Shentu, J.; Yang, X.; Baligar, V. C.; Zhang, T.; Stoffella, P. J. Heavy Metal Contamination of Soils: Sources, Indicators, and Assessment. *J. Environ. Indicators* **2015**, *9*, 17–18.

Hembrom, S.; Singh, B.; Gupta, S. K.; Nema, A. K. A Comprehensive Evaluation of Heavy Metal Contamination in Foodstuff and Associated Human Health Risk: A Global Perspective. In *Contemporary Environmental Issues and Challenges in Era of Climate Change*; Springer: Singapore, 2020; pp. 33–63.

http://cadmium.org (accessed on 08.04.2021).

http://wikipedia.org (accessed on 08.04.2021).

Hu, Z.; Gao, S. Upper Crustal Abundances of Trace Elements: A Revision and Update. *Chem. Geol.* **2008**, *253* (3), 205–221. DOI: 10.1016/j.chemgeo.200805.010

Irfan, M.; Hayat, S.; Ahmad, A.; Alyemeni, M. N. Soil Cadmium Enrichment: Allocation and Plant Physiological Manifestations. *Saudi J. Biol. Sci.* **2013,** *20* (1), 1–10.

Islam, M. S.; Hoque, M. F. Concentrations of Heavy Metals in Vegetables Around the Industrial Area of Dhaka City, Bangladesh and Health Risk Assessment. *Int. Food Res.* **2014,** *21*, 2121–2126.

Iwegbue, C. M. Chemical Fractionation and Mobility of Heavy Metals in Soils in the Vicinity of Asphalt Plants in Delta State, Nigeria. *Environ. Forensics* **2013,** *14*, 248–259.

Jaishankar, M.; Tseten, T.; Anbalagan, N.; Mathew, B. B.; Beeregowda, K. N. Toxicity, Mechanism and Health Effects of Some Heavy Metals. *Interdiscip. Toxicol.* **2014,** *7* (2), 60.

Jan, S.; Parray, J. A. Metal Tolerance Strategy in Plants. In: *Approaches to Heavy Metal Tolerance in Plants*; Springer: Singapore, 2016; pp 19–32.

Jarup, L. Hazards of heavy metal contamination. *Br Med Bull.***2003,***68* (1), 167–182.

Jarup, L.; Bergluud, M.; Elinder, C. G. Health Effect of Cadmium Exposure—A Review of the Literature and a Risk Estimate. *Sci. J. Work Environ. Health* **1998**, *24*, 1–51.

Jeffery, G. H.; Bassett, J.; Mendham, J.; Denny, R. C. *Vogel's Textbook of Quantitative Chemical Analysis*, 5th edn.; John Wiley and Sons Inc: New York, 1989.

Kabata-Pendias, A.; Pendias, H. *Trace Metals in Soils and Plants*, 2nd edn; CRC Press: Boca Raton, 2001.

Keshavarzi, B.; Tazarvi, Z.; Rajabzadeh, M.; Najmeddin, A. Chemical Speciation, Human Health Risk Assessment and Pollution Level of Selected Heavy Metals in Urban Street Dust of Shiraz. *Iran. Atmos. Environ.* **2015**, *119*, 1–10.

Khan S.; Cao Q.; Zheng, Y. M.; Huang, Y. Z.; Zhu Y. G. Health Risks of Heavy Metals in Contaminated Soils and Food Crops Irrigated with Wastewater in Beijing, China. *Environ. Pollut*. **2008,** *152* (3), 686–692.

Khan, I.; Ghani, A.; Rehman, A. U.; Awan, S. A.; Noreen, A.; Khalid, I. Comparative Analysis of Heavy Metal Profile of *Brassica campestris* (L.) and *Raphanus sativus* (L.) Irrigated with Municipal Waste Water of Sargodha City. *J. Clin. Toxicol*. **2016**, *6*, 1–4. DOI: 10.4172/2161-0495.1000307

Khan, S.; Cao, Q.; Zheng, Y. M.; Huang, Y. Z.; Zhu, Y. G. Health Risks of Heavy Metals in Contaminated Soils and Food Crops Irrigated with Wastewater in Beijing, China. *Environ. Pollut*. **2008,** *152*, 686–692. https://doi.org/10.1016/j.envpol.2007.06.056.

Kiran, B. R.; Prasad, M. N. V. Responses of Ricinus communis L. (Castor Bean, Phytoremediation Crop) Seedlings to Lead (Pb) Toxicity in Hydroponics. *Selcuk J. Agric. Food Sci*. **2017,** *31*, 73–80.

Koirtyohann, S. R. A History of Atomic Absorption Spectroscopy. *Spectrochim. Acta Part B* **1980,** *35*, 663–670.

Koropatnick, J.; Leibbrandt, M. E. I. Effects of Metal on Gene Expression. In *Handbook of Experimental Pharmacology, Toxicology of Metals, Biochemical Aspects*; Goyer, R. A., Cherian, M. G., Eds., Vol. 115; Springer, 1995; pp 93–120.

Krupa, Z.; Skorzynska, E.; Malesymiec, W.; Baszynski, E. Effect of Cd Treatment on Photosynthetic Apparatus and Its Photochemical Activities in Greening Radish Seedling. *Photosynthetica* **1987,** *21*, 156–164.

Kubier, A.; Wilkin, R. T.; Pichler, T. Cadmium in Soils and Groundwater: A Review. *Appl. Geochem.* **2019**. DOI: 10.1016/j.apgeochem.2019.104388

Lavado, S. R.; Porcelli, A. C.; Alvarez, R. Nutrient and Heavy Metal Concentration and Distribution in Corn, Soybean and Wheat as Effected by Different Tillage Systems in Argentine Pampas. *Soil Tillage Res*. **2001,** *62*, 55–60.

Li, H.; Ji, H.; Shi, C.; Gao, Y.; Zhang, Y.; Xu, X. Distribution of Heavy Metals and Metalloids in Bulk and Particle Size Fractions of Soils from Coal-Mine Brownfield and Implications on Human Health. *Chemosphere* **2017**, *172*, 505–515.

Li, H.; Zuo, X. J. Speciation and Size Distribution of Copper and Zinc in Urban Road Runoff. *Bull. Environ. Contamin. Toxicol.* **2013,** *90*, 471–476.

Lindsay, W. L.; Norvell, W. A. Development of a DTPA Soil Test for Zinc, Iron, Manganese, and Copper. *Soil Sci. Soc. Am. J.* **1978,** *42*, 421–428.

Lone, M. I.; He, Z. L.; Stoffella, P. J.; Yang, X. E. Phytoremediation of Heavy Metal Polluted Soils and Water: Progresses and Perspectives. *J. Zhejiang Univ.–Sci. B (Biomed. Biotechnol.)*. **2008**, *9*, 210–220. DOI:10.1631/jzus.B0710633

Loutfy, N.; Fuerhacker, M.; Tundo, P.; Raccanelli, S.; El Dien, A. G.; Ahmed, M. T. Dietary Intake of Dioxins and Dioxin–Like PCBs, Due to the Consumption of Dairy Products, Fish/Seafood and Meat from Ismailia City, Egypt. *Sci. Total Environ.* **2006,** *370*, 1–8.

Ma, J.; Chen, Y.; Antoniadis, V.; Wang, K.; Huang, Y.; Tian, H. Assessment of Heavy Metal (Loid)s Contamination Risk and Grain Nutritional Quality in Organic Waste-Amended Soil. *J. Hazard. Mater.* **2020,** *399*, 123095. https://doi.org/10.1016/j.jhazmat.2020.123095

Mandal, J.; Golui, D.; Raj, A.; Ganguly, P. Risk Assessment of Arsenic in Wheat and Maize Grown in Organic Matter Amended Soils of Indo-Gangetic Plain of Bihar, India. *Soil Sediment Contamin.* **2019**, *28* (8), 757–772.

Mao, C.; Song, Y.; Chen, L.; Ji, J.; Li, J.; Yuan, X.; Yang, Z.; Ayoko, G. A.; Frost, R. L. Theiss, F. Human Health Risks of Heavy Metals in Paddy Rice Based on Transfer Characteristics of Heavy Metals from Soil to Rice. *Catena* **2019,** *175*, 339–348.

Marshall, F. M.; Holden, J.; Ghose, C.; Chisala, B.; Kapungwe, E.; Volk, J.; Agrawal, M.; Agrawal, R.; Sharma, R. K.; Singh, R. P. Contaminated Irrigation Water and Food Safety for the Urban and Peri– Urban Poor: Appropriate Measures for Monitoring and Control from Field Research in India and Zambia, Inception Report DFID Enkar R8160, SPRU. University of Sussex, Rome, 2007.

McLaughlin, M. J.; Hamon, R. E.; McLaren, R. G.; Speir, T. W.; Rogers S. L. Review: A Bioavailability-Based Rationale for Controlling Metal and Metalloid Contamination of Agricultural Land in Australia and New Zealand. *Aust. J. Soil Res.* **2000**, *38* (6), 1037–1086.

McLaughlin, M. J.; Parker, D. R.; Clarke, J. M. Metals and Micronutrients—Food Safety Issues. *Field Crop Res*. **1999**, *60*, 143–163.

Mehlich, A. Mehlich 3 Soil Test Extractant: A Modification of Mehlich 2 Extractant. *Commun. Soil Sci. Plant Anal.* **1984,** *15*, 1409–1416.

Montaser, A.; Golightly, D. W. *Inductively Coupled Plasmas in Analytical Atomic Spectrometry*; VCH Publishers, Inc.: New York, 1992.

Muradoglu, F.; Gundogdu, M.; Ercisli, S.; Encu, T.; Balta, F.; Jaafar, H. Z. E.; Zia–Ul–Haq, M. Cadmium Toxicity Affects Chlorophyll a and b Content, Antioxidant Enzyme Activities and Mineral Nutrient Accumulation in Strawberry. *Biol. Res.* **2015**, *48*, 1–7. DOI: 10.1186/S40659–015–0001–3

Nagajyoti, P. C.; Lee, K. D.; Sreekanth, T. V. M. Heavy Metals, Occurrence and Toxicity for Plants: A Review. *Environ. Chem. Lett.* **2010,** *8*, 199–216. DOI: 10.1007/s10311–010–0297–8

Najeeb, U.; Ahmad, W.; Zia, M. H.; Malik, Z.; Zhou, W. Enhancing the Lead Phytostabilization in Wetland Plant *Juncus effusus* L. Through Somaclonal Manipulation and EDTA Enrichment. *Arab. J. Chem.* **2014.**

Naser, H. M.; Rahman, M. Z.; Sultana, S.; Quddus, M. A.; Hossain, M. A. Heavy Metal Accumulation in Leafy Vegetables Grown in Industrial Areas Under Varying Levels of Pollution. *Bangladesh J. Agric. Res.* **2018,** *43*, 39–51.

Nath, K.; Shyam, S.; Singh, D.; Shanna, Y. K. Effect of Chromium and Tannery Effluent Toxicity on Metabolism and Growth in Cowpea (*Vigna sinensis* L. Saviex Hassk) Seedling. *Res. Environ. Life Sci.* **2008,** *1*, 91–94.

Niassy, S.; Diarra, K. Effect of Organic Inputs in Urban Agriculture and Their Optimization for Poverty Alleviation in Senegal, West Africa. In *Organic Fertilizers: Types, Production and Environmental Impact*; Singh, R. P., Ed.; Nova Science Publisher: Hauppauge, NY, 2012; pp 1–22.

Nieć, J.; Baranowska, R.; Dziubanek, G.; Rogala, D. Narażenie środowiskowe dzieci na metale ciężkie zawarte w glebach z placów zabaw, boisk, piaskownic i terenówprzedszkoli z obszaru Górnego Śląska. *J. Ecol. Health* **2013**, *2*, 55–62. In polish.

Noda, M.; Kitagawa, M. A Quantitative Study of Iliac Bone Histopathology on 62 Cases with Itai-Itai Disease. *Calcif. Tissue Int.* **1990**, *47*, 66–74.

Obeng-Gyasi, E. Sources of Lead Exposure in Various Countries. *Rev. Environ. Health* **2019**, *34* (1), 25–34. DOI: 10.1515/revh-2018–0037

O'Brien, T. J.; Cerjak, S.; Patierno, S. R. Complexities of Chromium Carcinogenesis: Role of Cellular Response, Repair and Recovery Mechanisms. *Mutat. Res.* **2003,** *533*, 2–26.

Ogo, Y.; Kakei, Y.; Itai, R. N.; Kobayashi, T.; Nakanishi, H.; Takahashi, H.; Nakazono, M.; Nishizawa, N. K. Spatial Transcriptomes of Iron-Deficient and Cadmium-Stressed Rice. *New Phytol.* **2014,** *201*, 781–794.

Ogunlade, M. O.; Agbeniyi, S. O. Impact of Pesticides Use on Heavy Metals Pollution in Cocoa Soils of Cross-River State, Nigeria. *Afr. J. Agric. Res.* **2011**, *6*, 3725–3728.

Oliveira, H. Chromium as an Environmental Pollutant: Insights on Induced Plant Toxicity. *J. Bot.* **2012**, *2012*, 375843. DOI: 10.1155/2012/375843

Osakwe, S. A.; Okolie, L. P. Distribution of Different Fractions of Iron, Zinc, Chromium, Lead and Nickel in Soils Around Petrol Filling Stations in Selected Areas of Delta State, Nigeria. *J. Appl. Sci. Environ. Manage.* **2015,** *19*, 706–716.

Ouzounidou, G. Change in Chlorophyll Fluorescence as Result of Copper Treatment: Dose Responses Relations in Silene and Thlaspi. *Photosynthetica* **1993,** *29*, 455–462.

Pajević, S.; Arsenov, D.; Nikolić, N.; Borišev, M.; Orčić, D.; Župunski, M.; Mimica-Dukić, N. Heavy Metal Accumulation in Vegetable Species and Health Risk Assessment in Serbia. *Environ. Monit. Assess.* **2018,** *190*, 1–14.

Pourrut, B.; Shahid, M.; Douay, F.; Dumat, C.; Pinelli, E. Molecular Mechanism Involved in Lead Uptake, Toxicity and Detoxification in Higher Plants. In: *Heavy Metal Stress in Plants*; Gupta, D. K., Corpas, F. J., Palma, J. M., Eds.; Springer: Heidelberg, 2013; pp 121–147.

Rahimi, E. Lead and Cadmium Concentrations in Goat, Cow, Sheep, and Buffalo Milk from Different Regions of Iran. *Food Chem.* **2013,** *136* (2), 389–391.

Raju, K. V.; Somashekar, R. K.; Prakash, K. L. Spatio–Temporal Variation of Heavy Metals in Cauvery River Basin. *Proc. Int. Acad. Ecol. Environ. Sci.* **2013,** *3* (1), 59–75.

Saifullah, M. E.; Qadir, M.; de Caritat, P.; Tack, F. M. G.; Laing, G. D.; Zia, M. H. EDTA-Assisted Pb Phytoextraction. *Chemosphere* **2009,** *74*, 1279–1291.

Satarug, S.; Garrett, S. H.; Sens, M. A.; Sens, D. A. Cadmium, Environmental Exposure, and Health Outcomes. *Ciência Saúde Coletiva* **2011**, *16* (5), 2587–2602.

Seregin, I. V.; Shpigun, L. K.; Ivanov, V. B. Distribution and Toxic Effects of Cadmium and Lead on Maize Roots. *Russ. J. Plant Physiol.* **2004,** *51*, 525–533.

Shahid, M.; Dumat, C.; Khalid, S.; Schreck, E.; Xiong, T.; Niazi, N. K. Foliar Heavy Metal Uptake, Toxicity and Detoxification in Plants: A Comparison of Foliar and Root Metal Uptake. *J. Hazard. Mater.* **2017,** *325*, 36–58.

Shahid, M.; Shamshad, S.; Rafiq, M.; Khalid, S.; Bibi, I.; Niazi, N. K.; Dumat, C.; Rashid, M. I. Chromium Speciation, Bioavailability, Uptake, Toxicity and Detoxification in Soil-Plant System: a Review. *Chemosphere* **2017,** *178*, 513–533.

Sharma, R. K.; Agrawal, M. Biological Effects of Heavy Metals: An Overview. *J. Environ. Biol.* **2005,** *26* (2), 301–313.

Sheoran, I. S.; Single, H. R.; Singh, R. Effect of Cd and Ni on Photosynthesis and the Enzymes of Photosynthetic Carbon Reduction cycle in Pigeon Pea (Cajanus cajan). *Photosyn. Res.* **1990,** *23*, 345–811.

Shiyu, Q. I. N.; Hongen, L. I. U.; Zhaojun, N. I. E.; Rengel, Z.; Wei, G. A. O.; Chang, L. I.; Peng, Z. H. A. O. Toxicity of Cadmium and Its Competition with Mineral Nutrients for Uptake by Plants: A Review. *Pedosphere* **2020,** *30*, 168–180.

Shukla, O. P.; Dubey, S.; Rai, U. N. Preferential Accumulation of Cadmium and Chromium: Toxicity in Bacopa monnieri L. Under Mixed Metal Treatments. *Bull. Environ. Contamin. Toxicol.* **2007,** *78*, 252–257.

Singh, H. P.; Mahajan, P.; Kaur, S.; Batish, D. R.; Kohli, R. K. Chromium toxicity and tolerance in plants. *Environ. Chem. Lett.***2013,** *11*, 229–254.

Singh, P. K.; Tewari, S. K. Cadmium Toxicity Induced Changes in Plant Water Relations and Oxidative Metabolism of *Brassica juncea* L. Plants. *J. Environ. Biol.***2003,** *24*,107–117.

Singh, S.; Kumar, M. Heavy Metal Load of Soil, Water and Vegetables in Peri–Urban Delhi. *Environ. Monit. Assess.* **2004,** *120* (1–3), 79–91.

Sinha, R. K.; Herat, S.; Tandon, P. K. 14 Phytoremediation: Role of Plants in Contaminated Site Management. In: *Book of Environmental Bioremediation Technologies*; Singh, S. N., Tripathi, R. D., Eds.; Springer: Berlin, Germany, 2004; pp 315–330.

Skogerboe, R. K.; Coleman, G. N. Microwave Plasma Emission Spectrometry. *Anal. Chem.* **1976,** *48*, 611A–622.

Soltanpour, P. N.; Schwab, A. P. A New Soil Test for Simultaneous Extraction of Macro-and Micro-Nutrients in Alkaline Soils. *Commun. Soil Sci. Plant Anal.* **1977,** *8*, 195–207.

Song, B.; Lei, M.; Chen, T. B.; Zheng, Y. M.; Xie, Y. F.; Li, X. Y.; Gao, D. Assessing the Health Risk of Heavy Metals in Vegetables to the General Population in Beijing, China. *J. Environ. Sci.* **2009,** *21* (12), 1702–1709.

Srivastava, V.; Sarkar, A.; Singh, S.; Singh, P.; de Araujo, A. S. F.; Singh, R. P. Agroecological Responses of Heavy Metal Pollution with Special Emphasis on Soil Health and Plant Performances. *Front. Environ. Sci.* **2017**, *5*, 64. DOI: 10.3389/fenvs.2017.00064

Steenland, K.; Boffetta, P. Lead and Cancer in Humans: Where Are We Now? *Am. J. Ind. Med.* **2000,** *38*, 295–299.

Stohs, S. J.; Bagchi, D. Oxidative Mechanisms in the Toxicity of Metal Ions. *Free Radic. Biol. Med.* **1995,** *18* (2), 321–336.

Støving, C.; Jensen, H.; Gammelgaard, B.; Stürup, S. Development and Validation of an ICP-OES Method for Quantitation of Elemental Impurities in Tablets According to Coming US Pharmacopeia Chapters. *J. Pharm. Biomed. Anal.* **2013,** *84*, 209–214.

Su, C.; Jiang, L. Q.; Zhang, W. J. A Review on Heavy Metal Contamination in the Soil Worldwide: Situation, Impact and Remediation Techniques. *Environ. Skeptics Cri.* **2014**, *3* (2), 24–38.

Sumner, M. E. Beneficial Use of Effluents, Wastes, and Biosolids. *Commun. Soil Sci. Plant Anal.* **2000**, *31* (11–14), 1701–1715.

Sun, W. H.; Jiang, Y. X.; Li, X. Research of the Evaluation on Heavy-Metal Pollution in Rice by Sewage Irrigation. *Appl. Mech. Mater.* **2013**, *295–298*, 1594–1599. DOI: 10.4028/www.scientific.net/AMM.295–298.1594.

Taghipour, M.; Jalali, M. Impact of Some Industrial Solid Wastes on the Growth and Heavy Metal Uptake of Cucumber (*Cucumis sativus* L.) Under Salinity Stress. *Ecotoxicol. Environ. Saf.* **2019,** *182*, 109347. https://doi.org/10.1016/j.ecoenv.2019.06.030

Tangahu, B. V.; Sheikh, Abdullah, S. R.; Basri, H.; Idris, M.; Anuar, N.; Mukhlisin, M. A Review on Heavy Metals (As, Pb, and Hg) Uptake by Plants Through Phytoremediation. *Int. J. Chem. Eng.* **2011,** *2011*, 939161. https://doi.org/10.1155/2011/939161

Tasrina, R. C.; Rowshon, A.; Mustafizur, A. M. R.; Rafiqul, I.; Ali, M. P. Heavy Metals Contamination in Vegetables and Its Growing Soil. *J. Environ. Anal. Chem.* **2015,** *2*, 2380–2391.

Tóth, G.; Hermann, T.; Da Silva, M. R.; Montanarella, L. Heavy Metals in Agricultural Soils of the European Union with Implications for Food Safety. *Environ. Pollut.* **2016**, *88*, 299–309. DOI: 10.1016/j.envint.2015.12.017

Tripathy, B. C.; Mohanty, P. Zinc Inhibited Electron Transport of Photosynthesis in Isolated barley Chloroplasts. *Plant Physiol.* **1980,** *66*, 1174.

Uduma, A. U.; Jimoh, W. L. O. Sequential Extraction Procedure for Partitioning of Lead, Copper, Cadmium, Chromium and Zinc in Contaminated Arable Soils of Nigeria. *Am. J. Environ. Energy Power Res.* **2013,** *1*, 186–208.

Vodyanitskii, Y. N.; Vasil'ev, A. A.; Vlasov, M. N. Hydrogenic Heavy Metal Pollution of Alluvial Soils in the City of Perm. *Eur. Soil Sci.* **2008**, *41*, 1238–1246.

WHO. Air quality guidelines for europe. WHO Regional Publications, European Series No 23. World Health Organization officer for Europe, Copenhagen, 1987.

WHO. *Guideline for Drinking Water Quality, Vol. II, Health Criteria and Other Supporting Information*; World Health Organization: Geneva, 1984.

Właśniewski, S.; Hajduk, E. Akumulacja kadmu w glebach i wybranych warzywach uprawianych w ogrodach działkowych Rzeszowa. Roczniki Gleboznawcze—*Soil Sci. Annu* **2012**, *48*, 55–60. In polish.

Wuana, R. A.; Okieimen, F. E. Heavy Metals in Contaminated Soils: A Review of Sources, Chemistry, Risks and Best Available Strategies for Remediation. *Int. Scholar. Res. Netw. Ecol.* **2011**, 1–20. DOI:10.5402/2011/402647

Yang, Y. B.; Sun, L. B. Status and Control Countermeasures of Heavy Metal Pollution in Urban Soil. *Environ. Prot. Sci.* **2009,** *35* (4), 79–81.

Yongsheng, W.; Qihui, L.; Qian, T. Effect of Pb on Growth, Accumulation and Quality Component of Tea Plant. *Procedia Eng.* **2011**, *18*, 214–219.

Zhang, M. K.; Liu, Z. Y.; Wang, H. Use of Single Extraction Methods to Predict Bioavailability of Heavy Metals in Polluted Soils to Rice. *Commun. Soil Sci. Plant Anal.* **2010**, *41* (7), 820–831.

Zhang, W. J.; Jiang, F. B.; Ou, J. F. Global Pesticide Consumption and Pollution: With China as a Focus. *Proc. Int. Acad. Ecol. Environ. Sci.* **2011,** *1* (2), 125–144.

Zhao, X.; Jiang, T.; Du, B. Effect of Organic Matter and Calcium Carbonate on Behaviors of Cadmium Adsorption–Desorption on/from Purple Paddy Soils. *Chemosphere* **2014,** *99*, 41–48.

Zhou, H.; Yang, W. T.; Zhou, X.; Liu, L.; Gu, J. F.; Wang, W. L.; Zou, J. L.; Tian, T.; Peng, P. Q.; Liao, B. H. Accumulation of Heavy Metals in Vegetable Species Planted in Contaminated Soils and the Health Risk Assessment. *Int. J. Environ. Res. Public Health* **2016,** *13*, 289–301.

Zojaji, F.; Hassani, A. H.; Sayadi, M. H. Bioaccumulation of Chromium by Zea Mays in Wastewater– Irrigated Soil: An Experimental Study. *Proc. Int. Acad. Ecol. Environ. Sci.* **2014**, *4* (2), 62–67.

Zwolak, A.; Sarzyńska, M.; Szpyrka, E.; Stawarczyk, K. Sources of Soil Pollution by Heavy Metals and Their Accumulation in Vegetables: A Review. *Water Air Soil Pollut.* **2019**, *230*, 164. https://doi.org/10.1007/s11270–019–4221-y

CHAPTER 4

Origin, Distribution, Fate, and Remediation of Microplastics in Biowastes and Biowaste-Amended Soil

PRASHANT SRIVASTAVA[1], NANTHI BOLAN[2,3],
VERÔNICA CASAGRANDE[4], RAVINDU PERERA[5], and
HASINTHA WIJESEKARA[6]

[1]*The Commonwealth Scientific and Industrial Research Organisation (CSIRO), Environment, Waite Campus, Urrbrae, SA, Australia*

[2]*UWA School of Agriculture and Environment, The University of Western Australia, Perth, WA, Australia*

[3]*The UWA Institute of Agriculture, The University of Western Australia, Perth, WA, Australia*

[4]*Environmental Sanitation Institute, University of Caxias do Sul, Caxias do Sul, Rio Grande do Sul, Brazil*

[5]*Department of Livestock Production, Faculty of Agricultural Sciences, Sabaragamuwa University, Belihuloya, Sri Lanka*

[6]*Department of Natural Resources, Faculty of Applied Sciences, Sabaragamuwa University, Belihuloya, Sri Lanka*

ABSTRACT

Microplastics are an emerging class of contaminants that are ubiquitous in the environment and have been detected in terrestrial, marine, and freshwater ecosystems, and even at remote locations. These are becoming a huge challenge due to their adverse impact on human health and the

Environmental Contaminants: Impact, Assessment, and Remediation.
Pritam Ganguly, Jajati Mandal, M. Paramsivam, & Sandip Patra, (Eds.)

environment, particularly the marine environment. Biowastes, such as farmyard manure, composts, crop residues, and biosolids are land-applied due to the abundance of carbon and nutrients. However, due to the contamination of microplastics, the land application of biowastes inadvertently adds particulate plastic to agricultural soil, making them available for uptake by soil biota and eventually, the food chain. Due to their fine particle size and large surface area, microplastics (and nanoplastics) adsorb other contaminants on their surface and act as a vector; therefore, uptake of microplastics by soil biota, followed by humans and animals can lead to adverse effects due not only to microplastics but also to the contaminants transported by them. Unfortunately, due to the inherent properties of plastics and polymers, it is difficult to remediate microplastics in the environment, particularly in biowastes and biowaste-amended soil. There are limited physical, chemical, and biological remediation methods available to remove, treat, or degrade microplastics, which are described in this chapter.

4.1 INTRODUCTION

Microplastics are defined as plastic pieces smaller than 5 mm and are classified into primary and secondary microplastics, depending on how they were generated (Masura et al., 2015; Bradney et al., 2019). Nanoplastics is another term defined to represent plastic particles smaller than microplastics. However, no consistent definition is available in the literature defending and explaining the word "nanoplastics" (Gigault et al., 2018). According to a review published by Gigault et al. (2018), the particle size of nanoplastics range from 1 to 1000 nm (although some authors have set the upper size limit at 100 nm) resulting from the degradation of industrial plastic objects and can exhibit colloidal behavior.

Microplastics are ubiquitous and have contaminated terrestrial, marine, and freshwater ecosystems (Rochman, 2018), thereby, raising increasing concerns about their implications for animal and human health (Vethaak and Legler, 2021; Wright and Kelly, 2017). Alarmingly, microplastics have been detected in the air (Allen et al., 2019), rainwater (Brahney et al., 2020), wastewater (Conley et al., 2019), Arctic sub-surface waters (Kanhai et al, 2018), Antarctic sea ice (Kelly et al., 2020), Mount Everest (Napper et al., 2020), beer, sea salt, and tap water (Kosuth et al., 2018), fish (Smith et al., 2018), fruit, vegetables (Oliveri Conti et al., 2020), human feces (Schwabl et al., 2019), and even human placentas (Ragusa et al., 2021).

Recent estimates suggest that depending on age and sex and individual circumstances, humans are exposed to up to 200,000 microplastic particles annually (Cox et al., 2019). It is estimated that at the current rate of plastic usage and disposal, approximately 12 billion metric tons of plastic waste is likely to end up in landfills or the natural environment (Geyer et al., 2017).

As reviewed by Bucci et al. (2020), the very properties that make plastic a versatile useful household product may cause environmental damage (Andrady, 2015). Besides, these make plastics so durable that they can persist in the natural environment (e.g., in soil) for hundreds to thousands of years (Barnes et al., 2009). Their lightweight assists in the transportation of plastics via air and water (Barnes, 2002). The hydrophobic nature of plastics helps in the adsorption of especially organic contaminants on their surface, thereby making them a vector for a range of environmental contaminants (Ogata et al., 2009).

Once microplastic particles enter the body, they could release adsorbed contaminants (Teuten et al., 2009). Fine microplastic particles could be absorbed by cells in the lungs, gut, and digestive tract (Powell et al., 2010). Little is known about the impact of microplastics once they are inside the human body. Besides, limited information is available on microplastic contamination of major sources of nutrition, including grains, vegetables, beef, and poultry.

Based on the source, microplastics can be categorized into primary and secondary microplastics. Primary microplastics are manufactured and used primarily for industrial processes or in consumer products (van Wezel et al., 2016; Kumar, 2016). Conversely, secondary microplastics are produced due to the disintegration of larger (primary) plastic particles (Gouin et al., 2015), industrial abrasion process, erosion of synthetic paints, wearing of car tires (Lassen et al., 2015; Patel et al., 2019), and tearing of synthetic textiles (Napper and Thompson, 2016).

A significant amount of microplastics ends up in wastewater treatment plants (WWTPs) via human activities including but not limited to microbeads used in personal care products, the release of microfibers from washing, stormwater, incidental release, industrial waste, and atmospheric deposition (Sarkar et al., 2021). Besides, plastic mulch used for temperature, moisture, and weed control in agriculture contributes microplastics to terrestrial ecosystems (Steinmetz et al., 2016; Qi et al., 2018).

In this chapter, we have covered the origin, distribution, and fate of microplastics in solid biowastes and biowaste-amended soil. A limited number of remediation/management technologies for microplastics in biowastes have also been included.

4.2 BIOWASTES: SOURCES AND THEIR LAND APPLICATION

Biowastes are biological waste materials or organic amendments, which are used to improve soil quality, as they are rich in carbon and nutrients, and help in improving soil fertility and physical properties and conditioning. Biowastes include farmyard manure (FYM), composts, crop residues, and biosolids (Quilty and Cattle, 2011). Inorganic fertilizers are the primary source of plant nutrients and their demand is constantly increasing despite their adverse ecological impacts (Srivastava et al., 2009). With organic farming becoming more popular, biological waste materials are frequently being used as a nutrient source (Park et al., 2011). Apart from FYM, composts, crop residues, and biosolids, other biological waste materials, such as food waste composts (Beiyuan et al., 2018), paper and pulp sludges (Wang et al., 2019), and anaerobic digestates (Chen et al., 2017) are also land-applied. This section describes different types of biowastes along with their advantages and disadvantages.

4.2.1 FARMYARD MANURE

FYM is a decomposed mixture of animal excrements and crop residues. A large volume of FYM is generated through cattle and poultry excreta and flushing materials from confined poultry and animal industries. The FYM requires environmentally benign disposal or utilization as a resource (Shen et al., 2017). The enrichment of trace elements (e.g., Cu and Zn) derived from animal diets makes FYM a great nutritional source for promoting plant growth (Bolan et al., 2010). Excrements from chicken, swine, and cattle farms are the primary sources of FYM. The excrements are generally stored in pits, lagoons, or heaps to allow the formation of composted manure, which may take a few weeks to several months, during which there is a reduction in the amounts of organic matter and different microorganisms (Ben et al., 2013). The formation of FYM depends on factors like temperature, humidity, and redox potential (Tavazzi et al., 2018). Once formed, FYM is applied to agricultural fields as a carbon and nutrient source. The feasibility of using FYM with different physicochemical and biological properties directly as soil conditioners and fertilizers depends on several factors including the season, animal characteristics (type, sex, age, and size), bedding types, water usage, and manure management (e.g., collection, handling, and storage) (Petersen et al., 2007). Land disposal of FYM, which is related to weather conditions, must comply with the legally prescribed maximum content of organic-N in the manures to prevent aquatic eutrophication due to infiltration and surface runoff (Aga et al., 2005). Typically, before land application, FYM should be processed via composting, anaerobic digestion, granulation, and alum treatment (Wallace et al., 2018).

4.2.2 COMPOSTS

Composting is, in simple terms, an aerobic biological process that leads to the decomposition of organic materials, such as food waste, garden scraps, and garden waste into simpler nutrient-rich organic and inorganic compounds. It is a controlled process to stabilize organic wastes for recycling and disposal via aerobic biological decomposition (Chia et al., 2020). Microorganisms, such as bacteria and fungi are the main drivers of this process. The decomposing organic substances are transformed into an organically stable product that does not have a phytotoxic effect on the plant when used for soil amendment.

Biological waste materials generated from households often contain non-decomposable materials, such as pieces of plastic, which makes the composting process arduous. Therefore, the composting products, when applied to agricultural soil, may inadvertently contribute to plastic. Conventionally, composts are land-applied in layers on surface soil and incorporated into the topsoil. Similar to FYM, composts also improve soil fertility by providing nutrients, such as C and N (Benitez et al., 2003; Eghball, 2002). Composts also act as a soil conditioner, by improving the stability of soil aggregates (Annabi et al., 2011), lowering the soil bulk density (Somerville et al., 2018), reducing soil erosion, increasing the soil aeration, soil porosity, and water-holding capacity (Hargreaves et al., 2008).

4.2.3 CROP RESIDUES

Crop residues are postharvest organic remains, including stalks, stems, leaves, and seed pods. These contain a significant amount of plant nutrients that can be land-applied for crop growth (Liu et al., 2010), such as soil amendments (Erenstein, 2002) and animal feed (Karlen et al., 2009), thus enhancing agricultural productivity. Crop residues have also been used in building materials (Krausmann et al., 2008), and as industrial resources (Valbuena et al., 2012). In developing countries, the burning of crop residues is a common practice, which leads to a significant amount of greenhouse gas (GHG) emissions (Yokelson et al., 2011). Rice straw and corn straw are among the crop residues which are most commonly used as amendments in agricultural soil. Increasingly, crop residues are used as a feedstock material for the synthesis and soil application of biochar. Their application not only contributes to soil organic matter, but also improves the water-holding capacity (Ampofo, 2018), bulk density (Anik et al., 2017), aggregation (Yang et al., 2018), pesticide retention (Joshi et al., 2019), and microbial biomass and activity (Rong et al., 2018) of the amended soil, as well as the water use efficiency of plants (Yin et al., 2015).

4.2.4 BIOSOLIDS

Biosolids are organic solids that originate from wastewater and can be beneficially reused as a nutrient source in agricultural crops after suitable treatments such as anaerobic digestion and composting (Tchobanoglous, 2014). Biosolids are made up of suspended solid particles removed from the wastewater treatment process. A significant amount of biosolids is produced at the WWTPs (Letcher et al., 2020). Biosolids typically contain 40–70% organic matter (Torri et al., 2014), and are also rich in essential plant nutrients, such as N (~ 3.2%), P (~ 2.3%), and K (~ 0.3%). Due to these and many other properties, biosolids are beneficially reused as fertilizers for the revitalization or rehabilitation of agricultural soils (Haynes et al., 2009; Wijesekara et al., 2021), as they can increase the soil organic matter (Sharma et al., 2017), macronutrients (e.g., N, P, K), micronutrients (e.g., Zn, Cu, Fe, and Mn), air and water permeability (Samaras et al., 2010), aggregate stability (Roca-Pérez et al., 2009), porosity (Rostagno and Sosebee, 2001), water-holding capacity (Veeresh et al., 2003), cation exchange capacity, and microbial activity (Lloret et al., 2016) of agricultural soils, which subsequently enhance the soil fertility and crop yield (Walter et al., 2000), and inhibit soil erosion (Zerzghi et al., 2010).

4.2.5 RISKS ARISING FROM BIOWASTES

Despite the various benefits, land application of biowastes may pose a significant risk to agricultural soils due to the inadvertent introduction of various contaminants, such as metals and metalloids (e.g.; Zn, Cu, Ni, Cd, Co, As, and Pb) (Zhi et al., 2020), polychlorinated biphenyls (PCBs), disinfectants, detergents, pharmaceuticals and personal care products (PPCPs), and steroid hormones (Bourdat-Deschamps et al., 2017, Verlicchi and Zambello, 2015), pathogenic microorganisms (Jiang et al., 2020), per- and poly-fluoroalkyl substances (Lakshminarasimman et al., 2020; Wang et al., 2020), and microplastics (Wijesekara et al., 2018). In addition to introducing contaminants, land application of biowastes may potentially lead to nutrient imbalances in the amended soil and contaminate surface water with excess nutrients and further enhance GHG emission (Thangarajan et al., 2013). Therefore, it is paramount that biowastes undergo appropriate preprocessing and/or pretreatment, such as alkaline stabilization of biosolids and anaerobic digestion prior to their land application to agricultural soils.

4.3 DISTRIBUTION OF MICROPLASTICS IN BIOWASTES AND BIOWASTES-AMENDED SOILS

Microplastics accumulate in a range of biowaste types, including FYM, composts, biosolids, and municipal solid waste (Wijesekara et al., 2018; Wu et al., 2020). Apart from these main biowaste types, industrial sludge (e.g., digestate) has also been identified as a possible source of microplastics (Weithmann et al., 2018). Globally, all these biowastes are generated in significant quantities (Wijesekara et al., 2016). For example, the annual global production of biosolids was estimated at 100 million tons, of which the United States and Australia produced 5.6 and 0.4 million tons, respectively (Thangarajan et al., 2013). Similarly, the annual global municipal solid waste production rate was 1.7 billion tons in 2008 (Foo and Hameed, 2009), which is estimated to increase to 2.2 billion tons by 2025 (Hoornweg and Bhada-Tata, 2012). Due to the high amounts of nutrients and organic matter, land application of biowastes has been practiced in many countries, thereby resulting in microplastics accumulation in the soil environment (Nizzetto et al., 2016).

4.3.1 WASTEWATER EFFLUENTS AND SLUDGE

WWTPs have become a significant source of discharging microplastics to the environment. Microplastics in the WWTPs are derived from personal care products, cosmetics, and plastic items such as textiles, car tires, or street paints (Crossman et al., 2020). Microplastics originating from sludge primarily include polyolefin and acrylic fibers, polyethylene and polyamide films, alkyl resin flakes, and polystyrene spheres. Reuse of both wastewater and sludge causes the reintroduction of microplastics into the environment and incarnates an environmental threat (Gatidou et al., 2019), the function of WWTPs in relation to the removal of plastic fragments and fibers or their subsequent accumulation in the soil is still not thoroughly investigated by scientists. Therefore, a deliberation exists for clarifying the extent to which treatment and management of wastewater discharges lead to the accumulation of microplastics in environmental compartments (Carr et al., 2016). Nizzetto et al. (2016) reported that North American farm soils annually receive 300,000 tons of microplastics from biosolid applications. Further, repeated land application of biosolids creates a risk of accumulating microplastics on soil(Brodhagen et al., 2017; Corradini et al., 2019). Crossman et al. (2020) found that biosolids-amended soils had high concentrations of several polymers, such as polyethylene (4772 MP kg^{-1}), polypropylene (2394 MP kg^{-1}), acrylic (815 MP kg^{-1}), and polyester (2305 MP kg^{-1}). Besides, biosolids contain several toxic compounds, such as toxic trace metals, pharmaceuticals, antibiotics, endogenous hormones, and synthetic steroids (Sharma et al., 2017; Smith, 2009). Due to their property of hydrophobicity, microplastics can bind these toxic compounds and cause severe problems to biota. For example, microplastics can change the biophysical environment by interacting with soil fauna and also introducing toxic compounds to the food webs (Ashton et al., 2010; Papaioannou et al., 2020; Picó et al., 2020).

The utilization of biosolids as a fertilizer is a common practice in horticulture. In 2015, a significant amount (>50–77%) of sludge produced in Albania, Bulgaria, Ireland, Norway, Spain, and the UK was applied to agricultural fields, whereas only <5% of the sludge produced in Finland and Romania was land-applied. Contrarily, no sludge was used in the agricultural fields in Greece, Malta, Slovakia, and Slovenia (Hudcová et al., 2019). Biosolids-derived pollutants are subjected to regulations, thereby limiting their land application, mainly because these regulations are based on heavy metals and pathogens. However, plastics are not categorized as a potentially unwanted ingredient in biosolids by European Union (EU 86/278/EEC) or the U.S. (Code 503) regulation. Some studies related to the efficiency of WWTPs have reported that 90% of microplastic was removed from the wastewater and abided by biosolids (Carr et al., 2016; Mintenig et al., 2017). Plastic concentrations detected ranged from ~1500 to 24,000 particles per kg of biosolids (Mintenig et al., 2017) and 15,800 microplastic particles per kg of sludge (Mahon et al., 2017). Based on the application rate of biosolids, the annual microplastic input to agricultural land was estimated to be 63,000–430,000 tons in Europe and 44,000–300,000 tons in North America, respectively (Nizzetto et al., 2016). This study further claimed that the biosolids-derived microplastic load exceeded the estimated marine surface water-based microplastic emissions. Studies have also found high levels of microplastics in WWTP sludge (Mahon et al., 2017; Nizzetto et al., 2016). For example, Mahon et al. (2017) and Lassen et al. (2015) reported 4200–15,400 microplastic particles per kg of dry sludge and 1000–24,000 microplastic particles (≥10 μm) per kg of dry sludge in German WWTPs, respectively. The repetitive application of biosolids to agricultural land is, nevertheless, likely to be a significant input of microplastics into the soils (Nizzetto et al., 2016).

4.3.2 COMPOSTS

Given their origin from domestic and industrial wastes, composts often contain non-decomposable materials, such as pieces of plastic, which, when applied to agricultural soil, may inadvertently contribute to plastic. Polyethylene fibers and strands have been detected in the industrially produced compost in the USA. Film-shaped microplastics have been identified as a prominent contaminant in industrially produced compost. It is difficult to screen plastics and other contaminants from the waste material before the composting process phase due to the high moisture content of the feedstock material. It is especially difficult to remove film-shaped microplastics by screening because wet organic material sticks to the plastics, preventing physical and mechanical separation (David, 2013). Most industrial screeners use an 8–25 mm opening sieve. The smaller opening can differ by 6 mm, which means anything less than 6 mm remains in the compost, depending on the technology and the manufacturer. Because the size of microplastic particles is defined as less than 5 mm or smaller, the fragments of these particles may still be present in the compost. It has been concluded that removing or screening out the microplastic fragments from the finished compost is challenging (Brinton, 2005; Smith, 2018). Moreover, polystyrene and polyethylene plastics that are incorporated with compost to ameliorate the soil properties act as a source of microplastics to the soils (Hurley and Nizzetto, 2018). Approximately, 1.2 g plastics kg^{-1} organic compost was reported in Slovenia (Gajst, 2016). Based on an annual rate of compost application of 7 t ha^{-1}, approximately, 16–1200 g visible plastic particles ha^{-1} were estimated to be discharged into farmland annually (Bläsing and Amelung, 2018).

4.3.3 *FARMYARD MANURES*

Microplastics reach FYMs through direct excretion by animals and indirect accumulation resulting from manure collection and processing. The application of manure as a nutrient source for soils provides a possible pathway for microplastics to enter the soil environment. Microplastics may also exist in livestock manure from animal breeding and poultry production. For example, plastic bags are used to load feed and manure. Eventually, these bags can be degraded and mixed with the feeds and digested by animals, and then released into the environment. In agriculture, silage films are frequently used to wrap the silage balls. Moreover, cattle ear tags and animal feed packaging materials can act as sources for microplastics in manure. Beriot et al. (2020) have found microplastic debris in the corn silage brought to the animals. Ingesting them can cause microplastics contamination of the feces. Mekuanint et al. (2017) reported that plastic polyethylene bags are the most frequently found foreign body in ruminants' rumen and their frequency of occurrence and prevalence is 46.1%. Low-density microplastics have been found in sheep feces accounting for approximately 1000 particles per kg (Betriot et al., 2020). Because plastic materials are abundant on farms, accidental dismantling, and spreading of materials will introduce plastic waste into manure (Piehl et al., 2018). This situation can cause serious pollution in farmland after manure fertilization for resource production. As per a recent study in China, the average annual abundance of microplastics was 1250 ± 640 particles kg^{-1} in manure (Yang et al., 2021) They estimated the average accumulation of microplastics in the agricultural soil with long-term application of pig manure to be 3.50 ± 1.71 million particles ha^{-1} a^{-1}.

4.4 FATE OF MICROPLASTICS IN BIOWASTES AND BIOWASTE-AMENDED SOIL

Once biowastes are land-applied, microplastics (synthetic polymers) are distributed in all environmental compartments due to their long life span. UV-radiation and physical abrasion processes cause the degradation of microplastics (Song et al., 2017; Bläsing and Amelung, 2018). However, in soil, the degradation process occurs slowly due to a lack of driving forces (Krueger et al., 2015). A weight loss of polyethylene (0.1–0.4%) in the soil was observed after 800 days (Albertsson, 1980), and a weight reduction of polypropylene by 0.4% was observed following 1 year of soil incubation (Arkatkar et al., 2009). Whereas, there are instances where no degradation in polyvinyl chloride in soil was observed even after 10–35 years (Ali et al., 2014).

Earthworms have been shown to act as a vector of microplastics in soils, incorporating them into the soil, with the likely pathways being casts, burrows, egestion, and adherence to the earthworm skin (Rillig et al., 2017; Bläsing and Amelung, 2018). Vertical and lateral movements of polyethylene beads in sizes 710–2800 μm down to 10 cm caused by earthworms (*Lumbricus terrestris* L) were observed in 21 days (Rillig et al., 2017). Similar observations have been reported for 400 μm low-density polyethylene (LDPE) microplastics (Huerta Lwanga et al., 2017). Research suggests that fine microplastic particles promote bioturbation transport. Macrobenthos which live in biogenic structures buried in the sediment cause solid and liquid transport through their feeding and movement activities. Collembolan species, such as *Folsomia candida* and *Proisotoma minuta* have been reported to transport urea–formaldehyde (100–200 μm) microplastic (Maaß et al., 2017). The studies demonstrated that the macrobenthic species help to incorporate and transfer microplastics vertically in the soil. However, because these species do not usually travel to the groundwater, these processes reduce their groundwater contamination risk. As shown in Figure 4.1, algae, fungi, and bacteria form biofilms with microplastics due to their large specific surface area (Tu et al., 2020). Besides, microplastics can bind to the soil particles and make aggregates, changing the physical properties of the soil profiles (Rillig et al., 2019).

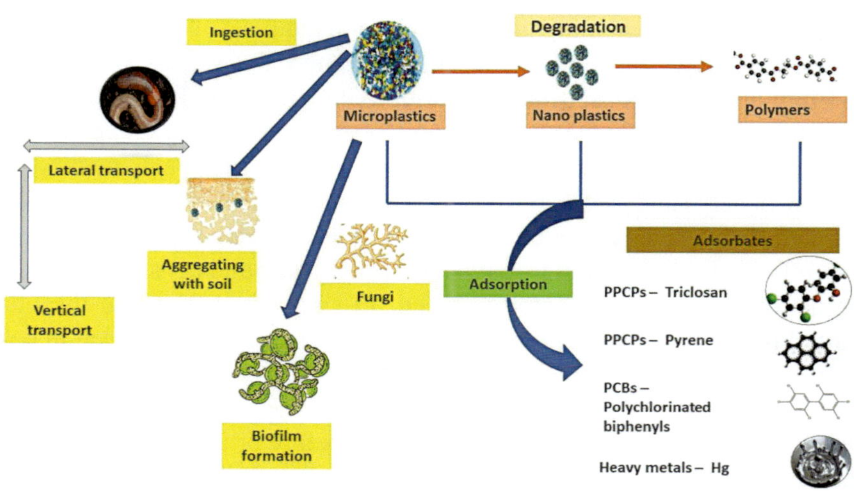

FIGURE 4.1 Schematic view for the fate of microplastics in biowastes and biowaste-amended soil.

4.5 REMEDIATION OF MICROPLASTICS IN BIOWASTES AND BIOWASTE-AMENDED SOIL

The remediation of soil contaminated with plastic materials constitutes a significant challenge due to the large-scale production and the high consumption of polymeric material – both are proportional to population growth and the lack of effective technologies applied to waste treatment. The treatment technologies must consider different types of polymers, the composition, and structure of molecular chains, and different additives, which add properties and characteristics relevant to commercial applications. The additives commonly used in polymers are related to dyes, plasticizers, which modify the material's rigidity/flexibility properties, flame retardants, antistatic, and antioxidants (Bläsing and Amelung, 2018; Pathan et al., 2020), catalysts of polymerization, initiators, and accelerators (da Costa et al., 2016).

Removing microplastics from the contaminated environment resulting from biowaste application is challenging due to the fine particle size and the complex constituents of microplastics, and their wide distribution in biowastes. Although several researchers have investigated the distribution of microplastics in the environment and their adverse effects on sensitive receptors, an appropriate strategy for the removal and/or degradation of microplastics is yet to be developed (Weithmann et al., 2018). The potential remediation strategies for microplastics require prior assessment and selection based on the contaminating material, for example, solid biowastes, such as sewage sludge, biosolid, compost, and manure (Kowalska et al., 2020), and liquid biowastes used for fertigation (Chojnacka et al., 2019). Both in situ and ex situ techniques may be developed for removing and/or treating microplastics from biowastes (Megharaj et al., 2014). Each method has specific benefits and associated costs. Ex situ techniques involve the physical extraction of the contaminated material from the impacted site and its relocation off-site for appropriate treatment, which can be expensive due to the extraction and transport costs involved. In situ remediation, on the other hand, involves on-site treatment of the contaminated media, which may be relatively cheaper (Kuppusamy et al., 2016).

Most treatment methods employed for sewage sludge are unable to efficiently remove microplastics from biosolids before or after their land application (Li et al., 2018; Mahon et al., 2016). Advanced treatment processes, such as thermal drying and stabilization with lime rather disintegrate microplastics into even finer particles (e.g., nanoplastics) due to alkaline conditions and heat hydrolysis, which eventually increases the number of micro or nanoplastics (Mahon et al., 2016). The available technologies for the removal of microplastics from solid and liquid biowastes are presented hereunder.

4.5.1 PHYSICAL REMOVAL

To the best of our knowledge, there are limited studies on the physical removal of microplastics from biowastes. The physical methods for microplastics removal depend on the type, shape, size, and mass of the plastic materials or particles (Lambert and Wagner, 2018; Poerio et al., 2019). Physical removal techniques such as sieving are generally applicable to larger plastics (macroplastics) in solid biowastes and liquid biowastes due to the possible application of flow and counterflow of the volume of waste in membranes and filters. For example, membrane technology, conventional membrane separation processes, and membrane bioreactor processes have been reported (Poerio et al., 2019). Similarly, dynamic membranes (DM) technology is a promising treatment option for removing microparticles during wastewater treatment, including the removal of low-density materials. This technology was used to remove microplastics from synthetic effluent (Li et al., 2018), and can potentially be used for the removal of microplastics in biowaste. The advantages of DM technology for removing microplastics are low resistance to filtration, low transmembrane pressure, easy operation, and the application of nonchemical treatment (Padervand et al., 2020).

4.5.2 CHEMICAL DEGRADATION AND TREATMENT

Chemical degradation of microplastics considers the formation of nanoplastics that can be generated as a by-product of chemical reactions (Lambert and Wagner, 2016). It must be noted that the degradation process does not minimize or eliminate the impacts related to particulate plastics on sensitive receptors. Unfortunately, there is hardly any technique available for the chemical degradation of microplastics in solid biowastes before and/or after their land application. However, there are chemical degradation techniques available for microplastics in liquid biowastes, such as effluent from WWTPs, used for fertigation. These techniques utilize coagulants to bind microplastic particles and form strong bonds between the waste particles. The efficiency of coagulation and agglomeration processes for the removal of microplastics depends on the type of coagulant, pH, and chemical composition of the medium and concentrations (Padervand et al., 2020). Both processes have been reliably used to facilitate the separation of microplastics, albeit they need to be coupled with additional filtration stages. The most commonly used coagulants include Fe and Al salts (Chorghe et al., 2017), with Al performing better than Fe (Ariza-Tarazona et al., 2019). Application of Fe and Al salt coagulants along with ultrafiltration has successfully removed polyethylene microplastics from effluents through gravity sedimentation and sieving processes. For small microplastic particles (with diameter <0.5 mm), polyacrylamide (PAM) was found to be a much better coagulant, as the removal of microplastics increased from 25.83% without PAM to 61.19% with 15 mg/L of PAM. Besides, the removal efficiency for small microplastics decreased with increasing pH (Ariza-Tarazona et al., 2019).

4.5.3 BIOREMEDIATION

Bioremediation is a cost-effective and environmentally friendly remediation technique involving degradation, attenuation, and/or detoxification of contaminants via activities of living organisms, such as microbes (e.g., bacteria), fungi, plants, and algae (e.g., microalgae) (Megharaj et al., 2014; Skipper, 1998), and plastic-eating insect (Yang et al., 2015, 2018). The available bioremediation techniques for microplastics, particularly applicable to biowastes and biowaste-applied soil are described hereunder.

4.5.3.1 PHYTOREMEDIATION OF MICROPLASTICS

Phytoremediation techniques involve the interception of contaminants with plant roots for their accumulation in the rhizosphere, absorption in the root, and further translocation to aboveground plant parts (Gardea-Torresdey et al., 2004). The phytoremediation approach for microplastics considers the ability of plants to uptake micro or nanoplastics from biowastes or biowaste-applied soils, and/or degrade polymers, or may act as a physical interceptor, where microplastics can be accumulated and localized. Once plant roots have become saturated with the contaminant of concern, the plant can be harvested and disposed of, such that it is safe for human health and the environment (Gardea-Torresdey et al., 2004). Potential phytoremediation techniques for biowastes or biowaste-applied soil include phytoextraction, phytostabilization, and phytofiltration.

Phytoextraction (also called phytoaccumulation) involves the uptake of microplastics from the biowaste or soil and their subsequent translocation to aboveground plant parts. In this process, microplastic or nanoplastic particles accumulate on plant roots through root interception and are subsequently absorbed by the root system for further translocation to areal plant parts. This phenomenon was demonstrated by Li et al. (2019), who noted that 0.2 μm microbeads of polystyrene (PS) were extracellularly trapped in the root cap of the lettuce plant. They also observed the absorption of two sizes of PS microspheres (0.2 and 1.0 μm) in the edible parts of lettuce. Fluorescent markers, used to track PS microspheres in plant tissues, provided evidence of the adherence, absorption, accumulation, and translocation of submicrometric microplastics within the edible plant parts (Li et al., 2019).

Phytostabilization involves the immobilization of microplastics within the rhizosphere which reduces their bioavailability and prevents their further movement, distribution, and leaching into groundwater. If microplastics have moved to the groundwater, or are present in wastewater, phytofiltration removes microplastics from groundwater and aqueous waste streams via absorption or adsorption of microplastics through a profuse root system.

Some plants have shown the ability to accumulate microplastics, thus serving as a potential phytoremediation agent for microplastics in terrestrial ecosystems (Enyoh et al., 2019). For example, the absorption of LDPE microplastics and starch-based biodegradable plastic mulch has been demonstrated by wheat (Qi et al., 2018) and corn (Zhang et al., 2015).

4.5.3.2 BACTERIAL DEGRADATION OF MICROPLASTICS

Microorganisms, like bacteria, have been shown to decompose plastics in soils (Peixoto et al., 2017), and compost (Leejarkpai et al., 2011; Mercier et al., 2017), and can degrade microplastics during sludge treatments (Mahon et al., 2016; Li et al., 2018). Bacterium *Enterobacter cloacae* AKS7 has been found to reduce the LDPE-derived plastic (Sarker et al., 2020). *E. cloacae* AKS7 exploits the hydrophobicity on the cell surface to develop a microbial biofilm covering the LDPE surface, which enhances LDPE degradation (Tribedi and Sil, 2013). Although this bacterium has been found to efficiently degrade LDPE, its efficacy for degradation of microparticles, smaller than 5 × 4 cm has not been confirmed (Sarker et al., 2020). Other bacteria, such as *Rhodococcus* spp., *Bacillus sphericus*, *Pseudomonas citronellolis* EMBS027, and *P. syringae* (DC3000 ATCC 10,862) have also been reported to degrade LDPE (Nanda and Sahu, 2010; Bhatia et al., 2014; Kyaw et al., 2012).

Anaerobic digestion of sludge caused by anaerobic microbial communities has been found to slowly degrade microplastics (Mahon et al., 2016), which is accelerated at relatively high temperatures by thermophilic bacteria presumably due to the cleavage of -C-C- bonds (Manzur et al., 2004; Chen et al., 2020). Composting increases the temperature and provides a niche environment for the activity of thermophilic bacteria, which help in the sludge treatment. However, as traditional sludge treatment technologies cannot efficiently remove microplastics, hyperthermophilic composting (hTC) technology was developed to utilize thermophilic bacteria for microplastic biodegradation at high temperatures (Chen et al., 2020). This study demonstrated that hTC technology significantly enhanced the removal efficiency of sludge-based microplastics, representing a promising strategy for in situ biodegradation of microplastics. A better understanding of the impact of composting and its thermophilic bacteria on microplastics is critical for developing efficient methods for the biodegradation of microplastics in biowastes.

4.5.3.3 FUNGAL DEGRADATION OF MICROPLASTICS

Fungal bioremediation is a promising technology capable of removing, reducing, or degrading microplastics (Sanchez, 2020). Fungi, in general, are better than bacteria at degrading polymers, such as polyurethane and PE (Pathak and Navneet, 2017; Muhonja et al., 2018). Several studies have been conducted to evaluate the fungal degradation of polymers such as PVC, PP, PE, polyethylene terephthalate (PET), and polyesters. For example, white-rot fungi have been found to degrade plastics (Ellouze and Sayadi, 2016), the fungus *Zalerion maritimumto* has been shown to degrade PE pellets (Paço et al., 2017), and fungus belonging to ascomycetes have been found to degrade PE (Sangale et al., 2019), PVC (Vivi et al., 2019), and LDPE (Munir et al., 2018; Muhonja et al., 2018).

The biodegradation and biodeterioration of plastics by fungi involves enzymatic processes. For example, polyurethane biodegradation with Monascus sp. strains was facilitated by esterase, lipase, and protease enzymes (El-Morsy et al., 2017), and degradation of LDPE by a consortium of ascomycete strains occurred due to the activities of enzymes laccase, manganese peroxidase, and lignin peroxidase (Ameen et al., 2015). The primary mechanism by which microplastics are degraded by fungal enzymes is the reduction in the length of PE chains (Restrepo-Flórez et al., 2014).

4.6 SUMMARY AND CONCLUSIONS

Microplastics have posed a significant challenge to the scientific community—both in terms of their impact on human health and the environment as well as difficulties in their remediation in various matrices. Biowastes are useful for agricultural crops due to the presence of organic matter as well as essential plant nutrients. However, the presence of particulate plastics in biowastes inadvertently leads to their addition to soil and subsequently to their potential uptake by plants, animals, and humans. The uptake of microplastics adversely impacts human health, not only due to the toxic effect of microplastics but also to the other contaminants transported by them. Limited remediation methods are available for physical removal and chemical degradation or treatment of microplastics from biowastes (primarily liquid biowastes). Similarly, a few bioremediation methods, including phytoremediation and fungal and bacterial degradation of microplastics have also been developed. There is a need to develop novel remediation technologies to completely remove and degrade microplastics from the biowaste and biowaste-applied soil.

KEYWORDS

- **microplastics**
- **ecosystem**
- **contamination**
- **risk**
- **biowaste**

REFERENCES

Aga, D. S.; O'Connor, S.; Ensley, S.; Payero, J. O.; Snow, D.; Tarkalson, D. Determination of the Persistence of Tetracycline Antibiotics and their Degradates in Manure-Amended Soil Using Enzyme-Linked Immunosorbent Assay and Liquid Chromatography–Mass Spectrometry. *J. Agric. Food Chem.* **2005,** *53*, 7165–7171.

Akhbarizadeh, R.; Moore, F.; Keshavarzi, B.; Moeinpour, A. Microplastics and Potentially Toxic Elements in Coastal Sediments of Iran's Main Oil Terminal (Khark Island). *Environ. Pollut.* **2017,** *220*, 720–731.

Albertsson, A. C. The Shape of the Biodegradation Curve for Low and High Density Polyethenes in Prolonged Series of Experiments. *Eur. Polym. J.* **1980,** *7*, 623–630.

Ali, M. I.; Ahmed, S.; Robson, G.; Javed, I.; Ali, N.; Atiq, N.; Hameed, A. Isolation and Molecular Characterization of Polyvinyl Chloride (PVC) Plastic Degrading Fungal Isolates. *J. Basic Microbiol.* **2014,** *1*, 18–27.

Allen, S.; Allen, D.; Moss, K.; Le Roux, G.; Phoenix, V.; Sonke, J. Examination of the Ocean as a Source for Atmospheric Microplastics. *PLoS One* **2020,** *15*, e0232746.

Allen, S.; Allen, D.; Phoenix, V.; Le Roux, G.; Durántez Jiménez, P.; Simonneau, A.; Binet, S.; Galop, D. Atmospheric Transport and Deposition of Microplastics in a Remote Mountain Catchment. *Nat. Geosci.* **2019,** *12*, 339–344.

Ameen, F.; Moslem, M.; Hadi, S.; Al-Sabri, A. E. Biodegradation of Low Density Polyethylene (LDPE) by Mangrove Fungi from the Red Sea Coast. *Prog. Rubber. Plast. Re.* **2015,** *31*, 125–144.

Ampofo, E. A. Influence of Organic Mulches on Soil Physico-Chemical Properties and Maize (*Zea mays* L.) Crop Performance. *J Agric. Stud.* **2018,** *6*, 1–16.

Andrady, A. Persistence of Plastic Litter in the Oceans. In *Marine Anthropogenic Litter*; Bergmann, M., Gutow, L., Klages, M., Ed.; Springer: Cham, Switzerland, 2015; pp 57–72.

Anik, M. F. A.; Rahman, M. M.; Rahman, G. M.; Alam, M. K.; Islam, M. S.; Khatun, M. F. Organic Amendments with Chemical Fertilizers Improve Soil Fertility and Microbial Biomass in Rice-Rice-Rice Triple Crops Cropping Systems. *J. Agric Stud.* **2017,** *7*, 87.

Annabi, M.; Le Bissonnais, Y.; Le Villio-Poitrenaud, M.; Houot, S. Improvement of Soil Aggregate Stability by Repeated Applications of Organic Amendments to a Cultivated Silty Loam Soil. *Agric. Ecosyst. Environ.* **2011,** *144*, 382–389.

Ariza-Tarazona, M. C.; Villarreal-Chiu, J. F.; Barbieri, V.; Siligardi, C.; Cedillo-González, E. I. New Strategy for Microplastic Degradation: Green Photocatalysis using a Protein-Based Porous N-TiO_2 Semiconductor. *Ceram. Int.* **2019,** *45*, 9618–9624.

Arkatkar, A.; Arutchelvi, J.; Bhaduri, S.; Uppara, P. V.; Doble, M. Degradation of Unpretreated and Thermally Pretreated Polypropylene by Soil Consortia. *Int. Biodeterior. Biodegrad.* **2009,** *1*, 106–111.

Aryal, N.; Wood, J.; Rijal, I.; Deng, D.; Jha, M. K.; Ofori-Boadu, A. Fate of Environmental Pollutants: A Review. *Water Environ. Res.* **2020,** *92*, 1587–1594.

Ashton, K.; Holmes, L.; Turner, A. Association of Metals with Plastic Production Pellets in the Marine Environment. *Mar. Pollut. Bull.* **2010,** *60*, 2050–2055.

Barnes, D. K. A. Invasions by Marine Life on Plastic Debris. *Nature* **2002,** *416*, 808–809.

Barnes, D. K. A.; Galgani, F.; Thompson, R. C.; Barlaz, M. Accumulation and Fragmentation of Plastic Debris in Global Environments. *Philos. Trans. Royal Soc.* B **2009,** *364*, 1985–1998.

Beiyuan, J.; Tsang, D. C.; Bolan, N. S.; Baek, K.; Ok, Y. S.; Li, X. D. Interactions of Food Waste Compost with Metals and Metal-Chelant Complexes During Soil Remediation. *J. Clean Prod.* **2018,** *192*, 199–206.

Ben, W.; Pan, X.; Qiang, Z. Occurrence and Partition of Antibiotics in the Liquid and Solid Phases of Swine Wastewater from Concentrated Animal Feeding Operations in Shandong Province. China. *Environ. Sci. Process. Impacts* **2013,** *15*, 870–875.

Benitez, C.; Tejada, M.; Gonzalez, J. Kinetics of the Mineralization of Nitrogen in a Pig Slurry Compost Applied to Soils. *Compost. Sci. Util.* **2003,** *11*, 72–80.

Beriot, N.; Peek, J.; Zornoza, R.; Geissen, V.; Lwanga, E. H. Low Density-Microplastics Detected in Sheep Faeces and Soil: A Case Study from the Intensive Vegetable Farming in Southeast Spain. *Sci. Total Environ.* **2020,** *755*, 142653.

Bhatia, M.; Girdhar, A.; Tiwari, A.; Nayarisseri, A. Implications of a Novel Pseudomonas Species on Low-Density Polyethylene Biodegradation: An In Vitro to in Silico Approach. *Springerplus* **2014,** *3*, 497.

Bläsing, M.; Amelung, W. Plastics in Soil: Analytical Methods and Possible Sources. *Sci. Total Environ.* **2018,** *612*, 422–435.

Bolan, N. S.; Szogi, A.; Chuasavathi, T.; Seshadri, B.; Rothrock Jr, M.; Panneerselvam, P. Uses and Management of Poultry Litter. *Worlds Poult. Sci. J.* **2010,** *66,* 673–698.

Bourdat-Deschamps, M.; Ferhi, S.; Bernet, N.; Feder, F.; Crouzet, O.; Patureau, D.; Montenach, D.; Moussard, G. D.; Mercier, V.; Benoit, P. Fate and Impacts of Pharmaceuticals and Personal Care Products after Repeated Applications of Organic Waste Products in Long-Term Field Experiments. *Sci. Total Environ.* **2017,** *607*, 271–280.

Brahney, J.; Hallerud, M.; Heim, E.; Hahnenberger, M.; Sukumaran, S. Plastic Rain in Protected Areas of the United States. *Science* **2020,** *368*, 1257–1260.

Bradney, L.; Wijesekara, H.; Bolan, N. S.; Kirkham, M. B. Sources of Particulate Plastics in Terrestrial Ecosystems. In *Particulate Plastics in Terrestrial and Aquatic Environments*; Bolan, N. S., Kirkham, M. B., Halsband, C., Nugegoda, D., Ok, Y. S., Eds.; CRC Press, 2019; pp 3–18.

Brinton, W. F. Characterization of Man-made Foreign Matter And its Presence in Multiple Size Fractions From Mixed Waste Composting. *Compost. Sci. Utili.* **2005,** *13,* 274–280.

Brodhagen, M.; Goldberger, J. R.; Hayes, D. G.; Inglis, D. A.; Marsh, T. L.; Miles, C. Policy Considerations for Limiting Unintended Residual Plastic in Agricultural Soils. *Environ. Sci. Pol.* **2017,** *69*, 81–84.

Brunner, I.; Fischer, M.; Rüthi, J.; Stierli, B.; Frey, B. Ability of Fungi Isolated from Plastic Debris Floating in the Shoreline of a Lake to Degrade Plastics. *PLoS One* **2018,** *13*, e0202047.

Bucci, K.; Tulio, M.; Rochman, C. M. What is Known and Unknown about the Effects of Plastic Pollution: A Meta-Analysis and Systematic Review. *Ecol. Applic.* **2020,** *30*, e02044.

Carr, S. A.; Liu, J.; Tesoro, A. G. Transport and Fate of Microplastic Particles in Wastewater Treatment Plants. *Water Res.* **2016,** *91*, 174–182.

Chen, G.; Guo, X.; Cheng, Z.; Yan, B.; Dan, Z.; Ma, W. Air Gasification of Biogas-Derived Digestate in a Downdraft Fixed Bed Gasifier. *Waste Manag.* **2017,** *69*, 162–169.

Chen, Z.; Zhao, W.; Xing, R.; Xie, S.; Yang, X.; Cui, P.; Lü, J.; Liao, H.; Yu, Z.; Wang, S.; Zhou, S. Enhanced In Situ Biodegradation of Microplastics in Sewage Sludge using Hyperthermophilic Composting Technology. *J. Hazard. Mater.* **2020,** *384*, 121271.

Chia, W. Y.; Chew, K. W.; Le, C. F.; Lam, S. S.; Chee, C. S. C.; Ooi, M. S. L.; Show, P. L. Sustainable Utilization of Biowaste Compost for Renewable Energy and Soil Amendments. *Environ. Pollut.* **2020,** *267*, 115662.

Chojnacka, K.; Moustakas, K.; Witek-Krowiak, A. Bio-Based Fertilizers: A Practical Approach Towards Circular Economy. *Biores. Techno.* **2019,** *295*, 122223.

Chorghe, D.; Sari, M. A.; Chellam, S. Boron Removal from Hydraulic Fracturing Wastewater by Aluminum and Iron Coagulation: Mechanisms and Limitations. *Water Res.* **2017,** *126*, 481–487.

Conley, K.; Clum, A.; Deepe, J.; Lane, H.; Beckingham, B. Wastewater Treatment Plants as a Source of Microplastics to an Urban Estuary: Removal Efficiencies and Loading Per Capita Over One Year. *Water Res. X* **2019,** *3,* 100030.

Corradini, F.; Meza, P.; Eguiluz, R.; Casado, F.; Huerta-Lwanga; E.; Geissen, V. Evidence of Microplastic Accumulation in Agricultural Soils from Sewage Sludge Disposal. *Sci. Total Environ.* **2019,** *671,* 411–420.

Cox, K.; Covernton, G.; Davies, H; Dower, J.; Juanes, F.; Dudas, S. Human Consumption of Microplastics. *Environ. Sci. Technol.* **2019,** *53*, 7068–7074.

Crossman, J.; Hurley, R. R.; Futter, M.; Nizzetto, L. Transfer and Transport of Microplastics from Biosolids to Agricultural Soils and the Wider Environment. *Sci. Total Environ.* **2020,** *724,* 138334.

da Costa, J. P.; Santos, P. S.; Duarte, A. C.; Rocha-Santos, T. (Nano) Plastics in the Environment - Sources, Fates and Effects. *Sci. Total Environ.* **2016,** *566*, 15–26.

Das, M. P.; Kumar S. Microbial Deterioration of Low Density Polyethylene by Aspergillus and Fusarium sp. *Int. J. Chem. Tech. Res.* **2014,** *6*, 299–305.

David, A. Technical Document on Municipal Solid Waste Organics Processing; Environment Canada, 2013.

de Souza Machado, A. A.; Lau, C. W.; Till, J.; Kloas, W.; Lehmann, A.; Becker, R.; Rillig, M. C. Impacts of Microplastics on the Soil Biophysical Environment. *Environ. Sci. Technol.* **2018,** *52*, 9656–9665.

Deepika, S.; Jaya, M. R. Biodegradation of Low Density Polyethylene by Microorganisms from Garbage Soil. *J. Exp. Biol. Agric. Sci.* **2015,** *3*, 15–21.

Dobaradaran, S.; Schmidt, T. C.; Nabipour, I.; Khajeahmadi, N.; Tajbakhsh, S.; Saeedi, R.; Javad Mohammadi, M.; Keshtkar, M.; Khorsand, M.; Faraji Ghasemi, F. Characterization of Plastic Debris and Association of Metals with Microplastics in Coastline Sediment Along the Persian Gulf. *Waste Manage.* **2018,** *78*, 649–658.

Eghball, B. Soil Properties as Influenced by Phosphorus-and Nitrogen-Based Manure and Compost Applications. *Agronomy J.* **2002,** *94*, 128–135.

Ellouze, M.; Sayadi, S. White-Rot Fungi and their Enzymes as a Biotechnological Tool for Xenobiotic Bioremediation, In *Management of Hazardous Wastes*; Saleh, H., Rehab, A. R., 2016. DOI: 10.5772/64145.

El-Morsy, E. M.; Hassan, H. M.; Ahmed, E. Biodegradative Activities of Fungal Isolates from Plastic Contaminated Soils. *Mycosphere* **2017,** *8*, 1071–1087.

Enyoh, C. E.; Verla, A. W.; Verla, E. N. Uptake of Microplastics by Plant: A Reason to Worry or to be Happy? *World Sci. News* **2019,** *131*, 256–267.

Erenstein, O. Crop Residue Mulching in Tropical and Semi-Tropical Countries: An Evaluation of Residue Availability and Other Technological Implications. *Soil Tillage Res.* **2002,** *67*, 115–133.

Foo, K. Y; Hameed, B. H. An Overview of Landfill Leachate Treatment via Activated Carbon Adsorption Process. *J. Hazard Mater.* **2009,** *171*, 54–60.

Gardea-Torresdey, J.; Guadalupe, R.; Peralta-videa, J. Use of Phytofiltration Technologies in the Removal of Heavy Metals: A Review. *Pure Appl. Chem.* **2004,** *76*, 801–813.

Gast, T. Analysis of Plastic Residues in Commercial Compost. Nova Gorica, University of Nova Gorica, Slovenia, 2016.

Gatidou, G.; Arvaniti, O. S.; Stasinakis, A. S. Review on the Occurrence and Fate of Microplastics in Sewage Treatment Plants. *J. Hazard. Mater.* **2019,** *367*, 504–512.

Geyer, R.; Jambeck, J.; Law, K. Production, Use, and Fate of all Plastics Ever Made. *Sci. Adv.* **2017,** *3*, e1700782.

Gigault, J.; Halle, A.; Baudrimont, M.; Pascal, P.; Gauffre, F.; Phi, T.; El Hadri, H.; Grassl, B.; Reynaud, S. Current Opinion: What is a Nanoplastic? *Environ. Pollut.* **2018,** *235*, 1030–1034.

Gouin, T.; Avalos, J.; Brunning, I.; Brzuska, K.; de Graaf, J.; Kaumanns, J.; Koning, T.; Meyberg, M.; Rettinger, K.; Schlatter, H.; Thomas, J.; van Welie, R.; Wolf, T. Use of Micro-Plastic Beads in Cosmetic Products in Europe and their Estimated Emissions to the North Sea Environment. *SOFW J.* **2015,** *3*, 40–46.

Gouma, S.; Fragoeiro, S.; Bastos, A. C.; Magan, N. Bacterial and Fungal Bioremediation Strategies. In *Microbial Biodegradation and Bioremediation*; Das, S., Elsevier, 2014; pp 301–323. https://doi.org/10.1016/C2013-0-13533-7.

Gupta, S. K.; Herren T.; Wenger K.; Krebs R.; Hari, T. In Situ Gentle Remediation Measures for Heavy Metal-Polluted Soils. In *Phytoremediation of Contaminated Soil and Water*; Terry, N., Banuelos, G., Eds.; Lewis Publishers: Boca Raton, FL, USA, 2000; pp 303–322.

Hargreaves, J.; Adl, M.; Warman, P. A Review of the Use of Composted Municipal Solid Waste in Agriculture. *Agric. Ecosyst. Environ.* **2008,** *123*, 1–14.

Haynes, R.; Murtaza, G.; Naidu, R. Inorganic and Organic Constituents and Contaminants of Biosolids: Implications for Land Application. *Adv. Agron.* **2009,** *104*, 165–267.

He, L; Gielen, G; Bolan, N. S; Zhang, X; Qin, H; Huang, H; Wnag, H. Contamination and Remediation of Phthalic Acid Esters in Agricultural Soils in China: A Review. *Agronomy Sustain. Develop.* **2015,** *35*, 519–534.

Hoornweg, D.; Bhada-Tata, P. What a Waste: A Global Review of Solid Waste Management; Washington, DC, USA, 2012.

Horton, A. A.; Walton, A.; Spurgeon, D. J.; Lahive, E.; Svendsen, C. Microplastics in Freshwater and Terrestrial Environments: Evaluating the Current Understanding to Identify the Knowledge Gaps and Future Research Priorities. *Sci. Total Environ.* **2017,** *586*, 127–141.

Hudcová, H.; Vymazal, J.; Rozkošný, M. Present Restrictions of Sewage Sludge Application in Agriculture Within the European Union. *Soil Water Res.* **2019,** *14*, 104–120.

Huerta Lwanga, E.; Gertsen, H.; Gooren, H.; Peters, P.; Salanki, T.; van der Ploeg, M.; Besseling, E.; Koelmans, A.A.; Geissen, V. Incorporation of Microplastics from Litter into Burrows of Lumbricus Terrestris. *Environ. Pollut.* **2017,** *220* (Pt A), 523–531.

Hurley, R. R.; Lusher, A. L.; Olsen, M.; Nizzetto, L. Validation of a Method for Extracting Microplastics from Complex, Organic-Rich, Environmental Matrices. *Environ. Sci. Technol.* **2018,** *52*, 7409–7417.

Hurley, R. R.; Nizzetto, L. Fate and Occurrence of Micro(nano)Plastics in Soils: Knowledge Gaps and Possible Risks. *Curr. Opin. Environ. Sci. Health.* **2018,** *1*, 6–11.

Husain, Q.; Qayyum, S. Biological and Enzymatic Treatment of Bisphenol A and Other Endocrine Disrupting Compounds: A Review. *Crit. Rev. Biotech.* **2013,** *33*, 260–292.

Jeong, C. B.; Won, E. J.; Kang, H. M.; Lee, M. C.; Hwang, D.; Hwang, U. K.; Zhou, B.; Souissi, S.; Lee, S. J.; Lee, J. S. Microplastic Size-Dependent Toxicity, Oxidative Stress Induction, and p-JNK and p-p38 Activation in the Monogonont Rotifer (Brachionus koreanus). *Environ. Sci. Technol.* **2016,** *50*, 8849–8857.

Jiang, H.; Zhang, Y.; Wang, H. Surface Reactions in Selective Modification: The Prerequisite for Plastic Flotation. *Environ. Sci. Technol.* **2020,** *54*, 9742–9756.

Joshi, V.; Suyal, A.; Srivastava, A.; Srivastava, P. C. Role of Organic Amendments in Reducing Leaching of Sulfosulfuron Through Wheat Crop Cultivated Soil. *Emerg. Contam.* **2019,** *5*, 4–8.

Kanhai, L.; Gårdfeldt, K.; Lyashevska, O.; Hassellöv, M.; Thompson, R.; O'Connor, I. Microplastics in Sub-Surface Waters of the Arctic Central Basin. *Marine Pollut. Bull.* **2018,** *130*, 8–18.

Kanhai, L. D. K.; Gardfeldt, K.; Krumpen, T.; Thompson, R. C.; O'Connor, I. Microplastics in Sea Ice and Seawater Beneath Ice Floes from the Arctic Ocean. *Sci. Rep.* **2020,** *10*, 5004.

Karlen, D. L.; Lal, R.; Follett, R. F.; Kimble, J. M.; Hatfield, J. L.; Miranowski, J. M.; Cambardella, C. A.; Manale, A.; Anex, R. P.; Rice, C. W. Crop Residues: The Rest of the Story. *Environ. Sci. Technol.* **2009,** *43*, 8011–8015.

Kasirajan, S.; Ngouajio, M. Polyethylene and Biodegradable Mulches for Agricultural Applications: A Review. *Agron. Sustain. Dev.* **2012,** *32*, 501–529.

Kelly, A.; Lannuzel, D.; Rodemann, T.; Meiners, K.; Auman, H. Microplastic Contamination in East Antarctic Sea Ice. *Marine Pollut. Bull.* **2020,** *154*, 111130.

Khan, S.; Nadir, S.; Shah, Z. U.; Shah, A. A.; Karunarathna, S. C.; Xu, J.; Khan, A.; Hasan, F. Biodegradation of Polyester Polyurethane by Aspergillus Tubingensis. *Environ. Pollut.* **2017,** *225*, 469–480.

Kosuth, M.; Mason, S. A.; Wattenberg, E. V. Anthropogenic Contamination of Tap Water, Beer, and Sea Salt. *PLoS One* **2018,** *13*, e0194970.

Kowalska, A; Grobelak, A; Almås, A. R; Singh, B. R. Effect of Biowastes on Soil Remediation, Plant Productivity and Soil Organic Carbon Sequestration: A Review. *Energies* **2020,** *13*, 5813.

Krausmann, F.; Erb, K. H.; Gingrich, S.; Lauk, C.; Haberl, H. Global Patterns of Socioeconomic Biomass Flows in the Year 2000: A Comprehensive Assessment of Supply, Consumption and Constraints. *Ecol. Econ.* **2008,** *65*, 471–487.

Krueger, M. C.; Hofmann, U.; Moeder, M. Schlosser, D. Potential of Wood-Rotting Fungi to Attack Polystyrene Sulfonate and its Depolymerisation by Gloeophyllum Trabeum via Hydroquinone-Driven Fenton Chemistry. *PLoS One* **2015,** *10*, e0131773.

Kumar, M. Understanding the Remobilization of Copper, Zinc, Cadmium and Lead Due to Ageing Through Sequential Extraction and Isotopic Exchangeability. *Environ. Monit. Assess.* **2016,** *188*, 381.

Kuppusamy, S.; Palanisami, T.; Megharaj, M.; Venkateswarlu, K.; Naidu, R. Ex-Situ Remediation Technologies for Environmental Pollutants: A Critical Perspective. *Rev. Environ. Contam. Toxic.* **2016,** *236*, 117–192.

Kyaw, B. M.; Champakalakshmi, R.; Sakharkar; M. K.; Lim, C. S.; Sakharkar, K. R. Biodegradation of Low Density Polythene (LDPE) by Pseudomonas species. *Ind. J. Microbiol.* **2012,** *52*, 411–419.

Lakshminarasimman, N.; Gewurtz, S. B.; Parker, W. J.; Smyth, S. A. Removal and Formation of Perfluoroalkyl Substances in Canadian Sludge Treatment Systems–A Mass Balance Approach. *Sci. Total Environ.* **2020,** *754*, 142431.

Lambert, S.; Wagner, M. Characterisation of Nanoplastics During the Degradation of Polystyrene. *Chemosphere* **2016,** *145*, 265–268.

Lambert, S.; Wagner, M. Microplastics Are Contaminants of Emerging Concern in Freshwater Environments: An Overview. In *Freshwater Microplastics. The Handbook of Environmental Chemistry*; Wagner, M., Lambert, S., Eds.; Springer: Cham, 2018; vol 58.

Lassen, C.; Hansen, S. F.; Magnusson, K.; Norén, F.; Hartmann, N. B.; Jensen, P. R.; Nielsen, T. G.; Brinch, A. Microplastics: Occurrence, Effects and Sources of Releases to the Environment in Denmark, Danish Environmental Protection Agency, Copenhagen K, Environmental Project No. 1793, 2015.

Leejarkpai, T.; Suwanmanee, U.; Rudeekit, Y.; Mungcharoen, T. Biodegradable Kinetics of Plastics Under Controlled Composting Conditions. *Waste Manage.* **2011,** *31*, 1153–1161.

Letcher, R. J.; Chu, S.; Smyth, S.-A.; Side-Chain Fluorinated Polymer Surfactants in Biosolids from Wastewater Treatment Plants. *J. Hazard Mater.* **2020,** *388*, 122044.

Li, X; Chen, L; Ji, Y; Li, M; Dong, B; Qian, G; Zhou, J; Dai, X. Effects of Chemical Pretreatments on Microplastic Extraction in Sewage Sludge and their Physicochemical Characteristics. *Water Res.* **2020,** *171*, 115379.

Li, L, Xu, G, Yu, H, Xing, J. Dynamic Membrane for Microparticle Removal in Wastewater Treatment: Performance and Influencing Factors. *Sci. Total Environ.* **2018,** *627*, 332–340.

Li, L; Zhou, Q; Yin, Na; Tu, C; Luo, Y. Uptake and Accumulation of Microplastics in an Edible Plant. *Chinese J.* **2019,** *64*, 928–934.

Liu, J.; Chen, Y.; Kushwaha, R. Effect of Tillage Speed and Straw Length on Soil and Straw Movement by a Sweep. *Soil Tillage Res.* **2010,** *109*, 9–17.

Lloret, E.; Pascual, J. A.; Brodie, E. L.; Bouskill, N. J.; Insam, H.; Juárez, M. F. D.; Goberna, M. Sewage Sludge Addition Modifies Soil Microbial Communities and Plant Performance Depending on the Sludge Stabilization Process. *Appl. Soil Ecol.* **2016,** *101*, 37–46.

Ma, C.; Yu, J.; Wang, B.; Song, Z.; Xiang, J.; Hu, S.; Su, S.; Sun, L. Chemical Recycling of Brominated Flame Retarded Plastics from Ewaste for Clean Fuels Production: A Review. *Renew. Sustain. Energy Rev.* **2016,** *61*, 433–450.

Maaß, S.; Daphi, D.; Lehmann, A.; Rillig, M. C. Transport of Microplastics by Two Collembolan Species. *Environ. Pollut.* **2017,** 456–459.

Mahon, A. M.; O Connell, B.; Healy, M. G.; O Connor, I.; Officer, R.; Nash, R.; Morrison, L. Microplastics in Sewage Sludge: Effects of Treatment. *Environ. Sci. Technol.* **2016,** *51*, 810–818.

Mallampati, S. R.; Lee, B. H.; Mitoma, Y.; Simion, C. Heterogeneous Nano-Fe/Ca/CaO Catalytic Ozonation for Selective Surface Hydrophilization of Plastics Containing Brominated and Chlorinated Flame Retardants (B/CFRs): Separation from Automobile Shredder Residue by Froth Flotation. *Environ. Sci. Pollut. Res.* **2017,** *24*, 4469–4479.

Manjate, E; Ramos, S; Almeida, M. R. Potential Interferences of Microplastics in the Phytoremediation of Cd and Cu by the Salt Marsh Plant Phragmites Australis. *J. Environ. Chem. Eng.* **2020,** *8*, 103658.

Manzur, A.; Limon-Gonzalez, M.; Favela-Torres, E. Biodegradation of Physicochemically Treated LDPE by a Consortium of Filamentous Fungi. *J. Appl. Polym. Sci.* **2004,** *92*, 265–271.

Masura, J.; Baker, J.; Foster, G.; Arthur, C. Laboratory Methods for the Analysis of Microplastics in the Marine Environment. NOAA Marine Debris Program, Technical Memorandum NOS-OR&R-48, 2015.

Megharaj, M.; Venkateswarlu, K.; Naidu, R. Bioremediation. *Encycl. Toxic.* **2014,** 485–489.

Mekuanint, S.; Alemneh, T.; Asredie, T. G. Indigestible Rumen Foreign Bodies - Causes of Rumen Impaction in Cattle, Sheep and Goats Slaughtered at Addis Ababa Abattoir Enterprise, Ethiopia, 2017.

Mercier, A.; Gravouil, K.; Aucher, W.; Brosset-Vincent, S.; Kadri, L.; Colas, J.; Bouchon, D.; Ferreira, T. Fate of Eight Different Polymers under Uncontrolled Composting Conditions:

Relationships Between Deterioration, Biofilm Formation, and the Material Surface Properties. *Environ. Sci. Technol.* **2017,** *51*, 1988–1997.

Mintenig, S. M.; Int-Veen, I.; Loder, M. G. J.; Primpke, S.; Gerdts, G. Identification of Microplastic in Effluents of Waste Water Treatment Plants Using Focal Plane Array-Based Micro-Fourier-Transform Infrared Imaging. *Water Res.* **2017,** 365–372.

Miyazaki, J.; Kuriyama, Y.; Miyamoto, A.; Tokumoto, H.; Konishi, Y.; Nomura, T. Adhesion and Internalization of Functionalized Polystyrene Latex Nanoparticles Toward the Yeast Saccharomyces cerevisiae. *Ad. Powder Technol.* **2014,** *25*, 1394–1397.

Muhonja, C. N.; Makonde, H.; Magoma, G.; Imbuga, M. Biodegradability of Polyethylene by Bacteria and Fungi from Dandora Dumpsite Nairobi-Kenya. *PLoS One* **2018,** *13*, e0198446.

Munir, E.; Harefa, R. S. M.; Priyani, N.; Suryanto, D. In *Plastic Degrading Fungi Trichoderma Viride and Aspergillus Nomius Isolated from Local Landfill Soil in Medan*; IOP Conference Series: Earth and Environmental Science, 2018; vol *126*, p 012145.

Nanda, S.; Sahu, S. S. Biodegradability of Polyethylene by Brevibacillus, Pseudomonas and Rhodococcus spp. *NY Sci. J.* **2010,** *3*, 95–98.

Napper, I.; Davies, B.; Clifford, H.; Elvin, S.; Koldewey, H.; Mayewski, P.; Miner, K.; Potocki, M.; Elmore, A.; Gajurel, A.; Thompson, R. Reaching New Heights in Plastic Pollution—Preliminary Findings of Microplastics on Mount Everest. *One Earth* **2020,** *3*, 621–630.

Napper, I. E.; Thompson, R. C. Release of Synthetic Microplastic Plastic Fibres from Domestic Washing Machines: Effects of Fabric Type and Washing Conditions. *Mar. Pollut. Bull.* **2016,** *112*, 39–45.

Nizzetto, L.; Futter, M.; Langaas, S. Are Agricultural Soils Dumps for Microplastics of Urban Origin? *Environ. Sci. Technol.* **2016,** *50*, 10777–10779.

Ogata, Y.; Takada, H.; Mizukawa, K.; Hirai, H.; Iwasa, S.; Endo, S.; Mato, Y.; Saha, M.; Okuda, K.; Nakashima, A.; Murakami, M.; Zurcher, N.; Booyatumanondo, R.; Zakaria, M.; Dung, L.; Gordon, M.; Miguez, C.; Suzuki, S.; Moore, C.; Karapanagioti, H.; Weerts, S.; McClurg, T.; Burres, E.; Smith, W.; Velkenburg, M.; Lang, J.; Lang, R.; Laursen, D.; Danner, B.; Stewardson, N.; Thompson, R. International Pellet Watch: Global Monitoring of Persistent Organic Pollutants (POPs) in Coastal Waters. 1. Initial Phase Data on PCBs, DDTs, and HCHs. *Mar. Pollut. Bull.* **2009,** *58*, 1437–1446.

Ojha, N.; Pradhan, N.; Singh, S.; Barla, A.; Shrivastava, A.; Khatua, P.; Rain, V.; Bose, S. Evaluation of HDPE and LDPE Degradation by Fungus, Implemented by Statistical Optimization. *Sci. Rep.* **2017,** *7*, 39515.

Oliveri Conti, G.; Ferrante, M.; Banni, M.; Favara, C.; Nicolosi, I.; Cristaldi, A.; Fiore, M.; Zuccarello, P. Micro- and Nano-plastics in Edible Fruit and Vegetables. The First Diet Risks Assessment for the General Population. *Environ. Res.* **2020,** *187*, 109677.

Paço, A.; Duarte, K.; da Costa, J. P.; Santos, P. S. M.; Pereira, R.; Pereira, M. E.; Freitas, A. C.; Duarte, A. C.; Rocha-Santos, T. A. P. Biodegradation of Polyethylenemicroplastics by the Marine Fungus Zalerion Maritimum. *Sci. Total Environ.* **2017,** *586*, 10–15.

Padervand, M.; Lichtfouse, E.; Robert, D.; Wang, C. Removal of Microplastics from the Environment. A Review. *Environ. Chem. Lett.* **2020,** *18*, 807–828.

Papaioannou, D.; Koukoulakis, P. H.; Papageorgiou, M.; Lambropoulou, D. A.; Kalavrouziotis, I. K. Investigation of Pharmaceutical and Personal Care Product Interactions of Soil and Beets (Beta vulgaris L.) under the Effect of Wastewater Reuse. *Chemosphere* **2020,** *238*, 124553.

Park, J. H.; Lamb, D.; Paneerselvam, P.; Choppala, G.; Bolan, N.; Chung, J. W. Role of Organic Amendments on Enhanced Bioremediation of Heavy Metal (loid) Contaminated Soils. *J. Hazard Mater*. **2011,** *185*, 549–574.

Patel, A. K.; Das, N.; Kumar, M. Multilayer Arsenic Mobilization and Multi-Metal Co-enrichment in the Alluvium (Brahmaputra) Plains of India: A Tale of Redox Domination along the Depth. *Chemosphere* **2019,** *224*, 140–150.

Pathak, V. M.; Navneet. Review on the Current Status of Polymer Degradation: A Microbial Approach. *Bioresour. Bioprocess.* **2017**, *4*, 15.

Pathan, S. I.; Arfaioli, P.; Bardelli, T.; Ceccherini, M. T.; Nannipieri, P.; Pietramellara, G. Soil Pollution from Micro- and Nanoplastic Debris: A Hidden and Unknown Biohazard. *Sustainability* **2020,** *12*, 7255.

Peixoto, J.; Silva, L. P.; Kr Ger, R. H.. Brazilian Cerrado Soil Reveals an Untapped Microbial Potential for Unpretreated Polyethylene Biodegradation. *J. Hazard. Mater*. **2017,** *324*, 634–644.

Petersen, S. O.; Sommer, S. G.; Béline, F.; Burton, C.; Dach, J.; Dourmad, J. Y.; Leip, A.; Misselbrook, T.; Nicholson, F.; Poulsen, H. D. Recycling of Livestock Manure in a Whole-Farm Perspective. *Livest. Sci.* **2007,** *112*, 180–191.

Picó, Y.; Alvarez-Ruiz, R.; Alfarhan, A. H.; El-Sheikh, M. A.; Alshahrani, H. O.; Barceló, D. Pharmaceuticals, Pesticides, Personal Care Products and Microplastics Contamination Assessment of Al-Hassa Irrigation Network (Saudi Arabia) and its Shallow Lakes. *Sci. Total Environ.* **2020,** *701*, 135021.

Piehl, S.; Leibner, A.; Löder, M. G. J.; Dris, R.; Bogner, C.; Laforsch, C. Identification and Quantification of Macro- and Microplastics on an Agricultural Farmland. *Sci. Rep.* **2018,** *8*, 17950.

Poerio, T.; Piacentini, E.; Mazzei, R. Membrane Processes for Microplastic Removal. *Molecules* **2019,** *24*, 4148.

Powell, J. J.; Faria, N.; Thomas-McKay, E.; Pele, L. C. Origin and Fate of Dietary Nanoparticles and Microparticles in the Gastrointestinal Tract. *J. Autoimmun.* **2010**, *34*, J226– J233

Qi, Y.; Yang, X.; Pelaez, A. M.; Lwanga, E. H.; Beriot, N.; Gertsen, H.; Garbeva, P.; Geissen, V. Macro-and Micro-Plastics in Soil-Plant System: Effects of Plastic Mulch Film Residues on Wheat (Triticum aestivum) Growth. *Sci. Total Environ.* **2018,** *645*, 1048–1056.

Quilty, J.; Cattle, S. Use and Understanding of Organic Amendments in Australian Agriculture: A Review. *Soil Res*. **2011,** *49*, 1–26.

Raghavendra, V. B.; Uzma, M.; Govindappa, M.; Vasantha, R. A.; Lokesh, S. Screening and Identification of Polyurethane (PU) and Low Density Polyethylene (LDPE) Degrading Soil Fungi Isolated from Municipal Solid Waste. *Int. J. Curr. Res*. **2016,** *8*, 34753–34761.

Ragusa, A.; Svelato, A.; Santacroce, C.; Catalano, P.; Notarstefano, V.; Carnevali, O.; Papa, F.; Rongioletti, M.; Baiocco, F.; Draghi, S.; D'Amore, E.; Rinaldo, D.; Matta, M.; Giorgini, E. Plasticenta: First Evidence of Microplastics in Human Placenta. *Environ. Int*. **2021,** *146*, 106274.

Restrepo-Flórez, J. M.; Bassi, A.; Thompson, M. R. Microbial Degradation and Deterioration of Polyethylene-A Review. *Int. Biodeterior. Biodegradation* **2014,** *88*, 83–90.

Rillig, M. C.; Lehmann, A.; Souza Machado, A. A.; Yang, G. Microplastic Effects on Plants. *New Phytol.* **2019,** *223*, 1066–1070.

Rillig, M. C.; Ziersch, L.; Hempel, S. Microplastic Transport in Soil by Earthworms. *Sci. Rep.* **2017,** *7*, 1362.

Roca-Pérez, L.; Martínez, C.; Marcilla, P.; Boluda, R. Composting Rice Straw with Sewage Sludge and Compost Effects on the Soil–Plant System. *Chemosphere* **2009,** *75*, 781–787.

Rochman, C. M. Microplastics Research - from Sink to Source. *Science* **2018,** *360*, 28–29.

Rong, Q. L.; Li, R. N.; Huang, S. W.; Tang, J. W.; Zhang, Y. C.; Wang, L. Y. Soil Microbial Characteristics and Yield Response to Partial Substitution of Chemical Fertilizer with Organic Amendments in Greenhouse Vegetable Production. *J. Integr. Agric.* **2018,** *17*, 1432–1444.

Rostagno, C. M.; Sosebee, R. E. Biosolids Application in the Chihuahuan Desert: Effects on Runoff Water Quality. *J. Environ. Qual.* **2001,** *30*, 160–170.

Saleem, M.; Alibardi, L.; Cossu, R.; Lavagnolo, M. C.; Spagni, A. Analysis of Fouling Development under Dynamic Membrane Filtration Operation. *Chem. Engin. J.* **2017,** *312*, 136–143.

Samaras, V.; Tsadilas, C. D.; Stamatiadis, S.; Nortcliff, S. Effects of Biosolids from Tomato Processing on Soil Fertility and Wheat Growth in a Greek Alfisol. *J. Plant Nutr. Soil Sci.* **2010,** *173*, 252–259.

Sánchez, C. Fungal Potential for the Degradation of Petroleum-Based Polymers: An Overview of Macro- and Microplastics Biodegradation. *Biotechnol. Adv.* **2020,** *40*, 107501.

Sangale, M. K.; Shahnawaz, M.; Ade, A. B. Potential of Fungi Isolated from the Dumping Sites Mangrove Rhizosphere Soil to Degrade Polythene. *Sci. Rep.* **2019,** *9*, 5390.

Sarkar, D. J.; Das; Sarkar S.; Mukherjee, S.; Das, B. K. Impact and Fate of Microplastics in the Riverine Ecosystem. In *Contaminants in Drinking and Wastewater Sources*. Kumar, M., Snow, D., Honda, R., Mukherjee, S., Eds.; Springer Transactions in Civil and Environmental Engineering, Springer: Singapore, 2021.

Sarker, R. K.; Chakraborty, P.; Paul, P.; Chatterjee, A.; Tribedi, P. Degradation of Low-Density Poly Ethylene (LDPE) by Enterobacter Cloacae AKS7: A Potential Step Towards Sustainable Environmental Remediation. *Arch. Microbiol.* **2020,** *202*, 2117–2125.

Schwabl, P.; Köppel, S.; Königshofer, P.; Bucsics, T.; Trauner, M.; Reiberger, T.; Liebmann, B. Detection of Various Microplastics in Human Stool. *Ann. Intern. Med.* **2019,** *171*, 453.

Sekhar, V. C.; Nampoothiri, K. M.; Mohan, A. J.; Nair, N. R.; Bhaskar, T.; Pandey, A. Microbial Degradation of High Impact Polystyrene (HIPS), An e-Plastic with Decabromodiphenyl Oxide and Antimony Trioxide. *J. Hazard. Mater.* **2016,** *318*, 347–354.

Sepperumal, U.; Markandan, M.; Palraja, I. Micromorphological and Chemical Changes During Biodegradation of Polyethylene Terephthalate (PET) by Penicillium sp. *J. Microbiol. Biotechnol. Res.* **2013,** *3*, 47–53.

Sharma, B.; Sarkar, A.; Singh, P.; Singh, R. P. Agricultural Utilization of Biosolids: A Review on Potential Effects on Soil and Plant Grown. *Waste Manag.* **2017,** *64*, 117–132.

Sharma, K. R.; Srivastava, P. C.; Srivastava, P.; Singh, V. P. Effect of Farmyard Manure Application on Boron Adsorption–Desorption Characteristics of Some Soils. *Chemosphere* **2006,** *65*, 769–777.

Shen, Y.; Tan, M. T. T.; Chong, C.; Xiao, W.; Wang, C. H. An Environmental Friendly Animal Waste Disposal Process with Ammonia Recovery and Energy Production: Experimental Study and Economic Analysis. *Waste Manag.* **2017,** *68*, 636–645.

Singh, J.; Gupta, K. C. Screening and Identification of Low Density Polyethylene (LDPE) Degrading Soil Fungi Isolated from Polythene Polluted Sites Around Gwalior City (M.P.). *Int. J. Curr. Microbiol. App. Sci.* **2014,** *3*, 443–448.

Skipper, H. Bioremediation of Contaminated Soils. In *Principles and Applications of Soil Microbiology;* Sylvia, D., et al., Eds.; Prentice Hall, Inc.: New Jersey, U.S.A., 1998.

Smith, S. R. Organic Contaminants in Sewage Sludge (Biosolids) and their Significance for Agricultural Recycling. *Philos. Trans. A Math. Phy. Eng. Sci.* **2009,** *367*, 4005–4041.

Smith, M. Do Microplastic Residuals in Municipal Compost Bioaccumulate in Plant Tissue? MSc Thesis, Royal Roads University, Victoria, British Columbia, Canada, 2018.

Smith, M.; Love, D.; Rochman, C.; Neff, R. Microplastics in Seafood and the Implications for Human Health. *Curr. Environ. Health Reports.* **2018,** *5*, 375–386.

Somerville, P. D.; May, P. B.; Livesley, S. J. Effects of Deep Tillage and Municipal Green Waste Compost Amendments on Soil Properties and Tree Growth in Compacted Urban Soils. *J. Environ. Manag.* **2018,** *227*, 365–374.

Song, Y. K.; Hong, S. H.; Jang, M.; Han, G. M.; Jung, S. W.; Shim. W. J. Combined Effects of UV Exposure Duration and Mechanical Abrasion on Microplastic Fragmentation by Polymer Type. *Environ. Sci. Technol.* **2017,** *51*, 4368–4376.

Song, C.; Sarpong, C. K.; He, J.; Shen, F.; Zhang, J.; Yang, G.; Long, L.; Tian, D.; Zhu, Y.; Deng, S. Accelerating Phytoremediation of Degraded Agricultural Soils Utilizing Rhizobacteria and Rndophytes: A Review. *Environ. Rev.* **2019,** *28*, 115–127.

Srivastava, P.; Srivastava, P. C.; Singh, U. S.; Shrivastava, M. Effect of Integrated and Balanced Nutrient Application on Soil Fertility, Yield and Quality of Basmati Rice. *Arch. Agron. Soil Sci.* **2009,** *55*, 265–284.

Steinmetz, Z.; Wollmann, C.; Schaefer, M.; Buchmann, C.; David, J.; Tröger, J.; Muñoz, K.; Froer, O.; Schaumann, G. E. Plastic Mulching in Agriculture. Trading Short-Term Agronomic Benefits for Long-Term Soil Degradation? *Sci. Total Environ.* **2016,** *550*, 690–705.

Takaya, C.; Cooper, I. R.; Berg, M.; Sarker, D. Offensive Waste Valorisation in the UK: Assessment of the Potentials for Absorbent Hygiene Product (AHP) Recycling. *Waste Manag.* **2019,** *88*, 56–70.

Tavazzi, S.; Mariani, G.; Skejø, H.; Comero, S.; Głowacka, N.; Gaduš, J.; Gawlik, B. Antimicrobial Agents and Other Compounds of Emerging Concern in Manure-Exposed Waters: Part 1–Pilot Sampling Campaign in Slovakia and First Findings, EUR 29065 EN, 2018.

Tchobanoglous, G. In *Wastewater Engineering*; 1st ed., McGraw-Hill: Boston, 2014.

Teuten, E.; Saquing, J.; Knappe, D.; Barlaz, M.; Jonsson, S.; Björn, A.; Rowland, S.; Thompson, R.; Galloway, T.; Yamashita, R.; Ochi, D.; Watanuki, Y.; Moore, C.; Viet, P.; Tana, T.; Prudente, M.; Boonyatumanond, R.; Zakaria, M.; Akkhavong, K.; Ogata, Y.; Hirai, H.; Iwasa, S.; Mizukawa, K.; Hagino, Y.; Imamura, A.; Saha, M.; Takada, H. Transport and Release of Chemicals from Plastics to the Environment and to Wildlife. *Philos. Trans. B Biol. Sci.* **2009,** *364*, 2027–2045.

Thangarajan, R.; Bolan, N. S.; Tian, G.; Naidu, R.; Kunhikrishnan, A. Role of Organic Amendment Application on Greenhouse Gas Emission from Soil. *Sci. Total Environ.* **2013,** *465*, 72–96.

Torri, S. I.; Corrêa, R. S.; Renella, G. Soil Carbon Sequestration Resulting from Biosolids Application. *Appl. Environ. Soil Sci.* **2014,** 821768.

Tribedi, P.; Sil, A. K. Low-Density Polyethylene Degradation by Pseudomonas sp. AKS2 Biofilm. *Environ. Sci. Pollut. Res. Int.* **2013,** *20*, 4146–4153.

Truu, J; Truu, M; Espenberg, M; Nõlvak, H; Juhanson, J. Phytoremediation and Plant-Assisted Bioremediation in Soil and Treatment Wetlands: A Review. *The Open Biotech. J.* **2015,** *9*, 85–92.

Tu, C.; Chen, T.; Zhou, Q.; Liu, Y.; Wei, J.; Waniek, J.; Luo, Y. Biofilm Formation and its Influences on the Properties of Microplastics as Affected by Exposure Time and Depth in the Seawater. *Sci. Total Environ.* **2020,** *734*, 139237.

Tu, C.; Zhou, Q.; Zhang, C.; Liu, Y.; Luo, Y. Biofilms of Microplastics. In *Microplastics in Terrestrial Environments. The Handbook of Environmental Chemistry*; He, D., Luo, Y., Eds.; Springer: Cham, 2020; vol 95. https://doi.org/10.1007/698_2020_461

Valbuena, D.; Erenstein, O.; Tui, S. H. K.; Abdoulaye, T.; Claessens, L.; Duncan, A. J.; Gérard, B.; Rufino, M. C.; Teufel, N.; van Rooyen, A. Conservation Agriculture in Mixed Crop–Livestock Systems: Scoping Crop Residue Trade-Offs in Sub-Saharan Africa and South Asia. *Field Crops Res.* **2012,** *132*, 175–184.

van Wezel, A.; Caris, I.; Kools, S. A. Release of Primary Microplastics from Consumer Products to Wastewater in The Netherlands. *Environ. Toxicol. Chem.* **2016,** *35*, 1627–1631.

Veeresh, H.; Tripathy, S.; Chaudhuri, D.; Ghosh, B.; Hart, B.; Powell, M. Changes in Physical and Chemical Properties of Three Soil Types in India as a Result of Amendment with Fly Ash and Sewage Sludge. *Environ. Geol.* **2003,** *43*, 513–520.

Verlicchi, P.; Zambello, E. Pharmaceuticals and Personal Care Products in Untreated and Treated Sewage Sludge: Occurrence and Environmental Risk in the Case of Application on Soil - A Critical Review. *Sci. Total Environ.* **2015,** *538*, 750–767.

Vethaak, A.; Legler, J. Microplastics and Human Health. *Science* **2021,** *371*, 672–674.

Vivi, V. K.; Martins-Franchetti, S. M.; Attili-Angelis, D. Biodegradation of PCL and PVC: Chaetomium Globosum (ATCC 16021) Activity. *Folia Microbiol.* **2019,** *64*, 1–7.

Wallace, J. S.; Garner, E.; Pruden, A.; Aga, D. S. Occurrence and Transformation of Veterinary Antibiotics and Antibiotic Resistance Genes in Dairy Manure Treated by Advanced Anaerobic Digestion and Conventional Treatment Methods. *Environ. Pollut.* **2018,** *236*, 764–772.

Waller, C.; Griffiths, H.; Waluda, C.; Thorpe, S.; Loaiza, I.; Moreno, B.; Pacherres, C.; Hughes, K. Microplastics in the Antarctic Marine System: An Emerging Area of Research. *Sci. Total Environ.* **2017,** *598*, 220–227.

Walter, I.; Cuevas, G.; Garcia, S.; Martinez, F. Biosolid Effects on Soil and Native Plant Production in a Degraded Semiarid Ecosystem in Central Spain. *Waste Manag. Res.* **2000,** *18*, 259–263.

Wang, X.; Dong, T.; Zhang, A.; Fang, Y.; Chen, D.; Zhao, C.; Luo, Q.; Yang, H. Isolation of Bacteria Capable of Hydrogen Production in Dark Fermentation and Intensification of Anaerobic Granular Sludge Activity. *Int. J. Hydrog. Energy* **2019,** *44*, 15853–15862.

Wang, W.; Rhodes, G.; Ge, J.; Yu, X.; Li, H. Uptake and Accumulation of Per-and Polyfluoroalkyl Substances in Plants. *Chemosphere* **2020,** *261*, 127584.

Wei, S.; Luo, H.; Zou, J.; Chen, J.; Pan, X.; Rousseau, D. P. L.; Li, J. Characteristics and Removal of Microplastics in Rural Domestic Wastewater Treatment Facilities of China. *Sci. Total Environ.* **2020,** 139935.

Weithmann, N.; Möller, J. N.; Löder, M. G. J.; Piehl, S.; Laforsch, C.; Freitag, R. Organic Fertilizer as a Vehicle for the Entry of Microplastic into the Environment. *Sci. Adv.* **2018,** *4*, eaap8060.

Wijesekara, H.; Bolan, N. S.; Bradney, L.; Obadamudalige, N.; Seshadri, B.; Kunhikrishnan, A.; Dharmarajan, R.; Ok, Y.; Rinklebe, J.; Kirkham, M.; Vithanage, M. Trace Element Dynamics of Biosolids-Derived Microbeads. *Chemosphere* **2018,** *199*, 331–339.

Wijesekara, H.; Bolan, N. S.; Vithanage, M.; Xu, Y.; Mandal, S.; Brown, S. L.; Hettiarachchi, G. M.; Pierzynski, G. M.; Huang, L.; Ok, Y. S.; Kirkham, M. B.; Saint, C. P.; Surapaneni, A. Utilization of Biowaste for Mine Spoil Rehabilitation. *Adv. Agron.* **2016,** *138*, 97–173.

Wijesekara, H.; Colyvas, K.; Rippon, P.; Hoang, S.; Bolan, N.; Manna, M.; Thangavel, R.; Seshadri, B.; Vithanage, M.; Awad, Y.; Surapaneni, A.; Saint, C.; Tian, G.; Torri, S.; Ok, Y.; Kirkham, M. Carbon Sequestration Value of Biosolids Applied to Soil: A Global Meta-Analysis. *J. Environ. Manage.* **2021,** *284*, 112008.

Wright, S. L.; Kelly, F. J. Plastic and Human Health: A Micro Issue? *Environ. Sci. Technol.* **2017,** *51*, 6634–6647.

Wu, R. T.; Cai, Y. F.; Xing, S. C.; Yang, Y. W.; Mi, J. D.; Liao, X. D. A Novel Method for Extraction of Polypropylene Microplastics in Swine Manure. *Environ. Sci. Pollut. Res.* **2020,** 1–10.

Yamada-Onodera, K.; Mukumoto, H.; Katsuyaya, Y.; Saiganji, A.; Tani, Y. Degradation of Polyethylene by a Fungus, Penicillium Simplicissimum YK. *Polym. Degrad. Stab.* **2001,** *72*, 323–327.

Yang, L.; Li, K.; Cui, S.; Kang, Y.; An, L.; Lei, K. Removal of Microplastics in Municipal Sewage from China's Largest Water Reclamation Plant. *Water Res.* **2019,** *155*, 175–181.

Yang, J.; Li, R.; Zhou, Q.; Li, L.; Li, Y.; Tu, C.; Zhao, X.; Xiong, K.; Christie, P.; Luo, Y. Abundance and Morphology of Microplastics in an Agricultural Soil Following Long-Term Repeated Application of Pig Manure. *Environ. Pollut.* **2021,** *272*, 116028.

Yang, S.; Wang, Y.; Liu, R.; Xing, L.; Yang, Z. Improved Crop Yield and Reduced Nitrate Nitrogen Leaching with Straw Return in a Rice-Wheat Rotation of Ningxia Irrigation District. *Sci. Rep.* **2018,** *8*, 1–8.

Yang, Y.; Yang, J.; Wu, W.; Zhao, J.; Song, Y.; Gao, L.; Yang, R.; Jiang, L. Biodegradation and Mineralization of Polystyrene by Plastic-Eating Mealworms: Part 2. Role of Gut Microorganisms. *Environ. Sci. Technol.* **2015,** *49*, 12087–12093.

Yin, W.; Yu, A.; Chai, Q.; Hu, F.; Feng, F.; Gan, Y. Wheat and Maize Relay-Planting with Straw Covering Increases Water use Efficiency up to 46%. *Agron Sustain. Dev.* **2015,** *35*, 815–825.

Yokelson, R. J.; Burling, I.; Urbanski, S.; Atlas, E.; Adachi, K.; Buseck, P.; Wiedinmyer, C.; Akagi, S.; Toohey, D.; Wold, C. Trace Gas and Particle Emissions from Open Biomass Burning in Mexico. *Atmospheric Chem. Phys.* **2011,** *11*, 6787–6808.

Yoshida, S.; Hiraga, K.; Takehana, T.; Taniguchi, I.; Yamaji, H.; Maeda, Y.; Toyohara, K.; Miyamoto, K.; Kimura, Y.; Oda, K. A Bacterium that Degrades and Assimilates Poly(ethylene terephthalate). *Science* **2016,** *351*, 1196–1199.

Zerzghi, H.; Gerba, C. P.; Brooks, J. P.; Pepper, I. L. Long-Term Effects of Land Application of Class B Biosolids on the Soil Microbial Populations, Pathogens, and Activity. *J. Environ. Qual.* **2010,** *39*, 402–408.

Zhang, Z.; Luo, X.; Fan, Y.; Wu, Q. Cumulative Effects of Powders of Degraded PE Mulching-Films on Chemical Properties of Soil. *Environ. Sci. Technol.* **2015,** *38*, 115–119.

Zhang, L.; Xu, Z. Separating and Recycling Plastic, Glass, and Gallium from Waste Solar Cell Modules by Nitrogen Pyrolysis and Vacuum Decomposition. *Environ. Sci. Technol.* **2016,** *50*, 9242–9250.

Zhi, L.; Zhipeng, R.; Minglong, L.; Rongjun, B.; Xiaoyu, L.; Haifei, L.; Kun, C.; Xuhui, Z.; Jufeng, Z.; Lianqing, L. Pyrolyzed Biowastes Deactivated Potentially Toxic Metals and Eliminated Antibiotic Resistant Genes for Healthy Vegetable Production. *J. Clean Prod.* **2020,** *276*, 124208.

CHAPTER 5

Herbicide Contamination in Ground Water and Their Probable Remediation

DIBAKAR ROY[1], DIBAKAR GHOSH[2], SUBHASH CHANDER[3], BEDABRATA SAHA[4], and DEBOJYOTI MOULICK[5]

[1]*ICAR-Directorate of Weed Research, Jabalpur, Madhya Pradesh, India*

[2]*ICAR-Indian Institute of Water Management, Bhubaneswar, Odisha, India*

[3]*ICAR-National Bureau of Plant Genetic Resources, New Delhi, Delhi, India*

[4]*Weed Research Department, Ne'we Yaar Research Centre, Agricultural Research Organization, Ramat Yishai, Israel*

[5]*Plant Stress Biology and Metabolomics Laboratory, Assam University, Silchar, Assam, India*

ABSTRACT

Herbicides become integral components in modern agriculture as it offers low cost solution to manage weed flora both crop and non-crop field. Currently, promotion of conservation agriculture (CA)-based crop cultivation practices solely depends on herbicides to counter weeds. Herbicides were dominant agricultural chemicals used for crop protection purpose and also used in higher quantities compared to others. There is large volume of research articles on persistence of herbicides in soil had been done published so far. But information on herbicide residues in ground water is scanty and majority of the studies were reported from USA, Canada, and European countries. Major citations on the presence of herbicide residues dominated by triazine (atrazine, simezine, and metribuzin), chlor acetanilides (alachlor, metalochlor, and acetachlor), phenoxy alkalonic acid (2,4-D), and organo

Environmental Contaminants: Impact, Assessment, and Remediation.
Pritam Ganguly, Jajati Mandal, M. Paramsivam, & Sandip Patra, (Eds.)

phosphate (glyphosate) group of herbicides and along with their degraded products. Spatial and temporal variation of herbicide residue content in ground water was well documented by many researchers. It was observed that herbicide residues present in ground water in ranged below or more than maximum acceptable concentration (MAC) set by European Union (0.1 μg L^{-1}). The presence of herbicide residues more than MAC level in drinking water may cause health hazard also reported from many countries like Peru, Argentina, and Sri Lanka, etc. There is necessity of strong regulatory body for monitoring of herbicide residues in ground water in each and every country. Thus, future research should be focused to minimize herbicide residue load in soil, water, and ecosystem. In this chapter, we have highlighted on the presence of herbicide residues in ground water reported from across the world, factors involved and way out pathways. Among the different methods of remediation of herbicide residue in ground water, preventive measures such as selection of appropriate herbicide, proper mixing & application procedure, prevention of runoff, and irrigation management are of sustainable approach. Phytoremediation is of widely used method amid different ground water remediation techniques. Bacterial (*Comamonadaceae* and *Sphingomonadales*) and algal (*Scenedesmus* species) degradation were also found effective for treatment of herbicidal ground water contamination. Apart from this, a new nano-remediation technique was found effective for remediation of atrazine and bromacil in ground water.

5.1 INTRODUCTION

Synthetic or natural chemicals used in agriculture to manage various pests like insect, disease, rodents, and weeds are generally called pesticides. A wide range of chemicals like insecticides, fungicides, acaricides, nematicides, rodenticides, and herbicides are categorized as pesticides. Worldwide, approximately 2 million tonnes of pesticides are used in agriculture; herbicide contributes 47.5% of total pesticide consumption followed by insecticide (29.5%) and fungicide (17.5%) (De et al., 2014). In terms of annual pesticide consumption, China ranks first followed by USA and Argentina and proportion of pesticides consumption is increasing day-by-day. It was estimated that global pesticide consumption will be near about 3.5 million tonnes by 2020 (Zhang et al., 2018). Whereas, in India, use for herbicide is significantly lower as compared to industrialized countries. The share of insecticide is about 65% and 16% for herbicide out of total pesticide consumption (Choudhary et al., 2016). From the beginning of herbicide use, the load of herbicides in Indian soils negligible in comparison to insecticides. As we know that, judicious applications of pesticides are beneficial for crop production vis-à-vis environmental safety. Pesticide intensive modern agriculture is gradually inviting known and unknown negative impacts of pesticide use by virtue of their bio-magnifications and bio-accumulation.

Weed is one for the major biotic constrains and causes more loss to crop produce than that caused by other pests (Choudhury et al., 2016). The degree of yield loss due to weeds may vary with several factors like nature of weed flora, its diversity, type of measures taken for its management, etc., (Ghosh et al., 2017a; Ghosh et al., 2017b; Ghosh et al., 2018; Kumar et al., 2018; Ghosh et al., 2020a). To manage weed flora in crop field, herbicides are extensively used in modern agriculture (Kumar et al., 2016; Banerjee et al., 2019; Kundu et al., 2020; Singh et al., 2018; Ghosh et al., 2016). Herbicides are specifically designed chemicals that used to kill weeds from crop as well as non-crop areas. The popularity of herbicide over other alternate weed control measures is increasing day-by-day. Managing weeds through herbicide application is easy; provide quick results and having longer phytotoxic effect on weeds (Ghosh et al., 2020b; Ghosh et al., 2021). The weed controlling efficiency of herbicides depends on several factors like dose, rate, timing, and purity of herbicide; soil properties, climatic situations, etc.

The continuous application of herbicides in modern agriculture leads to build up of herbicide residues in soil and water. There are numerous reports about presence of herbicide residues in soil from different parts of globe, especially from developed countries like USA and other European countries. But studies on herbicide residues in ground water are very few in number; mostly confined to countries like USA, Canada, UK, Germany, Italy, France, Spain, Brazil, etc. Herbicide residue concentration of soil and ground water depends on several factors like molecular nature of herbicide, dose of herbicide applied, adsorption behavior in soil, crop type, crop management factors, soil properties, and weather conditions. Herbicide molecules generally used before twenty first century were high dose low potency in nature. But introduction of low dose high potency herbicides cut down herbicide dose by several times, and it influences the herbicidal contamination in ground water. Molecular complexity or integrity also allows herbicide molecules to persist in environment against physical agents like sunlight and microbial attack. Crops with higher photosynthetic efficiency and root volume uptake higher amount of herbicides applied in foliage and soil, subsequently reduces herbicide load in environment. Soil properties like texture, type of clay minerals, organic matter content, moisture content, and pH play an important role in herbicide adsorption, degradation, and mobility of herbicide molecules. Soil with sandy texture and low in organic matter content increases the chances of herbicide leaching to ground water from soil. Crop management practices like use of soil applied herbicide followed by irrigation increases possibility of herbicide leaching (Fig. 5.1). Weather parameters like occurrence of rainfall within short period of herbicide application also serve as a potential carrier of herbicide molecules to surface and ground water. Information on persistence in soil and water system is required for environmental safety point view. In this chapter, we will highlight some studies that mainly discusses on ground water contamination with herbicide residues at global aspect.

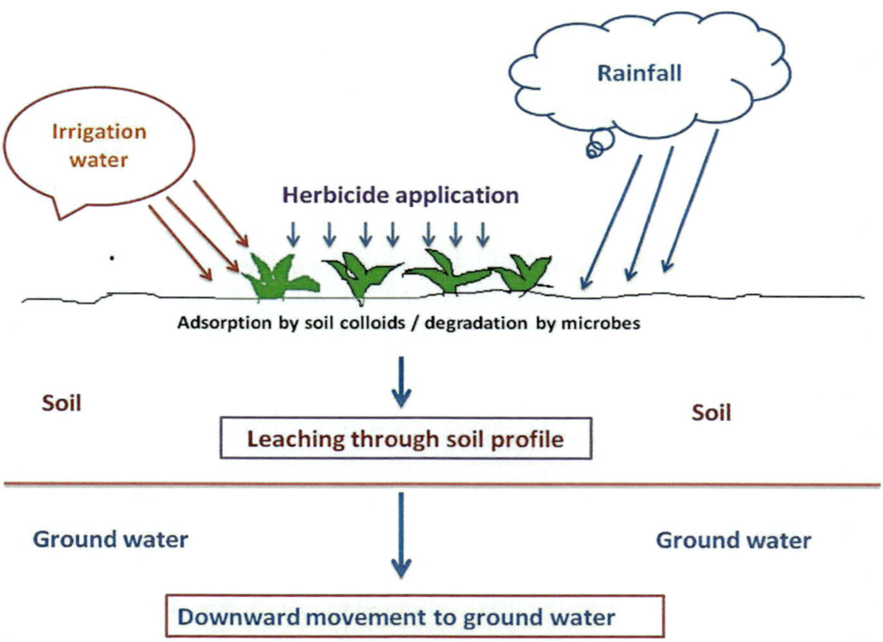

FIGURE 5.1 Schematic representation of herbicide movement from soil surface to ground water.

5.2 HERBICIDE RESIDUES IN GROUND WATER

5.2.1 TRIAZINE AND CHLORACETAMIDE

Atrazine, a triazine group of herbicide used for pre- and post-emergence control of broadleaf and grassy weeds in various crops like maize, sugarcane, sorghum, etc. It is one of the most widely used herbicides in USA and Australian. The documentation of herbicide residues in ground water was first cited in 1975. The presence of atrazine in ground water was first reported by Richard et al. (1975) from maize growing Iwoa region of USA. However, prime focus of their study was to determine organo-chlorin pesticides such as dichloro-diphenyl-trichloroethane (DDT), dichlorodiphenyldichloroethylene (DDE), DDE, aldrin, and dialdrin content in surface as well as ground water, but atrazine was taken in their study as atrazine was the "modern herbicide" at that time. They found the presence of atrazine was highest among all these pesticides in samples collected from maize growing areas. Later Junk et al. (1980) observed the presence of atrazine and alachlor in ground water samples collected from the USA. The concentration of atrazine residue varied from <0.01 to 88 $\mu g\ L^{-1}$ and maximum concentration found in shallow water table at the end of irrigated maize growing season; whereas alachlor concentration observed < 0.1 $\mu g\ L^{-1}$ in their study. Further, they reported atrazine and alachlor residues in ground water depend upon area of sampling, depth of ground water table, and time of sampling. In Europe, Berri et al. (1983) conducted a ground water study for atrazine and taken 300 bore water samples from rice and maize growing area of Italy. They found the presence of atrazine in 67% of the bore water sample with concentrations varying from 0.1 to 4 $\mu g\ L^{-1}$, and the concentration of atrazine reduces with the increment of sampling depth. Later Galassi and Leoni (1987) analyzed atrazine in 2000 bore water samples collected from public water supply purpose from Labordia, Itally. Results obtained in their study showed that atrazine concentration exceeded 0.1 $\mu g\ L^{-1}$ in 14% samples and 1.0 $\mu g\ L^{-1}$ in 1.4% samples indicating groundwater contamination with atrazine. The presence of atrazine and its residues like deethylatrazine along with other herbicides such as mesoprop and bentazone in ground water samples from Western Europe were reported by Leistra and Boesten, (1989). Atrazine concentration measured in series of ground water samples with varying depth was found below European safe limit (0.1$\mu g\ L^{-1}$) and maximum value obtained in their study was below 0.5 $\mu g\ L^{-1}$. The average concentration of bentazone detected in their study was almost 1.0 $\mu g\ L^{-1}$ and mesoprop was ranged up to 0.6 $\mu g\ L^{-1}$ in deep ground water samples. In ground water survey for herbicide residues done across Iowa region of USA in 1995, Kolpin et al. (1997a) detected presence of atrazine and deethylatrazine in 41 and 35% of ground water sample analyzed with maximum concentrations of 2.1 and 0.59 $\mu g\ L^{-1}$, respectively. The presence atrazine, alachlor, cyanazine, metolachlor, acetochlor, and prometon in ground water were found by Barbash et al. (2001) in the USA from 20 different sites under National Water Quality Assessment Programme. They found that, except atrazine concentration of other herbicide residues in ground water was less than 1 μg L^{-1}; and it was concluded that, atrazine residue was frequently found in ground water samples collected from agricultural areas which indicated that atrazine was extensively used in maize before sampling. Carabias-Mart´ınez et al. (2002) evaluated ground and surface water contamination with herbicides (atrazine, terbutryne, and chlorotoluron) and it secondary metabolites (desethylatrazine, desethyldesisopropylatrazine, 2-hydroxyatrazine, desethyl-2-hydroxyatrazine for atrazine; 2-hydroxyterbutylazine for terbutryne and 3-chloro-4-methylphenylurea for chlorotoluron) from agricultural zones in Spain using solid-phase extraction process. In irrigated maize growing areas of Spain, residues of atrazine and alachlor were observed in ground water by Sa´nchez-camazano et al. (2005) through enzyme-linked immunosorbent assays method. In this study, the concentration of atrazine varied from 0.04–3.45 μg L^{-1} and alachlor 0.05–4.85 μg L^{-1}. The metabolite of atrazine, that is, ethanesulfonate was also detected using HPLC and its concentration lies within 0.52–4.01 μg L^{-1}. Similarly, in maize growing areas of Greece, the residues of atrazine, metolachlor, and alachlor along with degradation product like deethylatrazine, deisopropylatrazine, and hydroxyatrazine was observed in ground water by Papastergiou and Papadopoulu-Mourkidou (2001). In most of the cases concentration of herbicide molecules were present < 0.1 μg L^{-1}. They observed the distribution pattern of herbicide follows atrazine > metolachlor > desethylatrazine > alachlor based on their concentration in ground water. Earlier, Ritter et al. (1994) also recorded atrazine, simazine, alachlor, and metolachlor in ground water in irrigated maize growing areas from Delaware, USA. Long-term application of atrazine and its impact on ground water safety in Jilin province of China was studied by Yue et al. (2013). The occurrence of atrazine and its degradation products like deethylatrazine, deisopropylatrazine, and hydroxyatrazine were monitored in ground water. These herbicide residues detected in 94% of the total sample and average residue concentration of deethylatrazine, deisopropylatrazine, and hydroxyatrazine were recorded 0.107, 0.0009, and 0.0003 μg L^{-1}. Their study highlighted the current scenario of herbicide residues in ground water lies within below hazardous level but continuous build up may cause health hazard in future. The occurrence of rain and application of irrigation water may enhance herbicide movement through soil specially atrazine. In India, Aslam et al. (2013) detected atrazine and simazine residues (0.017 and 0.040 μg L^{-1}) in ground water lower than below detectable limit

(<0.1 μg L^{-1}) collected from Delhi. Further, higher detection of atrazine was found from ground water samples collected from north Delhi compared to south and central Delhi. Their findings also showed that lateral movement of atrazine occurred in ground water, which carry herbicide residues from agricultural area to urban area.

It was observed that concentration of atrazine along with degradation product in ground water varied from with time and location. Sampling before cropping and post rainy season may lead to lower concentration of residues in ground water. Similarly, ground water sampling after herbicide application and during summer season reported to have higher concentration of atrazine residues. In rainy season, excess water comes into soil diluted the concentration of herbicide residues; in contrast in summer season due to evaporation and plants uptake there is decrease in total amount of ground water increases the concentration of herbicide residues in ground water. Temporal variation in atrazine along with different degradation products were reported by Guzella et al. (2006) from Italy (Fig. 5.2). Both spatial and temporal variation in atrazine concentration in ground water was monitored by Sánchez-Camazano et al. (2005) in Spain.

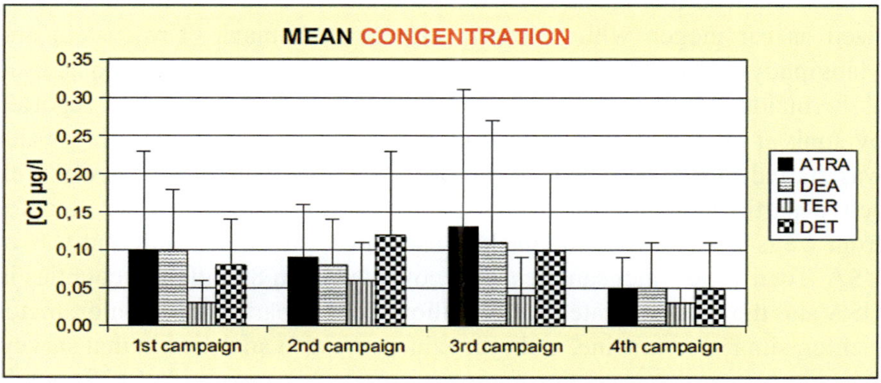

FIGURE 5.2 Temporal variation in atrazine (ATRA), deethylatrazine (DEA), terbuthylazine (TER), and diethyl-terbuthylazine (DET) concentration in ground water.

Source: Reprinted with permission from Guzella et al., 2006. Copyright 2006. Elsevier.

The studies conducted before 1990's mainly focused on the presence of primary herbicide molecules in ground water; whereas, studies done after 1990's also explored degradation products along with primary molecules. It was observed by various researchers that, sometimes total concentration of primary herbicide with degradation products exceeds the concentration of primary molecule. Furthermore, longer half-life of atrazine compared to other herbicides co-insides with higher frequency in detection in ground water. The presence of atrazine residues in ground water from maize growing areas had been widely cited across the world (Hallberg, 1989; Galasy and Leoni, 1987; Leistra and Boesten, 1989; Frank et al., 1990; Thurman et al., 1990; Bushway et al., 1991; Legrand et al., 1991; Gruessner et al., 1995; Kolpin et al., 1997a; Kolpin et al., 1997b; Wade et al., 1998; Garrido et al., 2000; Kolpin et al., 2000). The ground water contaminations with triazine and chloracetanilide herbicide in different countries are presented in Table 5.1.

Alachlor is a pre-emergence herbicide that belongs to chloracetanilide group of herbicide widely used in maize, soybean, sorghum, peanuts, cotton, vegetable, and forage crops. Among the various soil applied herbicide used, alachlor is moderately mobile in soil but less persistent is environment due to its lower half-life (Wauchope et al., 1992). Alachlor also used as mixed herbicide formulations with atrazine, trifluralin, and imazaquin, etc. The maximum contaminated level for alachlor in drinking water is 2 $\mu g\ L^{-1}$ was set by Environmental Protection Agency, USA. Studies conducted on animals showed that alachlor is slightly toxic for animals and also categorized as carcinogen with unknown effects on humans (Tirmenstein and Mangipudy, 2014). The European Union banned alachlor in 2006 to be used as herbicide. The presence of alachlor in ground water was first reported by Junk et al. (1980). An extensive survey on alachlor in drinking water was started in mid-1987 under National Alachlor Well Water Survey with active participation of Environmental Protection Agency. Water samples from wells were taken from alachlor used areas situated at different part of USA. Total 1430 water samples were collected from 89 different counties in USA and the data estimated for 6 million wells. Apart from alachlor, metolachlor, simazine, atrazine, and cyanazine were also analyzed in that survey. Results showed that detection frequency of alachlor was <1% (0.78), while detection frequency for metolachlor and simazine were 1.02% and 1.60%, respectively. But maximum detection frequency of 11.68% was obtained for atrazine, while the minimum 0.28% was detected for cyanazine in that survey. Concentration of alachlor was expected to be more than maximum contaminated level (2 $\mu g\ L^{-1}$) in only 0.02% of 6 million estimated wells. Later Potter and Carpenter, (1995) detected residues of alachlor along with 20 alachlor degradation compounds in ground water samples collected from maize growing areas of Massachusetts in USA. The total concentration of alachlor degradation products was ranged from 0.004 to 0.57 $\mu g\ L^{-1}$.

Similarly, in maize growing Iowa region of USA, atrazine, alachlor, metolachlor, prometon, acetochlor, and metribuzin along with their degradation products were detected in 70% of 106 municipal well water by Kolpin et al. (1997). The alachlor degradation product alachlor-ethanesulfonic acid was the most detected compound (65%) followed by atrazine (40.6%), deethylatrazine, and cyanazien amide (19.8%). The highest concentration was obtained for alachlor–ethanesulfonic acid (14.8 μg L^{-1}) followed by metolachlor (11.3 μg L^{-1}). The higher content of alachlor degradation product alachlor–ethanesulfonic acid in ground water than primary molecule was due to higher solubility of alachlor–ethanesulfonic acid in water over primary molecule (Thurman et al., 1996). Blanchard and Donald (1997) observed ground water contamination by herbicide molecules like atrazine, alachlor and along with their degradation products after 10 years of crop management where crop fields were treated with these herbicides. The well water samples collected quarterly during 1991 to 1996 for herbicide residue analysis. Atrazine and alachlor were detected in 7.2% and 0.4% samples with maximum concentration 0.12 and 0.14 μg L^{-1}, respectively. Variation in hydrology rather than herbicide application rate primarily governs concentration of herbicide residues in ground water.

Report generated on presence of herbicide residues (concentration >0.01 μg L^{-1}) in ground water in major hydrologic basins (20) in USA under US Geological Survey (USGS) – the National Water Quality Assessment (NAWQA) programme from 1993 to 1995 showed that atrazine, acetachlor, simazine, cyanazine, prometon, metalochlor, alachlor, and acetachlor were the major herbicide molecules detected in shallow and deep ground water in both agricultural, urban and mixed land use systems (Barbash,1999) (Fig. 5.3). Detection frequency was higher in shallow ground water compared to deep ground water; acetachlor was the least detected herbicide molecule among all in both shallow and deep aquifers.

A long-term study conducted (1993–2003) by United States Geological Survey to evaluate the presence of chloracetanilide herbicides along with their degradation products in groundwater (Scribner et al., 2004). Total 2420 ground water samples were analyzed for alachlor, acetochlor, flufenacet, dimethenamid, and metolachlor vis-à-vis their degradation products like ethanesulfonic acid, sulfinyl acetic acid, and oxanilic acid. Results showed that primary degradation products like ethanesulfonic acid and oxanilic acid of alachlor and metolachlor detected more frequently even at higher concentration than their parent compounds. Among various chloracetanilide herbicides, metolachlorwas most detected compound (12%) followed by

TABLE 5.1 Contamination of Ground Water by Triazine and Chloracetanilide Herbicide in Different Countries.

Metabolites detected	Concentration range	Average/maximum concentration	Analytical methodology	Country	Reference
Atrazine	0.1–4 μg L^{-1}	–	Gas chromatography (GC)	Italy	Berri et al. (1983)
Atrazine	0.1–1.0 μg L^{-1}		GC	Italy	Galassi and Leoni (1987)
Alachlor	0.004–0.57 μg L^{-1}	–	GC	USA	Potter and Carpenter, (1995)
Atrazine and alachlor	<0.01–88 μg L^{-1} and < 0.1 μg L^{-1}	–	GC	USA	Junk et al. (1980)
Atrazine and alachlor	–	0.12 μg L^{-1} and 0.14 μg L^{-1}	GC	Greece	Blanchard and Donald, (1997)
Atrazine and alachlor	0.04–3.45 μg L^{-1} and 0.05–4.85 μg L^{-1}	–	ELISA	Spain	Sa´nchez-camazano et al. (2005)
Atrazine, deethylatrazine, mesoprop, and bentazone	Atrazine conc. <0.1–0.5 μg L^{-1}	Bentazone – 1.0 μg L^{-1} Mesoprop–0.5 μg L^{-1}	GC	Western Europe	Leistra and Boesten, (1989)
Atrazine and deethylatrazine	–	2.1 and 0.59 μg L^{-1}	GC	USA	Kolpin et al. (1997a)
Attrazine, alachlor, cyanazine, metolachlor, acetochlor, and prometon	–	<1 μg L^{-1}	GC	USA	Barbash et al. (2001)
Attrazine, terbutryne and chlorotoluron, DEDIA, and DIHA	0.02–0.1 μg L^{-1}	–	Solid phase extraction (SPE)–GC	Spain	Carabias-Mart´ınez et al. (2002)

TABLE 5.1 *(Continued)*

Metabolites detected	Concentration range	Average/maximum concentration	Analytical methodology	Country	Reference
Aatrazine, metolachlor, alachlor, deethylatrazine, deisopropylatrazine, and hydroxyatrazine	< 0.1 $\mu g\ L^{-1}$	–	GC	Greece	Papastergiou and Papadopoulu-Mourkidou (2001)
Atrazine, deethylatrazine, deisopropylatrazine and hydroxyatrazine	< 0.1 $\mu g\ L^{-1}$	–	Liquid chromatography–mass spectrometry (LC–MS)/MS	China	Yue et al. (2013)
Alachlor, acetochlor, flufenacet, dimethenamid, metolachlor, ESA, SAA, and OXA	0.17–0.90 $\mu g\ L^{-1}$	–	GC	USA	Scribner et al. (2004)

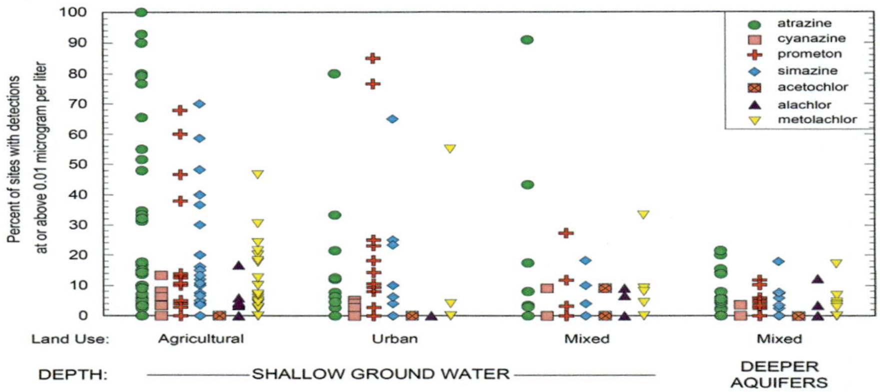

FIGURE 5.3 Detection frequency of triazine and chloracetanilide herbicide residues in ground water with varying land use and sampling depth in USA

Source: Reprinted with permission from Barbash, 1999. Copyright U.S. Geological Survey.

alachlor and acetochlor (2% each) and least was dimethenamid (0.18%). Among the various degradation products, metolachlor–ethanesulfonic acid was detected in 45% of the samples, followed by alachlor–ethanesulfonic acid and metolachlor–oxanilic acid found in 35% and 26% samples, respectively. The median concentration for metolachlor–ethanesulfonic acid and metolachlor–oxanilic acid were 0.97 and 0.70 µg L^{-1},, respectively; whereas, median concentration of parent metolachlor was 0.17 µg L^{-1}. Lee and Strahan, (2003) also found higher concentration of degradation product over parent compounds.

5.2.2 SULFONYLUREA

Sulfonylurea herbicides are one of the broad-spectrum herbicide with low chronic and acute toxicity to animals. This herbicide class having a half-life of <2 weeks and generally acts on acetolactase synthase enzyme which blocks synthesis of branched chain amino acids (Brown, 1990). Battaglin et al. (2000) reported the presence of sulfonylurea, sulfonamide, imidazolinone, and some other herbicides in surface water of rivers and reservoirs and ground water sample. To obtain this information, they collect ground water samples from 25 different locations in Midwestern United States. Results obtained in their study showed that imazathapyr was the most detected herbicide molecules (16% samples) followed by flumetsulam (12%), and nicosulfuronand imazaquin (8%). But the sum of total sulfonylurea, sulfonamide, and imidazolinone concentrations were lower than 0.01 μg L^{-1} except six samples. Among the other herbicide concerned atrazine was the most detected herbicide molecules in ground water.

5.2.3 PHENOXYACETIC ACID

Phenoxyacetic herbicides (synthetic 2,4-D) first introduced in 1940s, which were generally acts as broadleaf weed killer in both crop and non-crop areas. The 2,4-D was one of the oldest auxin herbicides widely used in USA and also used in Vietnam war to defoliate forest trees. Herbicides belongs to these class are highly water soluble in nature and adsorbed in soil weakly (Muszyński et al., 2020). Thus, these parent molecules along with their degradation products are readily transported to surface vis-à-vis ground water. McManus et al. (2014) detected 10 phenoxyacetic acid herbicides namely 2,4-D (2,4-di chlorophenoxyacetic acid), 2,3,6-trichlorobenzoic acid (TBA), mecoprop, mecoprop-p, MCPA, dicamba, bromoxynil, fluroxypr, triclopyr, and bentazone along with its six primary degradation products such as phenoxyacetic acid (PAC), 2,4-dichlorophenol (DCP), 2,4,5-trichloro-phenol(TCP), 4-chloro-2-methylphenol (4C2MP), 3,5-dibromo-4-hydroxybenzoic acid (BrAC), and 3,5,6-trichloro-2-pyridinol (T2P), and two benzonitrile degradation products such as 3,5-dichlorobenzoic acid (DBA) and 2,6-dichlorobenzamide (BAM) in Irish ground water which were never been detected earlier using HPLC coupled to tandem mass spectrometry. Average concentration of all compounds found below European Union permissible limit (0.1 μg L^{-1}) except 3,5-dichlorobenzoic acid (1.172 μg L^{-1}) and phenoxyacetic acid (0.456 μg L^{-1}). In alberta region of Canada, Munira et al. (2018) reported the presence of auxin herbicide residues specially 2,4-D and MCPA along with other pesticide residues in ground water samples collected from well and piezometer. The presence of phenoxyacetic herbicide residues in surface and ground water samples also been reported from different countries by different researchers from Poland (Ignatowicz-Owsieniuk and Skoczko, 2002), Spain (Köck et al., 2010), Norway (Roseth, 2013), Hungary (Székács et al., 2015), Denmark (McKnight et al., 2015), USA (DaSilva, 2016), and Australia (Schult, 2016).

5.2.4 GLYPHOSATE

Glyphosate, a non-selective herbicide widely used to eradicate weeds from crop as well as non-crop areas. Residues of glyphosate and its secondary metabolite aminomethylphosphonic acid (AMPA) had been widely detected in ground water samples across the globe. Residues of glyphosate in ground water depend on many factors like date of sampling, area of sampling, application of irrigation water, and occurrence of rainfall. Currently, various analytical methodology *viz.* gas chromatography, solid phase extraction, and enzyme-linked immunosorbent assay (ELISA) were used in estimation of glyphosate and its metabolite AMPA in ground and surface water. The glyphosate residue level in safe drinking water is varied among different countries. Based on 6 year review (1996–2001), United States Environmental protection Agency (USEPA) generated a detailed summery report in 2003 on ground water contamination. In this report, different agents in 16 states showed that out of 27,877 ground water samples analyzed glyphosate detected in only seven samples with maximum value 1.1 ug L^{-1} (EPA, 2003). However, United States Geological Survey conducted a survey for 6 years (2001–2006) in collaboration with various agencies found glyphosate, AMPA, and glufonsinate residue in ground water from 2135 surface and ground water samples. A relatively higher proportion of glyphosate residue was detected in surface water than ground water samples in this study. Glyphosate was detected in 7.78% of total ground water samples and maximum concentration of glyphosate detected in their study was 4.7 μg L^{-1}, whereas, AMPA detected in 15.23% of ground water samples with maximum concentration of 2.6 μg L^{-1}. Increased use of herbicide for both crop and non-crop areas favors enrichment of surface as well as ground water by these contaminants was highlighted in their study. Later, Battaglin et al. (2014) analyzed total 1171 ground water samples collected in between 2001 and 2010 to study occurrence of glyphosate and AMPA in 38 states. In 5.8% samples, glyphosate was detected with median concentration <0.02 μg L^{-1} and maximum concentration was 2.03 μg L^{-1}. AMPA detected in 14.3% samples with median concentration <0.02 μg L^{-1} and maximum concentration was 4.88 μg L^{-1}. Report generated by French Environmental Institute for 2001 and 2002 revealed that the occurrence of glyphosate was detected in surface and ground; and in 2002 out of 1389 ground water samples collected from 882 ground water monitoring sites only 3–4% samples showed the presence of glyphosate. In Germany, Ku¨ters and Gerhartz (2010) detected glyphosate and its secondary metabolite AMPA along with glufonsinate using solid phase extraction procedure in drinking water samples and the limit of detection values were 0.0012, 0.0014, and 0.0012 μg L^{-1}, respectively. Only AMPA detected in their study was very low (0.02 μg L^{-1}), while glufonsinate and glyphosate residues were not detected in their study. A multi-locational study was conducted in Hungery for residues of glyphosate and AMPA in ground and surface water by Mörtl et al. (2013) using ELISA method showed that residues present in ground water varies from sampling site and time of sampling. Average concentration of glyphosate was detected in their study was 0.12 μg L^{-1}. In their study, total 42 water samples were analyzed during 2010 and 2011, and 16 samples showed significantly higher level of glyphosate ranged from 0.54 to 0.76 μg L^{-1} during 2010, whereas, residues found in 2011 almost negligible indicating seasonal variation. In Spain, Sanchis et al. (2012) also reported the presence of glyphosate residues in 47% of ground water samples (total no. of sample was 139) collected from 11 sampling sites during 2007–2010. The average concentration of glyphosate residues detected from different site varied from 0.097 to 0.409 μg L^{-1}. At the initial period (2007) glyphosate residues were detected in 50% of sampling site; but in 2010, glyphosate was detected from all sampling sites with average concentration ranged from 0.097 to 0.560 μg L^{-1}. The highest concentration of glyphosate found in 2010 sampling is from site 1 (0.58 μg L^{-1}) where agriculture practices were done at moderate level. Their finding indicates that glyphosate applied for non-agricultural purpose contributed to ground water glyphosate residue build up by leaching and other pathways. Detection of higher level of glyphosate residues in these sites was mentioned due to sample taken immediately after glyphosate application and rainfall induced leaching. In Mexico, Ruiz-Toledo et al. (2014) detected glyphosate residues in ground water samples collected from intensive agricultural zone of southern Mexico using ELISA method. Glyphosate residues detected in their study ranged from <0.13 to 0.56 μg L^{-1} in rainy season and <0.13 to 18.43 μg L^{-1} in dry season, respectively. Lower glyphosate concentration in rainy season was due to dilution effect by precipitation. Later in soybean growing Campeche region of Mexico, glyphosate residues found in 90% of ground water samples collected and concentration ranged from 0.44 to 1.41 μg L^{-1} (Rendón-von Osten and Dzul-Caamal, 2017). It was documented that the presence of higher glyphosate residue level in ground water is associated with sampling from vicinity of agricultural field where glyphosate resistant soybean crop was widely cultivated coupled with gradual enrichment of glyphosate residues in ground water was a negative impact of introducing genetically modified (glyphosate resistant crop) crop in that area. In rice growing areas of Sri Lanka, glyphosate residues were detected in drinking waters from serving well vis-s-vis abandoned well with average concentration 0.7 and 3.5 μg L^{-1}, respectively, by Jayasumana et al. (2015) and further they concluded that the presence of glyphosate residue in drinking water may be a possible reason for chronic kidney failure disease of unknown etiology among paddy growing farmers in this region. Later, Gunarathna et al. (2018) reported the presence of trace levels glyphosate (1–4 μg L^{-1}) in all collected shallow groundwater samples and tracer amount of AMPA (2–11 μg L^{-1}) in few samples with vicinity of agricultural field in Sri Lanka. But the glyphosate concentrations detected in their study were much higher than maximum level of glyphosate residues prescribed by European council (0.1 μg L^{-1}). In a study conducted by Van Stempvoort et al. (2016) in Ontario region of Canada found that glyphosate and AMPA residues were present in 10.5% and 5% of ground water samples, respectively, and higher concentration of glyphosate (0.663 g μg L^{-1}) and AMPA (0.698 μg L^{-1}) were found from a shallow water depth as compared to deeper layer. Recently Lutri et al. (2020) detected glyphosate (1.3–2 μg L^{-1}) and AMPA in 15.8% of the total ground water samples collected from Córdoba, Argentina, and concluded that glyphosate distribution in ground water is related with hydrological cycle. The residues of different herbicides and their metabolites found in ground water are summarized in Table 5.2.

Most of the cases, contaminations of glyphosate in ground water in maize growing USA, Canada, Mexico, and European countries are positively related with volume of herbicide use, especially for both crop field and non-crop fields for weed control. Again more preferences toward conservation agriculture (CA) and glyphosate resistant transgenic crops such as maize which solely depends on herbicides to eradicate weeds also responsible for glyphosate residue build up. There is an argument on negative impact of glyphosate on human and animal health, but recent research findings from the University of Washington glyphosate increases risk of some cancers by 40% (Zhang et al., 2019). This aspect presence of glyphosate in ground water in those developed countries may pose serious health hazard, which should be taken care properly. Very few references had been found on glyphosate contamination in groundwater from developing countries from Asia and African continents is mainly due to scarce research in this aspect, insufficient funding and people unawareness. There is an urgent need to relook status of herbicide residue contamination in ground water across the world for peoples and environmental safety.

5.3 MANAGEMENT OR REMEDIATION OF HERBICIDE RESIDUE IN GROUND WATER

There are number of possible techniques to remediate or prevent ground water contamination with herbicides, includes microbial degradation,

TABLE 5.2 Residue of Auxin, Sulfonylurea, and Glyphosate Herbicides and Its Metabolites in Ground Water.

Herbicide and metabolites	Concentration range	Average/ maximum concentration	Analytical methodology	Country	Reference
Sulfonylurea, sulfonamide, imidazolinone, imazathapyr, flumetsulam, and imazaquin	–	<0.1 ug L^{-1}	High-performance liquid chromatography-mass spectrometry (HPLC–MS)	USA	Battaglin et al. (2000)
2,4-D (2,4-di chlorophenoxyacetic acid); 2,3,6-trichlorobenzoic acid (TBA); mecoprop; mecoprop-p; MCPA; dicamba; bromoxynil; fluroxypr; triclopyr; bentazone as herbicides Phenoxyacetic acid; 2,4-dichlorophenol; 2,4,5-trichloro-phenol; 4-chloro-2-methylphenol; 3,5-dibromo-4-hydroxybenzoic acid; and 3,5,6-trichloro-2-pyridinol as degradation product	–	<0.1 ug L^{-1}	Ultra high performance liquid chromatography - mass spectrometry (UHPLC–GC)	Ireland	McManus et al. (2014)
2,4-D and MCPA	–	<0.1 ug L^{-1} (Except)	Gas chromatography (GC)	Canada	Munira et al. (2018)
Glyphosate and Aminomethylphosphonic acid (AMPA)	–	4.5 and 2.6 µg L^{-1}	GC	USA	Scribner et al. (2007)
AMPA	–	0.02 µg L^{-1}	Solid phase extraction (SPE) followed by liquid chromatography (LC)	Germany	Ku¨ters and Gerhartz (2010)
Glyphosate	–	0.12 µg L^{-1}	ELISA	Hungary	Mörtl et al. (2013)

TABLE 5.2 *(Continued)*

Herbicide and metabolites	Concentration range	Average/ maximum concentration	Analytical methodology	Country	Reference
Glyphosate	0.097–0.409 µg L^{-1}	0.78 µg L^{-1}	ELISA	Spain	Sanchis et al., (2012)
Glyphosate and AMPA	–	2.03 and 4.88 µg L^{-1}	SPE and LC/mass spectrometry (MS)	USA	Battaglin et al. (2014)
Glyphosate	< 0.13–0.56 µg L^{-1}	–	ELISA	Mexico	Ruiz-Toledo et al. (2014)
Glyphosate and AMPA	–	0.7 µg L^{-1} and 3.5 µg L^{-1}	Inductively coupled plsma–mass spectroscopy(ICP–MS)	Sri Lanka	Jayasumana et al. (2015)
Glyphosate and AMPA	–	0.663 and 0.698 µg L^{-1}	Ion chromatography/ tandem mass spectrometry (IC/MS/ MS	Canada	Van Stempvoort et al. (2016)
Glyphosate	0.44–1.41 µg L^{-1}	–	ELISA	Mexico	Rendón-von Osten and Dzul-Caamal (2017)
Glyphosate and AMPA	1–4 and 2–11 µg L^{-1}	–	Liquid chromatography– mass spectrometry (LC–MS)	Sri Lanka	Gunarathna et al. (2018)
Glyphosate and AMPA	1.3–2 µg L^{-1}	–	UHPLC MS/MS	Argentina	Lutri et al. (2020)

phyto-remediation, nano-remediation, selection of appropriate herbicide, proper mixing and application procedure, prevention of runoff, irrigation management, microbial degradation, phyto-remediation, and nano-remediation.

5.3.1 SELECTING APPROPRIATE HERBICIDES, PROPER MIXING, AND APPLICATION

For safeguard water from contamination with herbicide need proper planning, monitoring, past history of herbicides (used) record will help to select the best control measure. When ground water is near to surface, hence there are more chances of leaching out of herbicides in through soil in ground water. Applicator must to read the label to know about the leaching potential of herbicide. The labels must have the "ground water advisory statement" whether the herbicides have high, medium, and low in leaching potential based on the depth of ground water. On the other hand, we have to focus more on the type of herbicide need to apply for controlling weeds *viz.* soil and foliage applied herbicide. To prevent herbicide contamination in ground water, foliage applied herbicides are more preferable than the soil active herbicide. Care should be taken during choosing herbicide molecules, that is, herbicide formulation. To reduce herbicide leaching through the soil system, water soluble formulation should be avoided. Otherwise, it may aggravate the chances of herbicide contamination in ground water if rain occurs after herbicide application.

Generally, herbicide spills during their measurement and mixing before the application. Select the mixing and loading site at least 100 feet away from pond, wells, lakes, streams, and other water sources. It should also be take in to account that the selected site should have impermeable surface such as cemented pad, which prevent soaking spill out chemicals. If is not possible to find the cemented pad, then select different site every time for mixing and loading to avoid chemical build-up at one site.

For proper application, calibrated equipment should be used; thus, we can prevent excess application of pesticides. Over application of herbicide is not only risky for the environment, but is also a violation of label directions and the law. As we know the herbicide application are generally over the surface of soil, hence proper dose and the site of application of herbicide should be away from wells, where depth of water is shallow and surface water is located. Apart from this, drift is another task to the applicator is to avoid drifting due to wind, this may cause contaminate water and cause other problems.

5.3.2 IRRIGATION MANAGEMENT

The irrigation management is of utmost important as heavy irrigation increases the chances of herbicides to migrate to groundwater. Irrigating soil more than the saturation or at higher rate which exceeds infiltration rate of soil supports runoff that can carry herbicide with it. In areas such as sandy soil where frequent irrigation is required due to the porous nature of soil which promotes the downward movement of water beyond the root zone of plants also supports the leaching of herbicides to ground water. Hence, appropriate irrigation management is needed to avoid the risk of herbicides moving to ground water.

5.3.3 PREVENTION OF RUNOFF

Soil compaction is the major reason for surface runoff of water. To avoid surface runoff of water, we need to adopt different techniques which directly or indirectly reduce compaction of soil. Apart from this, we can use planting cover crop, conservation tillage practices, enhancing crop residues, planting grasses to retard soil and water runoff, and include buffer strips to safeguard surface water bodies.

5.3.4 CROP ROTATION

Crop rotation with higher root volume can ameliorate water infiltration, which reduces runoff. Crop rotations with higher vegetative growth may also impart increased surface crop residue and may reduce the need for applications of specific herbicide repeatedly to a given field site.

5.3.5 INTEGRATED WEED MANAGEMENT

To reduce the herbicide load in the environment, we have to curtail down the herbicide use by integrating other non-chemical weed management approached suitable in different cropping systems.

5.3.6 MICROBIAL AND ALGAL DEGRADATION

Microbes from agricultural uses, landfills, septic tanks, or land disposal of sewage by products which is indicator of pollution are generally considered for the study of ground water microbiology (McCarty et al., 1984; Omenn, 1987; Wilson et al., 1983). Most of the ground water microbiology studies were related to aerobic degradation; however, few studies pointed out the significance of anaerobic bacteria in the transformation of some compounds (Rochkind-Dubinsky, 1987; Erickson et al., 1988). Few laboratory experiments, indicated aerobic bacteria totally degraded aromatic compounds for instances styrene, chlorobenzene, 1,4-dichlorobenzene, and naphthalene within 11 weeks, however, 1,3 dichlorobenzene was most difficult in degradation. During a field test (at Palo Alto, CA), where reclaimed municipal wastewater was infused into the water table, where heptaldehyde and naphthalene were degraded within 12 h and within 8 m of the wellbore. However, even after 3 months no reduction in trihalomethanes and chlorinated benzenes concentration was noticed (Robert et al., 1980). After injection ceased, anaerobic activity continued and concentrations of chloroform, tri-halomethanes, trichloroethylene, tetrachloroethylene, and 1, 1, 1-trichloroethane continue to decrease.

Feld et al. (2016) established natural bacterial degraders in a sand filter methodology for treating ground water contaminated with phenoxypropionate herbicides, that is, (RS)-2-(4-chloro-2-methylphenoxy) propanoic acid (MCPP) and (RS)-2-(2,4-dichloro- phenoxy) propanoic acid (DCPP). Ground water (anaerobic) was flowed through an aeration bay and afterward through a rapid sand filter, where water was kept for short period (with average flow of 0.30 m^3/h). The degradation of DCPP and MCPP in the sand filter was increased by 15–30% of initial concentration. *Comamonadaceae* and *Sphingomonadales* were potential degrading bacteria of DCPP and MCPP.

Mollamohammada et al. (2021) was found *Scenedesmus* species was effective in removing atrazine, nitrate, zinc, oxadiazon, triallate, phosphorus, and magnesium from ground water in a continuous flow reactor. During a laboratory-based experiment with synthetic ground water (prepared by 8.8 mg/L NO_3-N and 90 μg/L atrazine), where they performed at a hydraulic retention time of 7 days and the temperatures of 20°C and 35°C. The highest uptake of nitrate and atrazine was observed at 20°C (97% and 70%, respectively). However, during actual ground-water treatment, 92% of nitrate, 100% of magnesium, 99.9% of phosphorus, and 92% of zinc were successfully removed at the end of 29 days' treatment operations, whereas few herbicides diffused back into the solution toward the end of the treatment process. On the other hand, microbes also play an important role in degrading herbicide molecules in soil system which ultimately helps to reduce herbicide contamination in ground water through infiltration.

5.3.7 PHYTO-REMEDIATION

Due to runoff of herbicide, non-targeted crops and grasses are affected (Obrigawitch et al., 1998), however, few of them are able to sequester and metabolize those herbicides if they possess the capability (Hamilton, 1964; Harris and Sans, 1967; Lichtenstein, 1959; Lichtenstein and Schulz, 1960, 1965; Lichtenstein et al., 1965). Responses of various crops are summarized in Table 5.3.

TABLE 5.3 Plants Involved in the Remediation of Herbicides.

Name of plant	Herbicides	Reference
Poplar	Triazine	Nair et al. (1993)
Corbicula fluminea	Atrazine and glyphosphate	Roche et al. (2009)
Rice	Atrazine and glyphosphate	Roche et al. (2009)
Salix	2,4-D and 2,4,5-T	Field and Peel (1971a); Field and Peel (1971b); Field and Peel (1972); Shann et al. (1994)
Kochia	Atrazine	Perkovich et al. (1995)

It was found that about 70% of 2,4-D and 86% to 90% of trifluralin could be retained during runoff in a 25 m grass water-way and vegetative buffer strips, respectively (Asmussen et al., 1977; Rohde et al., 1980). Arora et al. (1996) reported that *Bromus inermis* could able to retain 11–100% of atrazine in natural rainfall and runoff events. Generally, the depletion in the proportion of herbicides as surface runoff is caused by infiltration into the root zone rather than by direct uptake of herbicide (Dosskey, 2002). Lowrance et al. (1997) found grass buffer strips of *Cynodon dactylon*, *Paspalum notatum*, and *Lolium perrene*, most effective, these buffer strips able to reduce concentration of two herbicides, atrazine, and alachlor. Zhao et al. (2003) found native prairie grasses *viz. Andropogon gerardii*, *Sorghastrum nutans*, and *Panicum virgatum* most effective in significant reduction of atrazine and metolachlor in soil, where indigenous atrazine or metolachlor degrading microorganism were not present.

5.3.8 NANOREMEDIATION

Gawel et al. (2020) evaluated two injectable reactive and sorption-active particle types for their applicability for in situ removal of herbicides (atrazine and bromacil). Carbo-Iron (zerovalent iron and colloidal activated carbon) reduced bromacil with increased activity compared to nanoscale zerovalent iron (nZVI). However, neither nZVI nor Carbo-Iron showed significant degradation of atrazine. As a novel subsurface treatment option, Trap-Ox® zeolite FeBEA35 was found effective to adsorb atrazine and bromacil.

5.4 CONCLUSION

Detection of herbicide residues in ground water mostly occurred in the developed countries positively correlated with volume of its use and showed reliance of modern agriculture on herbicides for controlling weeds in crop as well as non-crop fields. Frequency of herbicide residue determination was more when ground water samples were collected from near agricultural zone or crop field. To reduce the herbicide load, use of herbicide resistant transgenic crops like soybean and maize must be minimized. Herbicide application timing should be proper to avoid occurrence of rainfall or irrigation event, which facilitates leaching of herbicide in ground water especially in light textured soil. However, solubility of herbicide molecules and its half-life primarily determines the residence time in soil vis-à-vis ground water. The use of soil applied herbicides need to minimized for better environmental point of view. We have to focus more on developing new herbicide molecules with post emergence (foliage applied), less water solubility, faster plant absorption, and lower self-life in soil. Also negative impacts of new herbicide molecules on soil macro fauna like earthworms and termites; beneficial insects like butterfly, honey bees, wasp, and ants should be negligible. Further, detection of herbicides below safe limit is not always meaning that our ground water is safe for drinking purpose. Gradual accumulation of herbicides continuously for long term may affect animal and human health, which is not discussed in this chapter. Herbicide residues should be included in one of the parameters of good quality drinking water. Our study comes with the point that herbicide residues in ground water may not be a threat in today but possibly in tomorrow.

KEYWORDS

- **herbicide**
- **contamination**
- **residue**
- **ground water**
- **remediation**

REFERENCES

Arora, K.; Mickelson, S. K.; Baker, J. L.; Tierney, D. P. Herbicide Retention by Vegetative Buffer Strips from Runoff under Natural Rainfall. *Trans. ASAE* **1996,** *39 (6),* 2155–2162.

Aslam, M.; Alam, M.; Rais, S. Detection of Atrazine and Simazine in Ground Water of Delhi using High Performance Liquid Chromatography with Ultraviolet Detector. *Curr. World Environ.* **2013,** *8 (2)*, 323–329.

Asmussen, L. E.; White, A. W. Jr.; Hauser, E. W.; Sheridan, J. M. Reduction of 2,4-D load in Surface Runoff Down a Grassed Waterway. *J. Environ. Qual.* **1977,** *6 (2)*, 159–162.

Banerjee, H.; Garai, S.; Sarkar, S.; Ghosh, D.; Samanta, S.; Mahato, M. Efficacy of Herbicides Against Canary Grass and Wild Oat in Wheat and their Residual Effects on Succeeding Greengram in Coastal Bengal. *Indian J. Weed Sci.* **2019,** *51*, 246–251.

Barbash, J. E. Distribution of Major Herbicides in Ground Water of the United States; US Department of the Interior, US Geological Survey, 1999, *98* (4245).

Barbash, J. E.; Thelin, G. P.; Kolpin, D. W.; Gilliom, R. J. Major Herbicides in Water. *J. Environ. Qual.* **2001,** *30*, 831–845.

Battaglin, W. A.; Furlong, E. T.; Burkhardt, M. R.; Peter, C. J. Occurrence of Sulfonylurea, Sulfonamide, Imidazolinone, and Other Herbicides in Rivers, Reservoirs and Ground Water in the Midwestern United States 1998. *Sci. Total Environ.* **2000,** *248*, 123–133.

Battaglin, W. A., Meyer, M. T., Kuivila, K. M., Dietze, J. E. Glyphosate and its Degradation Product AMPA Occur Frequently and Widely in US Soils, Surface Water, Groundwater, and Precipitation. *J. Am. Water Resour. Assoc.* **2014,** *50*, 275–290.

Berri, A.; Allievi, C.; Burzi, F. Residui di atrazina e molinate nelle acque di falda di zone agricole sottoposte a diserbo. *Acqua-Aria* **1983,** *3*, 259–264.

Blanchard, P. E.; Donald, W. W. Herbicide Contamination of Groundwater Beneath Claypan Soils in North-Central Missouri. *J. Environ. Qual.* **1997,** *26 (6)*, 1612–1621.

Brown, H. M. Mode of Action, Crop Selectivity, and Soil Relations of the Sulfonylurea Herbicides. *Pestic. Sci.* **1990,** *29 (3)*, 263–281.

Bushway, R. J.; Perkins, B.; Fukal, L.; Harrinson, R. O.; Ferguson, B. S. Comparison of Enzyme-Linked Immunosorbent Assay and High-Peformance Liquid Chromatography for the Analysis of Atrazine in Water from Czechoslovakia. *Arch. Environ. Contam. Toxicol.* **1991,** *21*, 365–370.

Carabias-Martínez, R.; Rodríguez-Gonzalo, E.; Herrero-Hernández, E.; Sánchez-San Román, F. J.; Flores, M. G. P. Determination of Herbicides and Metabolites by Solid-Phase Extraction and Liquid Chromatography: Evaluation of Pollution Due to Herbicides in Surface and Groundwaters. *J. Chromatogr. A* **2002,** *950 (1-2)*, 157–166.

Choudhury, P. P.; Singh, R.; Ghosh, D.; Sharma, A. R. Herbicide Use in Indian Agriculture; ICAR - Directorate of Weed Research, Jabalpur, Madhya Pradesh, 2016, p 110.

DaSilva, A. Surface Water Monitoring for Pesticides in Agricultural Areas of Northern California, 2016. https://www.cdpr.ca.gov/docs/emon/pubs/ehapreps/report_306_dasilva.pdf.

De, A.; Bose, R.; Kumar, A.; Mozumdar, S. In *Targeted Delivery of Pesticides Using Biodegradable Polymeric Nanoparticles*; Springer India: New Delhi, 2014; pp 59–81.

Dosskey, M. G. Setting Priorities for Research on Pollution Reduction Functions of Agricultural Buffers. *Environ. Manage.* **2002**, *30* (5), 641–650.

EPA. EPA's Report on the Environment (2003 Draft), 2003.

Erickson, L. E.; Fan, L. T. Anaerobic Degradation of Toxic and Hazardous Wastes. In *Handbook on Anaerobic Fermentations*; Erickson, L. E., Fung, D. Y. C., Eds.; Marcel Dekker: New York, 1988; p 695.

Feld, L.; Nielsen, T. K.; Hansen, L. H.; Aamand, J.; Albers, C. N. Establishment of Bacterial Herbicide Degraders in a Rapid Sand Filter for Bioremediation of Phenoxypropionate-Polluted Groundwater. *Appl. Environ. Microbiol.* **2016,** *82*, 878–887.

Field, R. J.; Peel, A. J. The Metabolism and Radial Movement of Growth Regulators and Herbicides in Willow Stems, *New Phytol.* **1971a,** *70*, 743–749.

Field, R. J.; Peel, A. J. The Movement of Growth Regulators and Herbicides into the Sieve Elements of Willow, *New Phytol.* **1971b,** *70*, 997–1003.

Field, R. J.; Peel, A. J. The Longitudinal Mobility of Growth Regulators and Herbicides in Sieve Tubes of Willow. *New Phytol.* **1972,** *71*, 249–254.

Frank, R.; Clegg, B. S.; Sherman, C.; Chapman, N. D. Triazine, Cloroacetamide Herbicides in Sydenham River Water and Municipal Drinking Water, Dresden, Ontario, Canada, 1981–1987. *Arch. Environ. Contam. Toxicol.* **1990,** *19*, 319–324.

Galassi, S.; Leoni, V. In *The Problem of Atrazine in Drinking Water in Italy*, European Conference on Impact of Agriculture on Water Resources—Consequences and Perspective, 1987; pp 219–230.

Garrido, T.; Fraile, J.; Ninerola, J. M.; Figueras, M.; Ginebreda, A.; Olivella, L. Survey of Ground Water Pesticide Pollution in Rural Areas of Catalonia (Spain). *Int. J. Environ. Anal. Chem.* **2000,** *78*, 51–65.

Gawel, A.; Seiwert, B.; Sühnholz, S.; Schmitt-Jansen, M.; Mackenzie, K. In-situ Treatment of Herbicide-Contaminated Groundwater–Feasibility Study for the Cases Atrazine and Bromacil using Two Novel Nanoremediation-Type Materials. *J. Haz.Mat.* **2020,** *393*, 122470.

Ghosh, D.; Brahmachari, K.; Brestic, M.; Ondrisik, P.; Hossain, A.; Skalicky, M.; Sarkar, S.; Moulick, D.; Dinda, N. K.; Das, A.; et al. Integrated Weed and Nutrient Management Improve Yield, Nutrient Uptake and Economics of Maize in the Rice-Maize Cropping System of Eastern India. *Agronomy* **2020a,** *10*, 1906.

Ghosh, D.; Brahmachari, K.; Das, A.; Hassan, M. H.; Mukherjee, P. K.; Sarkar, S,; Dinda, N. K.; Pramanick, B.; Moulick, D.; Maitra, S.; Hossain, A. Assessment of Energy Budgeting and Its Indicator for Sustainable Nutrient and Weed Management in a Rice-Maize-Green Gram Cropping System. *Agronomy* **2021**. https://doi.org/10.3390/agronomy11010166

Ghosh, D.; Brahmachari, K.; Skalicky, M.; Hossain, A.; Sarkar, S.; Dinda, N. K.; Das, A.; Pramanick, B.; Moulick, D.; Brestic, M.; et al. Nutrients Supplementation through Organic Manures Influence the Growth of Weeds and Maize Productivity. *Molecules* **2020b,** *25*, 4924.

Ghosh, D.; Rathore, M.; Brahmachari, K.; Singh, R.; Kumar, B. Impact of Burial and Flooding Depths on Indian Weedy Rice. *Crop Prot.* **2017b,** *100*, 106–110.

Ghosh, D.; Singh, U. P.; Brahmachari, K.; Singh, N. K.; Das, A. An Integrated Approach to Weed Management Practices in Direct-Seeded Rice under Zero-Tilled Rice–Wheat Cropping System. *Int. J. Pest. Manag.* **2017a,** *63*, 37–46.

Ghosh, D.; Singh, R.; Chander, S. Effect of Nitrogen Fertilizer and Weed Management Practices on Weed Growth and Crop Yield of Zero-Till Transplanted Rice. *Indian J. Weed Sci.* **2018a,** *50*, 287–289.

Ghosh, D.; Singh, U. P.; Ray, K.; Das, A. Weed Management Through Herbicide Application in Direct-Seeded Rice and Yield Modeling by Artificial Neural Network. *Span. J. Agric. Res.* **2016,** *14*, e1003.

Gruessner, B.; Shambaugh, N. C.; Watzin, M. C. Comparison of an Enzyme Immunoassay and Gas Chromatography/Mass Spectrometry for the Detection of Atrazine in Surface Waters. *Environ. Sci. Technol.* **1995**, *29*, 251–254.

Gunarathna, S.; Gunawardana, B.; Jayaweera, M.; Manatunge, J.; Zoysa, K. Glyphosate and AMPA of Agricultural Soil, Surface Water, Groundwater and Sediments in Areas Prevalent with Chronic Kidney Disease of Unknown Etiology, Sri Lanka. *J. Environ. Sci. Health B* **2018,** *53* (11), 729–737.

Guzzella, L.; Pozzoni, F.; Giuliano, G. Herbicide Contamination of Surficial Groundwater in Northern Italy. *Environ. Pollut.* **2006,** *142* (2), 344–353.

Hallberg, G. R. Pesticides Pollution of Groundwater in the Humid United States. *Agric. Ecosys. Environ.* **1989,** *26* (3-4), 299–367.

Hamilton, R. H. Tolerance of Several Grass Species to 2-chloro-s-triazine Herbicides in Relation to Degradation and Content of Benzoxazinone Derivatives. *J. Agric. Food Chem.* **1964,** *12*, 14–17.

Harris, C. R.; Sans, W. W. Absorption of Organochlorine Insecticide Residues from Agricultural Soils by Root Crops. *J. Agric. Food Chem.* **1967**, *15*, 861–863.

Ignatowicz-Owsieniuk, K.; Skoczko, I. Dependence of Sorption of Phenoxyacetic Herbicides on their Physico-Chemical Properties. *Polish J. of Environ. Stud.* **2002,** *11* (4), 339–344.

Jayasumana, C.; Paranagama, P.; Agampodi, S.; Wijewardane, C.; Gunatilake, S.; Siribaddana, S. Drinking Well Water and Occupational Exposure to Herbicides is Associated with Chronic Kidney Disease, in Padavi-Sripura, Sri Lanka. *Environ. Health* **2015,** *14* (1), 6.

Junk, G. A.; Spalding, R. F.; Richard, J. J. Areal, Vertical, and Temporal Differences in Ground Water Chemistry: II. Organic Constituents. *J. Environ. Qual.* **1980,** *9* (3), 479–483.

Köck, M.; Farré, M.; Martínez, E.; Gajda-Schrantz, K.; Ginebreda, A.; Navarro, A.; De Alda, M. L.; Barceló, D. Integrated Ecotoxicological and Chemical Approach for the Assessment of Pesticide Pollution in the Ebro River Delta (Spain). *J. Hydrol.* **2010,** *383* (1-2), 73–82.

Kolpin, D. W.; Kalkhoff, S. J; Goolsby; D. A.; Sneck-Fahrer, D. A.; Thurman, E. M. Occurrence of Selected Herbicides and Herbicide Degradation Products in Iowa's Ground Water 1995. *Ground Water* **1997a,** *35*, 679–688.

Kolpin, D. W.; Sneck-Fahrer, D. A.; Hallberg, G. R.; Libra. R. D. Temporal Trends of Selected Agricultural Chemicals in Iowa's Groundwater, 1982–1995: Are Things Getting Better? *J. Environ. Qual.* **1997b,** *26*, 1007–1017.

Kolpin, D. W.; Thurman, E. M.; Linhart, S. M. Finding Minimal Herbicide Concentrations in Ground Water? Try Looking for their Degradates. *Sci. Total Environ.* **2000,** *248* (2-3), 115–122.

Kumar, M.; Ghosh, D.; Singh, R. Effect of Crop Establishment and Weed Management Practices on Growth and Yield of Wheat. *Indian J. Weed Sci.* **2018,** *50*, 129–132.

Kumar, A.; Nandan, R.; Singh, K. K.; Ghosh, D. Integrated Weed Management in Lentil (Lens culinaris) in Calcareous Alluvial Soils of Bihar. *Indian J. Agron.* **2016,** *61*, 75–78.

Kundu, R.; Mondal, M.; Garai, S.; Banerjee, H.; Ghosh, D.; Majumder, A.; Poddar, R. Efficacy of Herbicides on Weed Control, Rhizospheric Micro-Organisms, Soil Properties and Leaf Qualities in Tea Plantation. *Indian J. Weed Sci.* **2020,** *52*, 160–168.

Küsters, M.; Gerhartz, M. Enrichment and Low-Level Determination of Glyphosate, Aminomethylphosphonic Acid and Glufosinate in Drinking Water after Cleanup by Cation Exchange Resin. *J. Sep. Sci.* **2010,** *33* (8), 1139–1146.

Lee, E. A.; Strahan, A. P. *Methods of Analysis by the U.S. Geological Survey Organic Geochemistry Research Group—Determination of Acetamide Herbicides and their Degradation Products in Water using Online Solid-Phase Extraction and High-Performance Liquid Chromatography/Mass* Spectrometry: USGS Open-File Report, 2003, pp 03–173.

Legrand, M. F.; Costentin, E.; Bruchet, A. Ocurrence of 38 Pesticides in Various French Surface and Ground Waters. *Environ. Technol.* **1991,** *12*, 985–996.

Leistra, M.; Boesten, J. J. T. I. Pesticide Contamination of Groundwater in Western Europe. *Agric. Ecosys. Environ.* **1989,** *26* (3-4), 369–389.

Lichtenstein, E. P. Absorption of some Chlorinated Hydrocarbon Insecticides from Soils into Various Crops. *J. Agric. Food Chem.* **1959,** *7*, 430–433.

Lichtenstein, E. P.; Myrdal, G. R.; Schulz, K. R. Absorption of In-secticidal Residues from Contaminated Soils into Five Carrot Varieties. *J. Agric. Food Chem.* **1965,** *13*, 126–131.

Lichtenstein, E. P.; Schulz, K. R. Translocation of some Chlorinated Hydrocarbon Insecticides into the Aerial Parts of Pea Plants. *J. Agric. Food Chem.* **1960,** *8*, 452–456.

Lichtenstein, E. P.; Schulz, K. R. Residues of Aldrin and Heptachlor in Soils and their Translocation into Various Crops. *J. Agric. Food Chem.* **1965,** *13*, 57–63.

Lowrance, R.; Vellidis, G.; Wauchope, R. D.; Gay, P.; Bosch, D. D. Herbicide Transport in a Managed Riparian Forest Buffer System. *Trans. ASAE* **1997,** *40* (4), 1047–1057.

Lutri, V. F.; Matteoda, E.; Blarasin, M.; Aparicio, V.; Giacobone, D.; Maldonado, L.; Quinodoz, F. B.; Cabrera, A.; Albo, J. G. Hydrogeological Features Affecting Spatial Distribution of Glyphosate and AMPA in Groundwater and Surface Water in an Agroecosystem. Córdoba, Argentina. *Sci. Total Environ.* **2020,** *711*, 134557.

McCarty, P. L.; Rittmann, B. E.; Bouwer, E. J. Microbiological Processes Affecting Chemical Transformation in Groundwater. In *Groundwater Pollution Microbiology*; Bitton, B., Gerba, P. G., Eds.; Wiley: New York, 1984; p 89.

McKnight, U. S.; Rasmussen, J. J.; Kronvang, B.; Binning, P. J; Bjerg, P. L. Sources, Occurrence and Predicted Aquatic Impact of Legacy and Contemporary Pesticides in Streams. *Environ. Pollut.* **2015,** *200*, 64–76.

McManus, S. L.; Moloney, M.; Richards, K. G.; Coxon, C. E.; Danaher, M. Determination and Occurrence of Phenoxyacetic Acid Herbicides and their Transformation Products in Groundwater using Ultra High Performance Liquid Chromatography Coupled to Tandem Mass Spectrometry. *Molecules* **2014,** *19* (12), 20627–20649.

Mollamohammada, S.; Aly Hassan, A.; Dahab, M. Immobilized Algae-Based Treatment of Herbicide-Contaminated Groundwater. *Water Environ. Res.* **2021,** *93* (2), 263–273.

Mörtl, M.; Németh, G.; Juracsek, J.; Darvas, B.; Kamp, L.; Rubio, F.; Székács, A. Determination of Glyphosate Residues in Hungarian Water Samples by Immunoassay. *Microchem. J.* **2013,** *107*, 143–151.

Munira, S.; Farenhorst, A.; Sapkota, K.; Nilsson, D.; Sheedy, C. Auxin Herbicides and Pesticide Mixtures in Groundwater of a Canadian Prairie Province. *J. Environ. Qual.* **2018,** *47* (6), 1462–1467.

Muszyński, P.; Brodowska, M. S.; Paszko, T. Occurrence and Transformation of Phenoxy Acids in Aquatic Environment and Photochemical Methods of their Removal: A Review. *Environ. Sci. Pollut. Res*. **2020,** *27* (2), 1276–1293.

Nair, D. R.; Burken, J. G.; Licht, L. A.; Schnoor, J. L. Mineralization and Uptake of Triazine Pesticide in Soil-Plant Systems, *J. Environ. Eng.* **1993**, *119*, 842–854

Obrigawitch, T. T.; Cook, G.; Wetherington, J. Assessment of Effects on Non-target Plants from Sulfonylurea Herbicides using Field Approaches. *Pestic. Sci.* **1998,** *52*, 199–217.

Omenn, G. S. Ed.; In *Environmental Biotechnology*; Plenum Press: New York, 1987.

Papastergiou, A.; Papadopoulou-Mourkidou, E. Occurrence and Spatial and Temporal Distribution of Pesticide Residues in Groundwater of Major Corn-Growing Areas of Greece (1996−1997). *Environ. Sci. Technol.* **2001,** *35* (1), 63–69.

Perkovich, B. S.; Anderson, T. A.; Kruger, E. L.; Coats, J. R. Enhanced Mineralization of [14C]-Atrazine in Kochia Scopariarhizospheric Soil from a Pesticide-Contaminated Site. *Pesti. Sci.* **1995,** *46*, 391–396

Potter, T. L.; Carpenter, T. L. Occurrence of Alachlor Environmental Degradation Products in Groundwater. *Environ. Sci. Technol.* **1995,** *29* (6), 1557–1563.

Rendón-von Osten, J.; Dzul-Caamal, R. Glyphosate Residues in Groundwater, Drinking Water and Urine of Subsistence Farmers from Intensive Agriculture Localities: A Survey in Hopelchén, Campeche, Mexico. *Int. J. Environ. Res. Pub. Health* **2017,** *14* (6), 595.

Richard, J. J.; Junk, G. A.; Avery, M. J.; Nehring, N. L.; Fritz, J. S.; Svec, H. J. Analysis of Various Iowa Waters for Selected Pesticides:Atrazine, DDE, and Dieldrin-1974. *Pest. Monit. J.* **1975,** *9* (3), 117–123.

Ritter, W. F.; Scarborough, R. W.; Chirnside, A. E. M. Contamination of Groundwater by Triazines, Metolachlor and Alachlor. *J. Cont. Hydrol.* **1994,** *15* (1-2), 73–92.

Roberts, P. V.; McCarty, P. L.; Reinhard, M.; Schreiner, J. Organic Contaminant Behavior During Groundwater Recharge. *J. Water Pollut. Control Fed.* **1980**, *52*, 161.

Roche, H.; Vollaire, Y.; Martin, E.; Rouer, C.; Coulet, E.; Grillas, P.; Banas, D. Rice Fields Regulate Organochlorine Pesticides and PCBs in Lagoons of the Nature Reserve of Camargue. *Chemosphere* **2009,** *75*, 526–533.

Rochkind-Dubinsky, M. L.; Sayler, G. S.; Blackburn, J. W. In *Microbial Decom-Position of Chlorinated Aromatic Compounds*; Marcel Dekker: New York, 1987.

Rohde, W. A.; Asmussen, L. E.; Hauser, E. W.; Wauchope, R. D.; Allison, H. D. Trifluralin Movement in Runoff from a Small Agricultural Water- Shed. *J. Environ. Qual.* **1980,** *9* (10), 37–42.

Roseth, R. Plantevernmidlerigrunnvannijordbruksområder. Resultaterfraprøvetakingi 2010-2012; Bioforsk Rapport, 2013.

Ruiz-Toledo, J.; Castro, R.; Rivero-Pérez, N.; Bello-Mendoza, R.; Sánchez, D. Occurrence of Glyphosate in Water Bodies Derived from Intensive Agriculture in a Tropical Region of Southern Mexico. *Bullet. Environ. Contam. Toxicol.* **2014,** *93* (3), 289–293.

Sánchez-Camazano, M.; Lorenzo, L. F; Sánchez-Martín, M. J. Atrazine and Alachlor Inputs to Surface and Ground Waters in Irrigated Corn Cultivation Areas of Castilla-Leon Region, Spain. *Environ. Monitor. Assess.* **2005,** *105* (1), 11–24.

Sanchís, J., Kantiani, L.; Llorca, M.; Rubio, F.; Ginebreda, A.; Fraile, J.; Garrido, T.; Farré, M. Determination of Glyphosate in Groundwater Samples using an Ultrasensitive Immunoassay and Confirmation by On-Line Solid-Phase Extraction Followed by Liquid Chromatography Coupled to Tandem Mass Spectrometry. *Analyt. Bioanalyt. Chem.* **2012,** *402* (7), 2335–2345.

Schult, J. Pesticides and Nutrients in Groundwater of the Darwin Region; Northern Territory Department of Environment and Natural Resources, Report No. 21/2016D, Palmerston, 2016.

Scribner, E. A.; Battaglin, W. A.; Gilliom, R. J.; Meyer, M. T. Concentrations of Glyphosate, its Degradation Product, Aniomomethylphosphonic Acid, and Glufosinate in Ground and Surface-Water, Rainfall, and Soil Samples Collected in the United States, 2001–2006; US Department of Interior and US Geological Survey, Scientific Investigation Report 2007–5122.

Scribner, E. A.; Dietze, J. E.; Thurman, E. M. Acetamide Herbicides and their Degradation Products in Ground Water and Surface Water of the United States, 1993–2003; US Department of the Interior, US Geological Survey, 2004.

Shann J. R.; Boyle J. J. Influence of Plant Species on In Situ Rhizosphere Degradation. In *Bioremediation Through Rhizosphere Technology, ACS Symposium Series*; Anderson, T. A., Coats J.R., Eds.; American Chemical Society: Washington, DC, 1994; vol *563*, pp 70–81.

Singh, R.; Ghosh, D.; Dubey, R. P.; Singh, V. P. Weed Control in Sesame with Pre-emergence Herbicides. *Indian J. Weed Sci.* **2018,** *50*, 91–93.

Székács, A.; Mörtl, M.; Darvas, B. Monitoring Pesticide Residues in Surface and Ground Water in Hungary: Surveys in 1990–2015. *J. Chem.* **2015.** http://dx.doi.org/10.1155/2015/717948.

Thurman, E. M.; Goolsby, D. A.; Aga, D. S.; Pomes, M. L.; Meyer, M. T. Occurrence of Alachlor and its Sulfonated Metabolite in Rivers and Reservoirs of the Midwestern United States: The Importance of Sulfonation in the Transport of Chloroacetanilide Herbicides. *Environ. Sci. Technol.* **1996,** *30* (2), 569–574.

Thurman, E. M.; Meyer, M.; Pomes, M.; Prey, C. A.; Schwab, A. P. Enzyme-Linked Immunosorbent Assay Compared with Gas Chromatography/Mass Spectrometry for the Determination of Triazine Herbicides in Water. *Anal. Chem.* **1990,** *62*, 2043–2048.

Tirmenstein, A.; Mangipudy, R. In *Alachlor, Encyclopedia of Toxicology*, Third Ed.; Wexler, P., Ed.; Academic Press, 2014; pp 107–109.

Van Stempvoort, D. R.; Spoelstra, J.; Senger, N. D.; Brown, S. J.; Post, R.; Struger, J. Glyphosate Residues in Rural Groundwater, Nottawasaga River Watershed, Ontario, Canada. *Pest. Manag. Sci.* **2016**, *72* (10), 1862–1872.

Wade, H. F.; York, A. C.; Morey, A. E.; Padmore, J. M.; Rudo, K. M. The Impact of Pesticide use on Groundwater in North Carolina. *J. Environ. Qual.* **1998,** *7*, 1018–1026.

Wauchope, R. D.; Buttler, T. M.; Hornsby, A. G.; Augustijn-Beckers, P. W. M.; Burt, J. P. The SCS/ARS/CES Pesticide Properties Database for Environmental Decision-Making. In *Reviews of Environmental Contamination and Toxicology*; Springer: New York, NY 1992; pp 1–155.

Wilson, J. T.; McNabb, J. F.; Balkwill, D. L.; Ghiorse, W. C. Enumeration and Characterization of Bacteria Indigenous to a Shallow Water-Table Aquifer. *Ground Water* **1983,** *21*, 134.

Yue, G. E. N. G.; Jing, M. A.; Ran, J. I. A.; Xue, L.Q.; Tao, C. J.; Li, C. J.; Yan, L. I. N. Impact of Long-Term Atrazine use on Groundwater Safety in Jilin Province, China. *J. Int. Agri.* **2013,** *12* (2), 305–313.

Zhang, W. Global Pesticide Use: Profile, Trend, Cost/Benefit and More. *Proc. Int. Acad. Ecol. Environ. Sci.* **2018,** *8* (1), 1–27.

Zhang, L.; Rana, I.; Shaffer, R. M.; Taioli, E.; Sheppard, L. Exposure to Glyphosate-Based Herbicides and Risk for Non-Hodgkin Lymphoma: A Meta-Analysis and Supporting Evidence. *Rev. Mutat. Res.* **2019,** *781*, 186–206.

Zhao, S.; Arthur, E. L.; Coats, J. R. The Use of Native Prairie Grasses to Degrade Atrazine and Metolachlor in Soil. In *Environmental Fate and Ef- fects of Pesticides*; Coats, J. R.; Yamamoto, H., Eds.; ACS Symposium Series 853, ACS: Washington, DC, 2003; pp 157–167.

CHAPTER 6

Fluoride Contamination in Ecosystem and Mitigation Strategy

BISHWAJIT SINHA[1], PRAKASH RANJAN BEHERA[2], and BHOLANATH SAHA[3]

[1]*College of Agriculture, Odisha University of Agriculture and Technology, Bhawanipatna, Odisha, India*

[2]*Siksha 'O' Anusandhan Deemed to be University, Bhubaneswar, Odisha*

[3]*Department of Soil Science, Dr. Kalam Agricultural College, Bihar Agricultural University, Kishanganj, Bihar, India*

ABSTRACT

The groundwater and soil quality rely upon geogenic and anthropogenic circumstances. There are naturally occurring minerals in aquitards in different regions of India that control the concentration of pollutants, such as fluoride (F), arsenic (As), and nitrate (NO_3^-). Among these environmentally polluting elements, high concentrations of fluoride, typically much over the acceptable limit of 1 mg fluoride per liter, are a serious concern in the arid and semiarid tropics and subtropics of the world, and in some states of India too, where alluvial soil prevails. Fluorine in nature occurs as fluoride, and is beneficial to human health in trace amounts but hazardous in excess. Drinking groundwater in these areas has led to a mild form of dental fluorosis to debilitating skeleton fluorosis. There are several methods which are evolved for defluoridation of water like Nalgonda technique, reverse osmosis, IISC method, membrane filtration process, and electrochemical methods for removal of fluorides to provide safe drinking water. Although the concentration of F^- in groundwater is controlled by the local geological

Environmental Contaminants: Impact, Assessment, and Remediation.
Pritam Ganguly, Jajati Mandal, M. Paramsivam, & Sandip Patra, (Eds.)

setting, leaching through the soil profile, weathering of bedrock, and climatic condition of an area, the higher concentration in agricultural land might be due to the high pumping of irrigation water, long term drought, and longer residence time in the soil profile.

6.1 INTRODUCTION

Fluorine is the most electronegative element (3.98 at Pauling's scale) of the halide groups of the periodic table (group VII) and occupies 24th rank in universal abundance and 13th in terrestrial abundance (0.3 g kg^{-1} in earth's crust), discovered by Henri Moissan in 1886 for which he was awarded the Nobel prize in 1906. Fluorine is most reactive in nature, found mostly with different minerals like fluorspar, cryolite, etc., and discharged into the atmosphere through the phosphate fertilizer industry, brick, cement, and other industrial processes.

Fluoride ion has both beneficial and harmful impact on plants, animals, and humans. Intake in small quantities (<0.5 mg/L) reduces the dental problem, whereas excess amount (>1.5 mg/L) causes fluorosis, and excess availability in drinking water has biological hazardous effects on the environment. The weathering of rocks and minerals is a potential source of fluorine global danger to 200 million people, out of which 60 million people (Arlappa et al., 2013) in India are under this threat of fluorosis having fluorine concentration in drinking water >1.5 mg/L (Muralidharan et al., 2011). The most seriously saddened states of India are Tamil Nadu, Maharashtra, Gujarat, Punjab, Haryana, Andhra Pradesh, and Uttar Pradesh (Arlappa et al., 2013).

6.2 FLUORIDE DISTRIBUTION IN ECOSYSTEM: SOURCES AND FORMS

Fluorine enters the environment through different sources, like natural and anthropogenic sources.

6.2.1 SOIL

The total fluoride content in normal soil ranges from 150 to 400 mg/kg, whereas fluorine level in the clay soil can be up to 1000 mg/kg (Susheela, 1999). Nowadays fluoride contamination in agricultural lands is because of the application of different phosphatic fertilizers which have total 1.5% fluorine (Bombik et al., 2011). Soil contaminated with fluorine shows toxicity effects on plants and humans, and also contaminates the groundwater (Begum, 2012; Ericson et al., 2014). The primary minerals, after weathering, produce different minerals containing fluorine, such as muscovite or biotite, and their content depends on the mineralogical composition of the soil, inorganic fraction, clay percent, and pH.

6.2.2 WATER

Fluorine concentration in water for safety is below 1.0 mg/L, whereas marginal level is between 1.1 and 2.5 mg/L and when it goes above 2.6 mg/L, contaminates it (Susheela, 1999). The level of fluorine in groundwater is higher than surface water through leaching process. The factors such as nature of rocks, pH, temperature of soil, consistency of soil, chelating action of other elements, leakage of groundwater, and other geological factors determine the concentration of fluorine in the soil as well as groundwater (Li et al., 2014).

6.2.3 FORAGE, GRASSES, AND FOOD

The grasses and forages contain higher levels of fluorine than other areas, and it is due to ash and dust that are rich in fluorine. Plant contamination depends on factors like the amount of fluorine released into the atmosphere, type of vegetation, height of plants, source of fluorine to contaminated areas and seasons, etc. (NRC, 1960). The relationship between fluorine levels in soil and plants is crucial as each 100 ppm rise in soil fluorine can elevate plant fluorine content by 3 ppm up to 2200 ppm (Mascola et al., 1974). Due to its nonbiodegradable nature, excess fluorine application enters the food chain system and causes lingering toxicity. Crops such as wheat, cereal, maize, pulses, cabbage, cucumber, tomato, brinjal, lady's finger, beetroot, potato, onion, sweet potato, banana, grapes, mango, apple, coconut, groundnut, mustard, and food items such as mutton, beef, pork, fish, egg, etc., all have potential to carry high levels of fluorides.

6.2.4 VOLCANIC ACTIVITIES

Volcanic eruption affects animal and plant kingdoms by contributing high fluorine concentration mostly in the form of hydrogen fluoride. The geochemical cycles are altered due to the addition of fluoride ions, and after leaching, decaying causes casualty to domestic and wild animals (Bellomo et al., 2007).

6.2.5 FLUORIDES IN COSMETICS AND DRUGS

Smoking of cigarettes also serves as another significant route for fluoride intake by humans. The toothpaste also uses raw materials such as calcium carbonate, talc, and chalk that contain high amounts of fluoride (800–1000 ppm). Other fluoridated brands add 1000–4000 ppm of fluorides.

Continuous use of certain drugs has been known to lead to cause adverse effects of fluoride, for example, osteoporosis treatment by sodium fluoride, Niflumic acid for the cure of rheumatoid arthritis, fluoride mouthwash (Proflo) for the stronger tooth, etc.

TABLE 6.1 Percent Contamination and Habitations Affected by Different Contaminants.

Contaminants	No. of habitations	% contamination
Iron	42,788	41.66
Arsenic	3863	3.760
Fluoride	**29,065**	**28.30**
Nitrates	7296	7.100
Salinity	19,576	19.06
Sulfates	108	0.100
Total	102,696	100.0

Source: Reprinted with permission from Mahapatra, 2007.

6.3 GROUNDWATER CONTAMINATION

The water sources get contaminated by fluoride content by different variable climatic as well as other geological conditions, like arid climates, volcanic rocks, and granite basements, but the fluoride concentration lies below WHO guidelines. In general, the minerals like fluorite, fluorapatite, biotite, apatite, muscovite, hornblende, villiaumite, tremolite, sellaite, cryolite, topaz, fluocerite, yttrofluorite, gagarinite, bastnasite, microlite, sphene, epidote, amphibole, montmorillonite, kaolinite, pegmatite, mica, clays, villuanite, phosphorite, and some micas weathered from different rocks and minerals, release fluorides into groundwater (Adimalla et al., 2019).

The granite basement aquifers containing mica, apatite, or amphibole are more in India and Sri Lanka that elevate the fluoride concentrations in groundwater. In arid climate, the slow infiltration and flow rates favor the dissolution of fluorides present in rocks, as the interaction of rock with the solution is more.

High fluorine containing minerals present in alkaline volcanic areas are mixed to groundwater through fluoride geothermal solutions (Edmunds and Smedley, 2005). The high fluoride decreases calcium ion concentration and increases sodium bicarbonate that restricts precipitation of fluoride as fluorite. Moreover, the extent of fluoride in groundwater relies on the factors like geochemical structure of the rock, the contact time between the rock strata and water, pH or alkalinity of the soil/water, leaching of fluoride from the fluoride-containing rocks, abundance of rainfall, and excess utilization of the water.

6.4 CHEMISTRY IN SOIL SYSTEM

The fluoride ion concentration in surface soil (1 mg F g^{-1} soil) is lesser and has low solubility (<1 μg g^{-1}). The different contamination sources such as fluoride-releasing industries, phosphatic fertilizer applied sites, and coal-fired plants have high levels of fluoride (300 μg g^{-1}). Predominantly, pH, clay content, and organic carbon influence fluoride retention in the soil environment. Clayey soils and loamy soils, having Fe and Al, can deposit or retain >95% F. Factors such as soil types and concentrations of F control the amount of F sorbed in the soil. Fluoride uptake is nearly 10 times more in acidic soil than in alkaline soil, as Fe and Al have greater tendency to form a bond with it than clay. As it is highly electronegative, it forms a strong bond with clay by replacing OH^-/H_2O from hydroxides or oxyhydroxides compounds. A low pH will be favorable for bonding with Fe and Al. Retention is particularly higher at pH 5.5 and declines gradually above 6.5. Higher pH changes the electrostatic potential, hence declines the retention of fluoride in soil particles. In presence of higher amount of calcium carbonate as in calcareous soil, the fluoride is fixed with calcium forming calcium fluoride. Organic soil has lower F content. More than 90% of fluoride content in soil is tightly attached to different soil particles and hence it is insoluble in nature. Micas and other soil minerals play a key role in it. Total fluoride content in soil ranges from 20 to 1000 μg g^{-1}.

6.5 HARMFUL LIMITS TO HUMANS AND PLANTS

6.5.1 EFFECT ON PLANTS

Plants show symptoms due to toxicity of fluoride that depend on concentration, time of exposure, temperature, age of plant, and forms of uptake of ion. Vascular tissues act as entry point for fluorides and accumulate on upper edge of leaves causing necrosis, and proceed toward leaf base. Hydrogen fluoride exposure causes severe lesions, whereas higher fluoride concentration can cause acute injuries leading to marginal necrosis and proceed toward leaf base. In palisade cells, the chlorophyll synthesis is affected by the accumulation of higher concentration of delta-aminolevulinic acid, and it also damages leaf spongy mesophyll.

Fluorides are severely phytotoxic to plants through changing of metabolic pathways. They have negative impacts on germination of seeds, different enzymatic activities, photosynthesis, respiration, and reproduction so as yield of crops, as well as affect the carbohydrate and protein content of plants. Hydrogen fluoride gas accumulation also affects the plants even at lower concentrations and when enters the plants, affects different plant physiological processes. Thus, the fluoride concentration also affects animals and human beings, hence considered as a most phytotoxic air pollutant.

6.5.2 IMPACT ON HUMAN

In soil, water, and animals, the fluoride content is generally low. It is required for the solidification of bones and also lowers dental decay issues when present in lower concentrations, but fluoride uptake in excessive quantity has many bad effects. It may enter the body by the consumption of contaminated water, food, or through inhalation. Longer duration consumption of fluoride contaminated food or water may affect thyroid activities, skeletal development, and exhibit symptoms such as teeth decay, osteoporosis, fluorosis, etc. Environmental pollution due to fluoride contamination has serious effects causing syndromes such as asthenia, muscular atrophy, heartburn, joint pain, disfigured teeth as well as skeletal joint deformation, osteosclerosis, and calcification of ligaments, and finally leading to kyphosis, stiffness of the spines, and bony exostosis (Figs. 6.1a, 6.1b).

6.5.2.1 TYPES OF FLUOROSIS

Based on clinical observations, there are three types of fluorosis:

- **Dental fluorosis:** Teeth mottling, a yellowish white-glistening tooth that develops a yellow-white spot, a brown streak at the tip of the teeth, in the middle of the teeth, or the upper part of the teeth, indicates that a child has been exposed to high fluoride in food or water, or both, up to the age of 2 years, from the age of 2 to 4 years, and

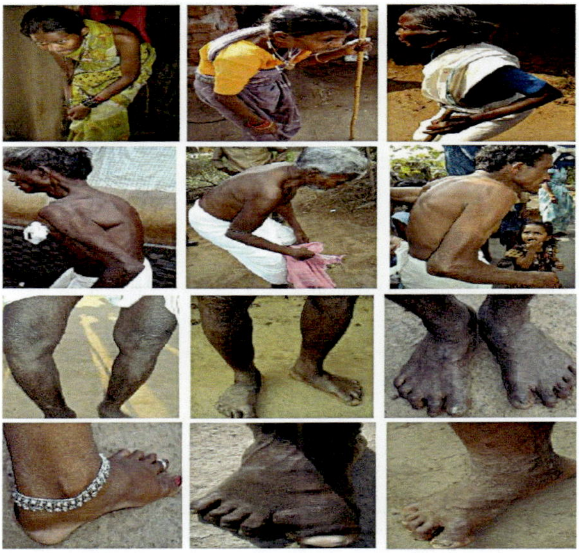

FIGURE 6.1A Adverse effects of fluoride on human beings.

Source: Reprinted from Dash et al., 2016.

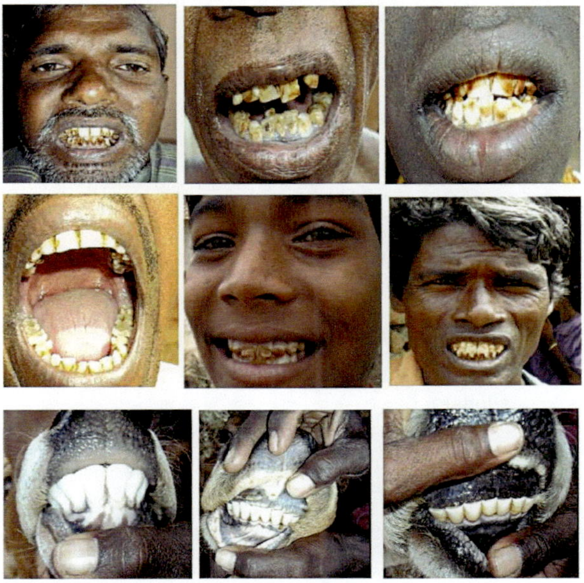

FIGURE 6.1B Adverse effects of fluoride on teeth.

Source: Reprinted from Dash et al., 2016.

TABLE 6.2 Drinking Water Standards for Fluoride Prescribed by Various Authorities.

Sl. No.	Authorities	Acceptable fluoride content (mg/L)
1	WHO (international standard for drinking water)	0.5
2	US public health standard	0.7–1.2
3	BIS (IS 10500)	1.0–1.5
4	Indian Council of Medical Research	1.0–2.0
5	CPHEEO	1.0–1.5

Source: Adapted from Kanthe, 2014

TABLE 6.3 Standards of Fluoride Intake.

Standard	Age group	Intake	Source
Adequate intake (AI)	0–6 months	0.01 mg/kg/day	Food and Nutrition Board of the USA (SCSEDRI, 1997)
	> 6 months	0.05 mg/kg/day	
	4–8 years	1 mg/day	(Based on assessment of requirement for caries prevention)
	Adults male	4 mg/day	
	Adults female	3 mg/day	
Tolerable upper intake level (UL)	0–8 years	0.1 mg/day	SCSEDRI, 1997
	> 8 years	10 mg/day	
Tolerable upper intake level (UL)	1–3 years	1.5 mg/day	European Food Safety Authority
	4–8 years	2.5 mg/day	EFSA, 2011
	9–14 years	5 mg/day	
	> 15 years	7 mg/day	
Dental fluorosis			US Department of Health and Human Services
NOAEL[a]	0.06 mg/kg/day		USDHHS, 2003
LOAEL[b]	0.12 mg/kg/day (8 mg/day for adults)		
Skeletal fluorosis			
NOAEL[a]	0.15 mg/kg/day		
LOAEL[b]	0.25 mg/kg/day (17.5 mg/day for adults)		

[a]No observed adverse effect level.

[b]Lowest observed adverse effect level (5% test population level).

Source: Adapted from Bader et al., 2015

from the age of 4 to 6 years and after, respectively. Teeth become black in final phases, resulting in loss of teeth at an early age.

- **Skeletal fluorosis:** There is a possibility of skeletal fluorosis developing when the bioaccumulation of fluoride in bone increases with age. Skeletal fluorosis causes lifelong bone and joint abnormalities. It causes severe pain and stiffness in backbone, joints, and rigidity in hip region. It increases thickening and density of bones as well as calcification of ligaments. It results in deformities in limbs, hunch back, paralysis, muscular wasting, and premature ageing.
- **Nonskeletal fluorosis:** Fluorosis not only affects bones and teeth, but it may also lead to neurological and gastrointestinal disorders. These disorders are manifested in various forms such as neurological, muscular, allergic, gastrointestinal, or urinary tract manifestations.

The health impact of fluorosis is different for different age groups and further can be divided into primary, secondary, and tertiary fluorosis based on its general symptoms. Primary fluorosis takes place when one individual has ingested an excess amount of F by contaminated water or vegetation for a prolonged period of 10–15 years, generally shown in the age group of 40–70. Secondary fluorosis occurs when a person uptakes higher F contaminated food or water since his/her birth. More severely, reported in the age group of 20–40. The victims of tertiary fluorosis are mostly of 1–20 years of age group; especially 8% of school going children bear a hump on their back and suffer from malaise.

Other than fluorosis, F affects human system in various other ways. When fluoride is ingested in human system, it accumulates on erythrocytes resulting in loss of calcium content and formation of echinocytes of lesser life span than normal. In turn, it causes early destruction of RBCs ascribed to anemia. Soft tissues like ligaments harden due to calcification and result in blockage of blood vessels (Siddiqui, 1970). Fluorides also stimulate oxygen consumption by white blood cells; hence affect the phagocytosis and antibody formation processes in blood. Fluorides are potential carcinogenic for human body, as they are known to cause chromosome damage (mainly due to interference with protein synthesis rather than direct interaction) as well as mutations. Fluorides are known to promote development of bone cancer. Moreover, fluorides inhibit iodine uptake; hence depress thyroid activity and promote thyroid enlargement. Excessive fluoride intake decreases 5-hydroxy-indole acetic acid and increases norepinephrine in the brain; hence, in early childhood, reduces mental work capacity (FHED, 2019).

6.6 DETECTION AND QUANTIFICATION

Fluoride detection and characterization are critical steps in understanding the advantages and possible toxicity of fluoride in biotic and abiotic natural sources, which may have a favorable or negative influence on human health. The presence of these ions in soil has the potential to dramatically diminish the size of the microbial population. Fluoride can be harmful to human health at high doses (>1.5 mg/L), hence it must be identified and quantified. Given these negative repercussions, considerable effort has gone into developing improved analytical techniques for fluoride sensing. This chapter discusses the analytical methods used to determine fluoride, such as inductively coupled plasma mass spectrometry (ICP–MS), atomic absorption spectrometry (AAS), molecular absorption spectrometry (MAS), and electrochemical approaches. The most often used methods are fluoride-sensing methods, including fluorescence and colorimetric methods, which have been studied because of their high sensitivities, low detection limits, and intracellular fluoride monitoring capacity.

6.6.1 MASS SPECTROMETRY

The approach is based on the formation of gas-phase ions that may be separated electrically or magnetically based on their mass to charge ratio (m/z). A mass analyzer separates ions according to their mass to charge ratio. A mass spectrometer analysis may provide a wealth of information about the analytes, including their structure, purity, and composition. Pagliano et al. coupled direct headspace (HS) or GC–MS with triethyloxonium tetrachloroferrate (III) to derivatize fluoride into fluoroethane. The approach exhibited a detection limit of 3.2 g/L. Derivatization using the suggested reagent exhibited a one-step reaction without the need of organic solvents. Because the suggested derivatizing reagent was nonvolatile, it was safer for fluoride detection. Fluoride detection is hampered by common fluoride-complexing agents such as Al(III) and Fe(III). As a result, using the approach of standard additions, the interference of these ions has been eradicated.

6.6.2 ICP–MS

ICP–MS is a multielement approach that needs atomization of the sample using a plasma source. Ions are subsequently identified using a mass spectrometer, where they are separated based on their mass to charge ratio. The approach has excellent detection limits in the ppt (parts per thousand) range. Multiple detectors are used in MC–ICP–MS to gather each isotope of a single element; however, the technique needs all isotopes in a restricted mass range, preventing them from being used for routine analysis. The approach has also been claimed to have reduced detection sensitivity, for example, the detection limit for negative ion mode is 110 g/L.

Bayón et al. used drinking and seawater samples to show a unique fluoride detection method. The scientists employed an excess of Al(III) to create an aluminum monofluoride complex, which was then separated into two species (AlF^{2+} and Al^{3+}) in a tiny ion-exchange column. The approach revealed interferences from Mg(II) and Zn(II) ions, but iron interference was reduced by adding EDTA to the sample solution. Using 0.45 M HNO_3 as an eluent, the approach exhibited a detection limit of 0.1 ng/mL. The ICP–MS approach outperformed FISE in terms of recovery and fluoride detection, even at extremely low concentrations.

6.6.3 AAS AND MAS

AAS is an accepted method for qualitative and quantitative analysis of fluoride in both organic and inorganic mixture. The MAS approach incorporates nonmetal determination in flames and graphite furnaces, which AAS cannot do. Because fluorine is the most electronegative element, it has a high ionization potential (17.42 eV), and its resonance lines are located in the vacuum-UV region at less than 100 nm. As a result, ICP–MS is deemed ineffective for fluoride sensing due to inadequate argon ionization potential for the ionization of this fluoride, except when the process of electrothermal vaporization is used for "dry" sample introduction. Therefore, an alternative for fluoride ion detection is MAS in flames and furnaces.

6.6.4 ELECTROCHEMICAL TECHNIQUES

Ion-selective electrodes (ISEs) give a quantitative and precise level of concentration of a variety of ionic surfactants. The ionophore used in ISEs serves as a detection platform for the selective measurement of a single ion of interest. FISE has been the most extensively used analytical technique for extracting fluoride ion concentrations in drinking water, fruit juice, bottled water, toothpaste, and other products. The surface of the electrode in the FISE is constructed of LaF_3 crystal with a small quantity of europium fluoride (EuF_2). The presence of EuF_2 in the crystal aids in the production of a response to external fluoride ion concentration. The electrochemical sensors and their binding with anions were studied and investigated by cyclic and square-wave voltammetry. With a reduction in current, this binding characteristic culminated in charges related to the anions specific in reduction potential. For biological samples, the high sensitivity and fast detection electrodes with reversible solid state type membrane references are made that can detect very much low concentrations in minimal time.

6.6.5 FLUORESCENCE TECHNIQUE FOR FLUORIDE SENSING

The fluorescent probes with receptors of signaling units for fluoride produce many significant physical changes, and these are proven to have high sensitivity and ease of handling. Recently, a technique for fluoride sensing in water that uses hydrogel as a supporter in a reaction involving a water-insoluble FP was developed. The researchers created N-(3-(benzo[d]thiazol-2-yl)-4-(tertbutyl diphenyl silyloxy) phenyl) acetamide (BTBPA) as the FP, which displayed a significant fluorescent color change from blue to green following the reaction with fluoride. The BTBPA/PVP probe is colorless and has a single absorption spectra band at 345 nm. When a neutral pH sodium fluoride solution was added, the peak shifted to 370 nm, and the fluorescence spectra showed a bathochromic shift from 406 to 494 nm, with an isoemissive point at 468 nm. Changes in color from blue-violet to green fluorescence suggested that the BTBPA/PVP is an efficient naked eye indicator. The properties of fluoride, like nucleophilic and other basic properties, have been utilized for sensor development providing dual-mode strategy like visible naked eye and a "fluorescence-on" detection. Another pathway, like Baylis–Hillman type, gives an idea to study chromogenic oligomerization reactions as it involves the addition of fluoride as a nucleophile on α,β-unsaturated system; this form an enolate that could start the polymerization reaction with the addition of nucleophilic compounds. When fluoride attack is there, then deprotonation of the α-position of amide compound and bathochromic shift of red-colored compounds occur.

6.7 DEFLUORIDATION OF WATER

Defluoridation is widely tested and conventional method for removal of fluorides to provide safe drinking water. It is the downward adjustment process of fluoride content in drinking water to an optimal level (Khaimar et al., 2015). This process can be achieved through two methods, that is, central treatment that is carried out at the source under the supervision of some skilled personnel and practiced in developed countries; the other method is point use treatment practiced in developing countries (Iyenger, 2003).

Defluoridation technique has components like membrane filtration, ion exchange, nanofiltration, and adsorption. Recent trends, like biosorption techniques, are more effective using low-cost adsorbents such as rice husk, red mud, sawdust, etc.

6.7.1 NALGONDA TECHNIQUE

The Nalgonda approach was one of the first to be developed and used in India by the National Environmental Engineering Research Institute (NEERI). The two most essential chemicals used in the Nalgonda technique are alum (aluminum sulfate or calcium aluminum sulfate) and lime (calcium oxide). The mixture is then swiftly combined with fluoride-contaminated water, followed by gentle stirring, after which flocs (aluminum hydroxides) form and are removed by settling. Lime expansion causes fluoride to precipitate as an insoluble calcium fluoride compound, raising the pH to 11–12. As the lime leaves a residual amount of 8.0 mg F^-/L, it is constantly treated with alum to the best possible guarantee for fluoride removal. Initially, precipitation happens by lime which is followed by a second step in which alum is added to come up with coagulation. When alum is added to water, generally two reactions happen. Alum reacts with a portion to produce insoluble aluminum hydroxide [Al(OH)3], and then with fluoride ions in water in the second reaction. The best pH range for fluoride elimination is between 5.5 and 7.5 (Razbe et al., 2013). Due to the sorption, the main contents of the fluoride are removed along with the flocs developed and ion exchange with hydroxide groups. In the popular Nalgonda technique, alum and lime compounds are widely used for defluoridation of drinking water (Nawlakhe et al., 1993).

Estimation of required dosage (Raw): The estimation of the amount of alum needed is based on the formula given by Dahi et al. (1995) using the Freundlich technique as,

$$A = (Fr - Ft) * V/a * Ft^1/b$$

where A is the amount of alum required (g);
Fr is the amount of fluoride concentration in the raw water (mg/L);
Ft is the residual fluoride concentration in the treated water (mg/L);
V is the volume of water to be treated in batch, (L);

a is the constant for sorption capacity, L (1–1/b) mg 2/b g^{-1};
b is the sorption intensity constant.

This method appears to be efficient in terms of larger quantities, simplicity of design, ease of construction and operation, ease of maintenance, less reliance on highly skilled workers, nonregeneration of media, low water waste, minimal mechanical and electrical equipment, minimal disposal issues, uniform quality defluoridated water, and the reuse of alum from generated sludge (Suneetha et al., 2015).

A prevalent criticism among researchers is that the Nalgonda approach necessitates an extensive monitoring and everyday operators require sufficient training and consistent operation (Ayoob et al., 2008). The wastewater flavor is disliked by customers since the maintenance requirements are expensive and the automatic systems do not perform, as evidenced by a report on the Nalgonda technique (Maheswari, 2006). To improve the sustainability of the Nalgonda systems, a reliable monitoring system, better fluoride removal capability, and community management capacity building are all essential. The Nalgonda system's fluoride removal method has been described as a coprecipitation in which the primary ingredient (fluoride) is removed as flocs by settling due to a combination of sorption and ion exchange with the hydroxide groups developed. Furthermore, coprecipitation procedures for aluminum salts have a few points of interest and limitations:

Interest

1. A commonly used approach.
2. When compared to other defluoridation techniques, this one is more practicable.
3. The technique is simple to grasp.

Restriction

1. The chemical doses required are high ($Al(OH)_3$ up to 700–1200 mg/L).
2. The sludge transmission problem.
3. Cannot estimate the passable reaches of fluoride so far.
4. Prerequisite of talented labor.
5. Aluminum releases in treated water, which may cause Alzheimer's disease.

6. The final fluoride concentration in treated water is highly dependent on the quantity of precipitated fluoride, calcium, and aluminum salt dissolved.
7. The use of aluminum sulfate as coagulant increases the sulfate ion concentration, which shows cathartic impacts in human.

6.7.2 REVERSE OSMOSIS

Due to its high waste removal capacity, reverse osmosis was regarded as an alternative to traditional waste treatment in the 1980s (Prihasto et al., 2009). By flowing across a semipermeable membrane, RO provides pressure to directly remove anions. RO operates at higher pressure, resulting in a significant rejection of suspended salts. In reverse osmosis, hydraulic pressure is applied to one side of the semipermeable membrane, forcing water through the membrane and leaving wastes behind. The reverse osmosis process involves the mechanical removal of compounds by the use of size exclusion, adsorption, charge repulsion, and diffusion. The membrane's porosity aids in nanofiltration, whereas a nonporous membrane follows the solution diffusion model's transport mechanism, and the solutes dissolved in it diffuse through the membrane down by concentration gradient (Richards et al., 2010).

Because the membranes are particularly sensitive to pH and temperature, the removal of fluoride by reverse osmosis varies from 45 to 90% as the pH of the water increases from 5.5 to 7. To demineralize water, reverse osmosis requires high energy, poor or insignificant selectivity for monovalent ions, and high pressure (Schneiter and Middlebrooks, 1983). Ndiaye et al. (2005) used a RO separation method for defluoridation of industrial wastewater and discovered that the rejection of fluoride ion was consistently greater than 98%, because the RO membrane was totally recovered after each arrangement of analyses. BerhanuAssefa et al. (2006) found that the fluoride retention of RO membranes in the Ethiopian Rift region ranged from 94 to 99%.

Diawara et al. (2011) used low-pressure reverse osmosis to remove fluoride and salt from brackish groundwater in a Senegal hamlet, with fluoride rejection rates ranging from 97 to 98.9%. Gedam et al. (2012) discovered that utilizing a polyamide RO membrane, 95–98% of fluoride was removed from groundwater in Moradgaon village, Chandrapur district.

Schoeman et al. (2010) found that fluoride can be eliminated with low-pressure RO in the feedwater concentration range of 10–17 mg/L to more or less 0.2 mg/L in the RO permeate, and with high-pressure RO from about 17 mg/L in the feedwater to approximately 0.2 mg/L in the RO permeate.

In southern Brazil, Briao et al. (2014) employed reverse osmosis to desalinate water from the Guarani aquifer system for drinking purposes. At 2 MPa pressure and 1.61 m/s cross section flow velocity, RO was able to reject 100% fluoride, 97% total dissolved solids (TDS), and 94% sulfate ions. Blended water, which was formed by blending groundwater and the permeate, was able to recover 93% of the drinking water.

6.7.3 IISC METHOD

Fluoride-contaminated groundwater with fluoride ion concentrations ranging from 1.5 to 7 mg/L is treated using the suggested approach. Magnesium oxide removes fluoride ions via a chemisorption method. Magnesium oxide hydrates to brucite [$Mg(OH)_2$] in the presence of water samples as:

$$MgO + H_2O \rightarrow Mg(OH)_2 \tag{6.1}$$

Fluoride ions (F^-) in polluted water samples are absorbed into the brucite lattice during brucite synthesis by isomorphous replacement of hydroxyl ions by fluoride ions. The $Mg(OH)_{2-y}$.Fy phase is formed as a result of the isomorphous substitution process. In addition to magnesium oxide, the raw water contains calcium chloride solution, sodium bisulfate solution, and calcium hydroxide (as solids). The addition of magnesium oxide raises the pH of water from 10.1 to 10.3. The reaction: Sodium bisulfate solution lowers the pH of magnesium oxide treated water to within drinkable water standards (6.5–8.5).

$$HSO_4^- + OH^- \Rightarrow SO_4^{2-} + H_2O \tag{6.2}$$

Bicarbonate ions present in the fluoride-bearing water samples convert to carbonate ions in the alkaline pH created by magnesium oxide addition according to the reaction:

$$HCO_3^- + OH^- \Rightarrow CO_3^{2-} + H_2O \tag{6.3}$$

The carbonate ions formed in reaction 6.3, in turn, consume additional sodium bisulfate as:

$$CO_3^{2-} + HSO_4^- \Rightarrow HCO_3^- + SO_4^{2-} \tag{6.4}$$

Calcium hydroxide (hydrated lime) and calcium chloride solution minimize bicarbonate ion interference with sodium bisulfate intake by converting soluble bicarbonates to insoluble calcium carbonates as:

$$HCO_3^- + Ca^{2+} + OH^- \Rightarrow CaCO_3 + H_2O \quad (6.5)$$

6.7.4 MEMBRANE FILTRATION PROCESS

Two membrane filtering technologies that can be used to remove fluoride are reverse osmosis (RO) and electrodialysis. Similarly, RO membranes have been claimed to be used to remove fluoride from polluted water sources. In reverse osmosis (RO), hydraulic pressure is applied to one side of a semi-permeable membrane, forcing water through the membrane while leaving salts behind. The size of the contaminants left behind is proportional to the pressure applied to the membrane. RO has been proven to be successful in decreasing traced concentrations of these pollutants in recent research by Fox (1981) and Huxstep (1981). The water treatment process has become more cost-effective and dependable because of advancements in membrane designs and materials (Fox, 1981).

The membranes in electrodialysis enable ions to flow but not water. The ions are propelled across the membranes by a driving force. Fluoride removal in the RO process has been reported to range from 45 to 90% as the pH of the water is adjusted from 5.5 to 7. The pH and temperature of the membranes are extremely sensitive. The various systems are also prone to fouling by particle contaminants, chemical assaults, and high waste concentrations. The waste quantities are significantly greater than those generated by the ion-exchange process. Pretreatment needs might be rather severe at times. Electrodialysis consumes a lot of energy and is costly (Huxstep, 1981).

Limitations (Kumar et al., 2000)

1. The membranes are sensitive to temperature and pH, and maintenance problem arises because of plugging, fouling by particulate matter, and concentrated with a large quantity of wastes.
2. Both processes are expensive and very complicated.
3. Removes all the ions present in water including some essential minerals for proper growth, and hence remineralization is required after treatment.
4. High energy consumption and large amount of water gets wasted as brine.

6.7.5 ELECTROCHEMICAL METHODS

Electrocoagulation (EC) utilizes an electrolytic process to generate a coagulant in situ by oxidation of an appropriate anodic material, and the coagulant ions then react with the target pollutant ions, initiating coagulation (Ayoob et al., 2008b). The defluoridation of water by EC using aluminum electrodes was introduced by Ming et al. (1983). With the electric current passing through the aluminum electrodes, an anodic reaction releases Al(III) ions, which then react with hydroxide ions produced at the cathode and with fluoride ions in solution. This particular method has been applied successfully by a group of engineers and spectacular scientists (Padmasiri et al., 2012) in the rural villages of Sri Lanka, where people are suffering from chronic kidney disease of unknown etiology (CKDu), to provide drinking water with less fluoride and hardness.

6.7.6 MOFS FOR THE TREATMENT OF FLUORIDE-CONTAMINATED DRINKING WATER

Easy design, facile operation, and low cost make adsorption process superior than other available techniques for fluoride removal. One of the most limiting factors is lower adsorption. Apart from arsenic, MOFs are also considered as potential adsorbents for the removal of fluoride in the recent years. In the pH range of 2–10, MOF-801 was evidenced with potential 40 mg/g of fluoride removal with high ionic potential as well as good anti-interference capability at stable operating temperatures in the range of 293–323 K (Zhu et al., 2018). Further, aluminum fumarate (AlFu MOF), with substantial surface area of 1156 m^2/g consisting of porous 3D structure ($Al(OH)(O_2C–CH]CH–CO_2)$) is recognized as an excellent adsorbent material for fluoride removal from groundwater with an adsorption capacity of 600 mg/g at 293 K (Karmakar et al., 2016). It is also reported that due to the powder-like nature, AlFu-based MOF is not recommended to use directly in water for fluoride removal. In view of that, MMM using cellulose acetate phthalate and AlFu-based MOF was prepared, which showed adsorption capacities in the range of 107–179 mg/g for fluoride removal in the presence of 2–10% (w/w) of MOF dosage (Karmakar et al., 2017). The significant adsorption capacity of a zirconium-based metal organic framework (Zr-MOFs) for fluoride with a specific surface area of 740.28 m^2/g is studied when used in membrane materials (He et al., 2016). Zr-MOFs exhibited 102.40 mg/g adsorption capacity at an initial fluoride concentration of 200 mg/L (He et al., 2016). Moreover, typical aluminum-based MOF, namely $Al_{12}O(OH)_{18}$ $(-H_2O)_3(Al_2(OH)_4)(BTC)_6.24H_2O$, also referred as MIL-96, attributes toward substantial defluoridation of drinking water with an adsorption capacity of 31.7 mg/g at 298 K (Zhang et al., 2014). In year 2014, a group of researchers investigated the stability of metal organic frameworks (MOFs) in fluoride solutions considering 11 water-stable MOFs (MIL-53 (Fe, Cr, Al), MIL-68 (Al), CAU-1, CAU-6, UiO-66 (Zr, Hf), and ZIFs-7, 8, 9). Among these, UiO-66 (Zr) exhibited higher fluoride removal efficiency of 41.36 mg/g (Zhao et al., 2014). Different factors including the central metal activity, coordination numbers, pore topology, and the number of –OH groups show significant impact on the stability of MOFs in fluoride solution (Zhao et al., 2014). However, further analysis of MOFs behavior in fluoride solution is required for better understanding.

TABLE 6.4 MOFs as Conventional Adsorbent for Fluoride Removal.

MOFs	pH	Adsorption capacity (mg/g)	References
MOF-801	2–10	40	Zhu et al. (2018)
Zr-MOF	7	102.4	He et al. (2016)
AlFu MOF	7	600	Karmakar et al. (2016)
UiO-66(Zr)	9	41.3	Zhao et al. (2014)
UiO-66-NH_2	7	58.8	Lin et al. (2016)
MIL-96	7	31.7	Zhang et al. (2014)

Source: Adapted from Haldar et al., 2020

6.7.7 BIOSORBENTS FOR FLUORIDE REMOVAL

Biosorbents include natural materials and chemically-modified natural materials, which are used for the adsorption process. Substantial research has been carried out on biosorbents specifically for water purification (Gadd, 2009). Biomaterials from agricultural, household, and industrial wastes, which are cheap and easily available, have been extensively used for defluoridation because of their comprehensive properties in the removal process (Waghmare and Arfin, 2015). Moreover, the presence of specific functional groups on these biomaterials is suitable for surface modification to obtain better removal efficiency.

TABLE 6.5 Applications of Different Biosorbents for Fluoride Removal.

Biosorbents	Mechanism and sorption efficiency	Reference
Polyalthia longifolia	The dried leaves removed fluoride up to 77% from initial concentration at the natural pH of the aqueous solution at a lower temperature. The kinetic study revealed that the fluoride adsorbed by the leaves was chemically intraparticle diffusion and physisorption.	Bharali and Bhattacharyya, 2014
Ailanthus exelsa	Charcoal obtained from *Ailanthus exelsa* showed a removal efficiency of about 94% at the initial fluoride concentration of 10 mg/L. Effective biosorption was achieved within 180 min of contact time, with pH 5 showing maximum adsorption.	Yadav et al., 2014
Azadirachta indica (neem) leaves	Effective fluoride removal of about 80% in the pH range of 5–7 was seen. The presence of several functional groups such as hydroxyl, carbonyl, carboxylic acid, ether, etc. formed a complex with fluoride and was adsorbed on the adsorbent.	Bharali and Bhattacharyya, 2015
Azolla pinnata	The presence of sulfhydryl and hydroxyl groups on the surface of the *Azolla* formed a strong interaction with fluoride ion. This material has exhibited to be a good biosorbent for fluoride removal from wastewater along with the water samples from oil and petroleum refining industry.	Parikh and Mazumder, 2015
Delonix regia pods	The pods were pyrolyzed at 800°C for 1 h to obtain activated carbon with the high surface area and porous structure, which are the prerequisites for fluoride removal. The material could show a maximum removal capacity up to 98.4% at pH 2 with an adsorbent dosage of 1.5 g/100 mL and the initial fluoride concentration of 10 mg/L within the contact time of 300 min.	Ajisha and Rajagopal, 2015
Eichhornia crassipes	The activated carbon obtained from the water hyacinth stem had a surface area of 97.68 m^2/g and a pore volume of 0.5185 cm^3/g. From the investigations, the material removed fluoride up to 70% with an initial concentration of 10 mg/L at pH 5.	Halder et al., 2015

TABLE 6.5 *(Continued)*

Biosorbents	Mechanism and sorption efficiency	Reference
Pithacelobiumducle	Activated carbon was extracted and the fluoride adsorption was purely physisorption, as the activated carbon contains a large pore size, and thus lead to intraparticle diffusion. The material removed about 81% of fluoride from an initial concentration of 5 mg/L.	Emmanuel et al., 2015
Wheat straw (*Triticum aestivum*)	The fluoride sorption on wheat straw might be due to van der Waals force or hydrogen bonding interaction without the release of the −OH group. Also, the formation of precipitates including fluoride and different elements present in wheat straw resulted in the formation of complexes such as CaF_2 or TiF.	Romar-Gasalla et al., 2017
Deoiled *Pongammia pinnata* seed cake	The obtained biochar was rich in carbonaceous material, which enhanced the fluoride removal. The material removed fluoride up to 98.5% and followed the Langmuir adsorption isotherm model.	De et al., 2018
Banana peel	The studies revealed that the biosorbent showed an efficient removal, as it has a surface area of 23.98 m^2/g and porosity of 178.26%. The reported material could remove 86.5% of the fluoride in the water sample containing the initial concentration of 10 mg/L at pH 5.	Mondal and Roy, 2018
Syzygium cumini (jamun leaf ash)	It contains calcium carbonate and calcium oxide/hydroxide, hence showed excellent fluoride removal of 77.8% with an uptake capacity of 4.56 mg/g, within 60 min of contact time.	Tirkey et al., 2018

6.8 CONCLUSION

In general, the concentration of F in groundwater is influenced by the geological context of the area, the leaching and weathering of bedrocks, and the climatic conditions. It's also worth noting that high evaporation rates, prolonged residence times, and comparably low rainfall in the aquifer zone, as well as extensive and long-term irrigation pumping, all contribute to fluoride concentrations. Therefore, to maintain the quality of groundwater, it is recommended for continuous monitoring of physicochemical parameters of water and can be used for drinking and other uses only after prior treatment.

Suitable measures, such as defluorinating the groundwater before use and recharging the groundwater by rainwater harvesting, need to be practiced to improve the groundwater quality. The challenges in fighting the fluoride menace can be met only through a well-planned strategy carried out diligently by the government and other stakeholders in close coordination with each other.

KEYWORDS

- **fluoride**
- **contamination**
- **detection**
- **mitigation**
- **defluoridation**

REFERENCES

Adimalla, N.; Venkatayogi, S.; Das, S. V. G. Assessment of Fluoride Contamination and Distribution: A Case Study from a Rural Part of Andhra Pradesh, India. *Appl. Water Sci.* **2019,** *9 (4)*, 94.

Ajisha, M. A. T.; Rajagopal, K. Fluoride Removal Study using Pyrolyzed *Delonix regia* Pod, An Unconventional Adsorbent. *Int. J. Environ. Sci. Technol.* **2015,** *12*, 223–236.

Anzenbacher, P.; Palacios, M. A.; Jursíková, K.; Marquez, M. Simple Electrooptical Sensors for Inorganic Anions. *Org. Lett.* **2005**, *7 (22)*, 5027–5030.

Arlappa, N.; Qureshi, A. A. T. I. F.; Srinivas, R. Fluorosis in India: An Overview. *Int. J. Res. Dev. Health* **2013,** *1 (2)*, 97–102.

Assefa, B. Defluoridation of Ethiopian Rift Valley Region Water using Reverse Osmosis Membranes. *J. Eur. Econ. Assoc*. **2006,** *23*, 1–6.

Ayoob, S.; Gupta, A. K.; Bhakat, P. B.; Bhat, V. T. Investigations on the Kinetics and Mechanisms of Sorptive Removal of Fuoride from Water using Alumina Cement Granules. *Chem. Eng. J.* **2008,** 140, 6–14. https://doi.org/10.1016/j.cej.2007.08.029

Bayón, M. M.; Garcia, A. R.; Alonso, J. I. G.; Sanz-Medel, A. Indirect Determination of Trace Amounts of Fluoride in Natural Waters by Ion Chromatography: A Comparison of On-Line Post-Column Fluorimetry and ICP-MS Detectors. *Analyst* **1999,** *124 (1)*, 27–31.

Begum, A. Soil Profiles and Fluoride Adsorption in Intensely Cultivated Areas of Mysore District, Karnataka, India. *Chem. Sci. Trans.* **2012,** *1*, 410–414. https://goo.gl/d1YEll

Bellomo, S.; Aiuppa, A.; D'Alessandro, W., Parello, F. Environmental Impact of Magmatic Fluorine Emission in the Mt. Etna Area. *J. Volcanol. Geoth. Res*. **2007,** *165*, 87–101. https://goo.gl/vi0hp9

Bharali, R. K.; Bhattacharyya, K. G. Kinetic and Thermodynamic Studies on Fluoride Biosorption by Devdaru (*Polyalthia longifolia*) Leaf Powder. *Octa J. Environ. Res.* **2014,** *2*, 22–31.

Bharali, R. K.; Bhattacharyya, K. G. Biosorption of Fluoride on Neem (*Azadirachta indica*) Leaf Powder. *J. Environ. Chem. Eng.* **2015,** *3*, 662–669.

Bombik, E.; Bombik, A., Gorski, K.; Saba, L.; Bombik, T.; Rymuza, K. Effect of Environmental Contamination by Fluoride compounds on Selected Horse Tissues. *Polish J. Environ. Stud.* **2011,** *20*, 37–43. https://goo.gl/Ev5Ob6

Brião, V. B.; Magoga, J.; Hemkemeier, M.; Brião, E. B.; Girardelli, L.; Sbeghen, L.; Favaretto, D. P. C. Reverse Osmosis for Desalination of Water from the Guarani Aquifer System to Produce Drinking Water in Southern Brazil. *Desalination* **2014,** *344*, 402–411.

Dahi, E.; Orio, L.; Bregnhj, H. In *Sorption Isotherms of Fluoride on Flocculated Alumina*, Proceedings of the 4th Workshop on Fluorosis and Defluoridation of Water, Ngurdoto, Tanzania, 1995.

Dash, Nilakantha. Fluoride Pollution in Kalahandi District Orissa. **2016.** http://hdl.handle.net/10603/472188

De, D.; Santosha, S.; Aniya, V.; Sreeramoju, A.; Satyavathi, B. Assessing the Applicability of an Agro-Industrial Waste to Engineered Bio-char as a Dynamic Adsorbent for Fluoride Sorption. *J. Environ. Chem. Eng.* **2018,** *6*, 2998–3009.

Diawara, C. K.; Diop, S. N.; Diallo, M. A.; Farcy, M. A. Determination Performance of Nanofiltration (NF) and Low Pressure Reverse Osmosis (LPRM) Membranes in the Removal of Fluorine and Salinity from Brakish Drinking Water. *J. Water Resour. Prot.* **2011,** *3*, 912–917.

D'Ulivo, A.; Pagliano, E.; Onor, M.; Pitzalis, E.; Zamboni, R. Vapor Generation of Inorganic Anionic Species after Aqueous Phase Alkylation with Trialkyloxonium Tetrafluoroborates. *Anal. Chem.* **2009,** *81 (15)*, 6399–6406.

Edmunds, W. M.; Smedley, P. L. Fluoride in Natural Waters. In *Essentials of Medical Geology*; Selinus, O., Ed.; Elsevier Academic Press: London, 2005; pp 301–329.

Emmanuel, K.; Veerabhadrarao, A.; Nagalakshmi, T.; Reddy, M. G.; Diwakar, P.; Sureshbabu, C. Factors Influencing the Removal of Fluoride from Aqueous Solution by *Pithecellobium dulce* Carbon. *Der. Pharma. Chem.* **2015,** *7*, 225–236.

Ericson, B.; Hanrahan, D.; Kong, V. The World's Worst Pollution Problems: The Top Ten of the Toxic Twenty; Blacksmith Institute and Green Cross, 2008. https://goo.gl/4saMAJ

Fluoride Health Effects Database- Research. Fluoride: Protected Pollutant or Panacea?, Calgary Real Estate Link. com, Selling Calgary. pro, 2019. www.fluoridation.com/fluorideresearch.htm. Fluoride Action Network

Fox, K. R. Removal of Inorganic Contaminants from Drinking Water by Reverse Osmosis EPA-600/2-81-115, Back to Cited Text No. 13. Cincinnati, OH: Municipal Environmental Research Laboratory, U.S. Environmental Protection Agency, 1981.

Gadd, G. M. Biosorption: Critical Review of Scientific Rationale, Environmental Importance and Significance for Pollution Treatment. *J. Chem. Technol. Biotechnol. Int. Res. Proc. Environ. Clean Technol.* **2009,** *84*, 13–28.

Gedam, V. V.; Patil, J. L., Kagne, S., Sirsam R. S.; Labhasetwar, P. Performance Evaluation of Polyamide Reverse Osmosis Membrane for Removal of Contaminants in Ground Water Collected from Chandrapur District. *J. Membr. Sci. Techno.* **2012,** *2 (3)*, 1–5.

Halder, G.; Sinha, K.; Dhawane, S. Defluoridation of Wastewater using Powdered Activated Carbon Developed from *Eichhornia crassipes* Stem: Optimization by Response Surface Methodology. *Desalin. Water Treat.* **2015,** *56*, 953–966.

He, J.; Cai, X.; Chen, K.; Li, Y.; Zhang, K.; Jin, Z.; Meng, F.; Liu, N.; Wang, X.; Kong, L.; Huang, X.; Liu, J. Performance of a Novelly-Defined Zirconium Metal-Organic Frameworks Adsorption Membrane in Fuoride Removal. *J. Colloid Interface Sci.* **2016,** *484*, 162–172.

Heitmann, U.; Becker-Ross, H.; Florek, S.; Huang, M. D.; Okruss, M. Determination of Non-metals via Molecular Absorption using High-Resolution Continuum Source Absorption Spectrometry and Graphite Furnace Atomization. *J. Anal. At. Spectrom.* **2006,** *21 (11)*, 1314–1320.

Huang, M. D.; Becker-Ross, H.; Florek, S.; Heitmann, U.; Okruss, M. Determination of Halogens via Molecules in the Air–Acetylene Flame using High-Resolution Continuum Source Absorption Spectrometry: Part I. Fluorine. *Spectrochim. Acta Part B At. Spectrosc*. **2006,** *61* (5), 572–578.

Huxstep, M. R. Inorganic Contaminants Removal from Drinking Water by Reverse Osmosis EPA-600/2-81. Cincinnati, OH: Office of Research and Development, U.S. Environmental Protection Agency, 1981.

Iyenger L. Technologies for Fluoride Removal. ; In *Small Community Water Supplies: Technology, People and Partnerships*; Smet, J., van Wijk, C., Ed.; IRC, Delft: Netherlands, 2003; pp 499–514.

Kang, Y. J.; Lutz, C.; Hong, S. A.; Sung, D. Y.; Lee, J. S.; Shin, J. H.; Nam, H. H.; Cha, G. S. Meyerhoff, M. E. Development of a Fluoride-Selective Electrode Based on Scandium (III) Octaethylporphyrin in a Plasticized Polymeric Membrane. *Bull. Korean Chem. Soc.* **2010,** *31 (6)*, 1601–1608.

Kanthe, V. N. Water Quality Assessment and Health Survey in Dhundi Village – A Case Study. *Int. J. Res. Advent Technol.* **2014,** *2* (1), 38–42.

Karmakar, S.; Bhattacharjee, S.; De, S. Experimental and Modeling of Fuoride Removal Using Aluminum Fumarate (AlFu) Metal Organic Framework Incorpo-Rated Cellulose Acetate Phthalate Mixed Matrix Membrane. *J. Environ. Chem. Eng.* **2017,** *5*, 6087–6097.

Khairnar, M. R.; Dodamani, A. S.; Jadhav, H. C.; Naik, R. G.; Deshmukh, M. A. Mitigation of Fluorosis - A Review. *J. Clin. Diagnostic Res.* **2015,** *9 (6)*, ZE05–ZE09. https://doi.org/10.7860/JCDR/2015/13261.6085

Kumar, S.; Gopal, K. A Review on Fluorosis and its Preventive Strategies. *Indian J. Environ. Prot.* **2000,** *20 (6)*, 430–436.

Li, C.; Gao, X.; Wang, Y. Hydrogeochemistry of high-Fluoride Groundwater at Yuncheng Basin, Northern China. *Sci. Total. Environ*. **2014,** *508C*, 155–165. https://goo.gl/FA4STx

Lin, K. Y. A.; Liu, Y. T.; Chen, S. Y. Adsorption of Fuoride to UiO-66-NH_2 in Water: Stability, Kinetic, Isotherm and Thermodynamic Studies. *J. Colloid Interface Sci.* **2016,** *461*, 79–87.

Mahapatra, M. K. Fluoride Menace in Orissa. **2007**. https://rcdcindia.org/PbDocument/067eff649dd180e-eae3-4a14-b69f-41a0d87be6c5fluoride%20menace%20in%20orissa.pdf

Maheshwari, R. C. Fluoride in Drinking Water and its Removal. *J. Hazard. Mater.* **2006,** *137 (1)*, 456–463.

Mansilla, J.; Saá, J. M. Enantioselective, Organocatalytic Morita-Baylis-Hillman and Aza-Morita-Baylis-Hillman Reactions: Stereochemical Issues. *Molecules* **2010,** *15* (2), 709–734.

Mascola, J. J.; Barth, K. M.; McLaren, J. B. Fluoride Intake of Cattle Grazing Fluoride-Contaminated Forage, as Determined by Esophageal-Fistulated Steers. *J. Anim. Sci.* **1974,** *38*, 1298–1303. https://goo.gl/xdgvvv

Ming, L.; Yi, S. R.; Hua, Z. J.; Lei, B. Y. W.; Ping, L.; Fuwa, K. C. Elimination of Excess Fluoride in Potable Water with Coacervation by Electrolysis using Aluminium Anode. *Fluoride* **1983,** *20*, 54–63.

Mondal, N. K.; Roy, A. Potentiality of a Fruit Peel (Banana Peel) Toward Abatement of Fluoride from Synthetic and Underground Water Samples Collected from Fluoride Affected Villages of Birbhum District. *Appl. Water Sci.* **2018,** *8*, 90.

Muralidharan, D.; Rangrajan, R.; Shankar, G. B. K. Vicious Cycle of Fluoride in Semi-Arid India – A Health Concern. *Curr. Sci.* **2011,** *100 (5)*, 638–640.

Razbe, N.; Kumar, R.; Pratima; Kumar, R. Various Options for Removal of Fluoride from Drinking Water. *IOSR J. Appl. Phys.* **2013,** *3 (2)*, 40–47.

Nagy, K.; Ekstrand, J. Determination of Fluoride in Biological Fluids with a Reversible Solid-State Contact Combination Fluoride Electrode. *Fresenius' Zeitschrift für analytische Chemie.* **1983,** *315 (2)*, 138–140.

Nawlakhe, W. G.; Paramasivam, R. Defuoridation of Potable Water by Nalgonda Technique. *Curr. Sci.* **1993,** *65 (10)*, 743–748.

Ndiaye, P. I.; Moulln, P.; Dominguez, L.; Millet, J. C.; Charbit, F. Removal of Fluoride from Electronic Industrial Effluent by RO Membrane Separation. *Desalination* **2005,** *173*, 25–32.

NRC. *The Fluorosis Problem in Livestock Production*; A Report of the NRC Committee on Animal Nutrition; National Research Council: Washington. Publication 824, 1960. https://goo.gl/EFXplv

Padmasiri, J. P.; Jayawardhene, W. M.; Dissanayake, C. B. In *Water Quality Improvements in the Dry Zone Areas in Sri Lanka*, International Symposium on Water Quality and Human Health: Challenges Ahead; Ileperuma, O. A.; Priyantha, N.; Navaratne, A.; Yatigammana, S. K.; Weragoda, S. K., Eds.; Symposium Proceedings, PGIS: Peradeniya, 2012.

Parikh, P. S.; Mazumder, S. K. Capacity of *Azolla pinnata* var. Imbricata to Absorb Heavy Metals and Fluorides from the Wastewater of Oil and Petroleum Refining Industry at Vadodara. *Int. J. All. Pract. Res.* **2015,** *2*, 37–43.

Prihasto, N.; Liu, Q. F.; Kim, S. H. Pre-treatment Strategies for Seawater Desalination by Reverse Osmosis System. *Desalination* **2009,** *249*, 308–316. https://doi.org/10.1016/j.desal.2008.09.01

Richards, L. A.; Vuachère, M.; Schäfer, A. I. Impact of pH on the Removal of Fuoride, Nitrate and Boron by Nanofltration/Reverse Osmosis. *Desalination* **2010,** *261*, 331–337. https://doi.org/10.1016/j. desal.2010.06.025

Romar-Gasalla, A.; Coelho, G. F.; Nóvoa-Muñoz, J. C.; Arias-Estévez, M.; Fernández-Sanjurjo, M. J.; Álvarez-Rodríguez, E.; Núñez-Delgado, A. Wheat Straw as a Bio-Sorbent for Arsenate, Chromate, Fluoride, and Nickel. *Water* **2017,** *9 (9)*, 690.

Schneiter, R. W.; Middlebrooks, E. J. Arsenic and Fuoride Removal from Groundwater by Reverse Osmosis. *Environ. Int.* **1983,** *9*, 289–291. https://doi.org/10.1016/0160-4120(83)90087-9

Schoeman, J. J. *Water Defluoridation, Water Denitrification and Water Desalination in Rural Areas in South Africa*, In Proceedings of the third IASTED African Conference, Power and Energy System (AfricaPES 2010), Gaborone, Botswana, Sept 6–8, 2010; pp 244–247.

Siddiqui, A. H. Fluorosis in Areas of India with a High Natural Content of Water Fluoride. Fluorides and Human Health. *WHO Monograph.* **1970,** *59*, 284.

Suneetha, M.; Syama Sundar, B.; Ravindhranath, K. Studies on Defuoridation Techniques: A Critical Review. *Int. J. Chem. Tech. Res.* **2015,** *8*, 295–309. https://doi.org/10.1186/s40543-014-0042-1

Susheela, A. K. Fluorosis Management Programme in India. *Curr. Sci.* **1999,** *77 (10)*, 1250–1256. https://goo.gl/V4VjHH

Tirkey, P.; Bhattacharya, T.; Chakraborty, S. Optimization of Fluoride Removal from Aqueous Solution Using Jamun (*Syzygium cumini*) Leaf Ash. *Process Saf. Environ. Protect.* **2018,** *115*, 125–138.

Tsunoda, K. I.; Haraguchi, H.; Fuwa, K. Spectrochemical Analysis of Fluorine by Aluminum Monofluoride Absorption in High Temperature. In *Studies in Environmental Science*; Elsevier, 1986; vol 27, pp 15–23.

Waghmare, S.; Arfin, T. Fluoride Removal by Industrial, Agricultural, and Biomass Wastes as Adsorbents: Review. *Int. J. Adv. Res.* **2015,** *IE1*, 628–653.

Xiong, L.; Feng, J.; Hu, R.; Wang, S.; Li, S.; Li, Y.; Yang, G. Sensing in 15 s for Aqueous Fluoride Anion by Water-Insoluble Fluorescent Probe Incorporating Hydrogel. *Anal. Chem.* **2013,** *85* (8), 4113–4119.

Yadav, B.; Garg, A.; Santra, P.; Santra, S.; Pathodiya, P. Use of Aralu (*Ailanthus exelsa*) Stem Charcoal for Defluoridation of Drinking Water. *Int. J. Bioassays*. **2014,** *3*, 1884–1888.

Zewge, F. Combined Aluminum Sulfate/Hydroxide Process for Fluoride Removal from Drinking Water. *Bull. Chem. Soc. Ethiop.* **2016,** *30 (3)*, 391–401.

Zhang, N.; Yang, X.; Yu, X.; Jia, Y.; Wang, J.; Kong, L.; Jin, Z.; Sun, B.; Luo, T.; Liu, J. Al-1,3,5-benzenetricarboxylic Metal-Organic Frameworks: A Promising Adsorbent for Defluoridation of Water with pH Insensitivity and Low Aluminum Residual. *Chem. Eng. J.* **2014,** *252*, 220–229.

Zhao, X.; Liu, D.; Huang, H.; Zhang, W.; Yang, Q.; Zhong, C. The Stability and Defluoridation Performance of MOFs in Fluoride Solutions. *Micropor. Mesopor. Mater*. **2014,** *185*, 72–78.

Zhu, X. H.; Yang, C. X.; Yan, X. P. Metal-Organic Framework-801 for Efficient Removal of Fluoride from Water *Micropor. Mesopor. Mater*. **2018,** *259*, 163–170.

CHAPTER 7

Impact of Pharmaceutical Contaminants in the Environment

KONAPPAN JAYALAKSHMI[1], ANANDA BASKARAN SANGEETHA[2], MANICKAM SASIKALA[3], CHELLAMUTHU SELVI[4], and MARIAPPAN PARAMASIVAM[4]

[1]*Department of Veterinary Medicine, Veterinary College and Research Institute, Orathanadu, Thanjavur, Tamil Nadu Veterinary and Animal Sciences University, Chennai, India*

[2]*Department of Veterinary Public Health and Epidemiology, Veterinary College and Research Institute, Orathanadu, Tamil Nadu Veterinary and Animal Sciences University, Chennai, Tamil Nadu, India*

[3]*Department of Veterinary Pathology, Veterinary College and Research Institute, Namakkal, Tamil Nadu Veterinary and Animal Sciences University, Chennai, Tamil Nadu, India*

[4]*Pesticide Toxicology Laboratory, Department of Agricultural Entomology, Tamil Nadu Agricultural University, Coimbatore, Tamil Nadu, India*

ABSTRACT

Pharmaceutical compounds are essential for human and animal health. However, their widespread use in hospitals, domestic residences, animal husbandry, aquaculture, agriculture and manufacturing industries has increased their discharge into water, soil, air, and food. Pharmaceutical contaminants are detected in ground waters, surface waters (lakes, rivers, and streams), sea water, wastewater treatment plants, soils, and sludges. These compounds have become environmental pollutants in recent years

Environmental Contaminants: Impact, Assessment, and Remediation.
Pritam Ganguly, Jajati Mandal, M. Paramsivam, & Sandip Patra, (Eds.)

causing both chronic and acute harmful effects on natural flora and fauna. The pharmaceutical contaminants in the environment are most commonly extracted by using solid-phase extraction and pressurized liquid extraction and analyzed by liquid chromatography-mass spectrometry. The pharmaceutical pollutions in the ecosystem may be minimized by the implementation of programme such as take-back strategies, pyrolysis-gasification waste treatment technology, manure composting, site-specific disposal/destruction of waste from hospital and nursing homes, and development of sustainable green pharmacy in the future.

7.1 INTRODUCTION

Pharmaceutical compounds are a wide range of chemicals that include therapeutic drugs, phytotherapeutics, biotechnological products, and veterinary drugs. These compounds are widely used in the human and animal health sector for the prevention and treatment of diseases (Kümmerer, 2010). The pharmaceutical compounds and their metabolites excreted from humans and animals may finally end up in the aquatic environment and eventually in drinking water supplies if sewage is not treated completely (Kümmerer, 2010). In addition, several pharmaceutical production facilities also act as sources to the environment (Kapoor, 2015). Once pharmaceutical contaminants are released into the environment, it will be transported and distributed to air, water, soil, or sediment. It has been revealed to be present in all kinds of environmental matrices that include surface water (lakes, rivers, stream, estuaries, and seawater) (Patel et al., 2019; Smith et al., 2007; Seifrtova et al., 2008), ground water (Fick et al., 2009; Patel et al., 2019), sludge, and wastewater treatment plant effluents (Lapworth et al., 2012; Homem and Santos, 2011). At present, they appear to be broadly present in the geosphere, biosphere (Kim et al., 2011), and earth's most uncontaminated polar regions (Kallenborn et al., 2008). The distribution is affected by physicochemical properties of compounds and receiving environment. The pharmaceutical compounds that reach the environment are considered as pharmaceutical contaminants or pollutants and may have unintended effects on animals and microorganisms in the environment. The pharmaceutical contaminants vary widely with regard to its concentration, solubilities, volatilities, absorbabilities, polarities, stabilities, and biodegradabilities in the environment (Ziylan and Ince, 2011).

Pharmaceutical contaminants are less persistent; however, they are being added continuously in small concentrations from several sources making them "pseudo-persistent" (Zwiener, 2007; Nikolaou et al., 2007) leading to the formation of a complex pharmaceutical pool in environmental matrices. Pharmaceutical contaminates persisting in water and soil get transferred into plants like cabbage, cucumbers, carrots, lettuces, and other vegetables (Boxall et al., 2006; Dolliver et al., 2007; Shenker et al., 2011). Waste treatment plants were planned to degrade moderately degradable organic compounds and are not designed to remove pharmaceutical contaminants completely. Remediation efficiency in wastewater treatment plants for pharmaceuticals (carbamazepine, atenolol, acetylsalicylic acid, diclofenac, mefenamic acid, propranolol, atenolol, clofibric acid, and lincomycin) was reported to be less than 10% (Patel et al., 2019).

Pharmaceutical contaminants can be very active at very low concentration (ng/L to μg/L). It has been shown that continuous exposure to low concentrations can cause unexpected effects on nontarget organisms and induce undesirable effects on humans and ecosystem. The presence of pharmaceuticals in the environment has been associated with the development of antimicrobial resistance in microorganisms that is recognized by the United Nation environment as one of the biggest public health problems globally. Therefore, it is essential to adopt strategies to mitigate their environmental impact by reducing and managing pharmaceuticals uses without compromising their effectiveness, availability, and affordability.

7.2 PHARMACEUTICALS CONTAMINANTS VERSUS OTHER CONTAMINANTS

The pharmaceutical compounds such as pills or liquids consist of active ingredients, excipients, additives, and inorganic or organic substances such as sugars, scents, pigments, and dyes. Active pharmaceutical ingredients (API) are compounds that are active pharmacologically, resist to degrade, are persistent in aquatic environment, and exhibit harmful effects on organisms and humans. Pharmaceutical compounds include a broad range of compounds with varied physical, chemical, structural, and biological properties; hence, these cannot be grouped as homogeneous compounds like chloroflurocarbons (CFCs) or poly-aromatic hydrocarbons (PAHs) (Taylor and Senac, 2014).

Most of the pharmaceuticals are polar in nature developed for a specific activity in target species. According to Rivera-Utrilla et al. (2013), pharmaceutical contaminants differ from other contaminants. The pharmaceutical contaminants are usually low molecular masses (<500 Da), chemical complexity, polar compounds, degree of ionization, lipophilic properties, and ability to persist and accumulate in nature and remain biologically active. For example, naproxen, sulfamethoxazole, and erythromycin can persist for almost one year, while clofibric acid can remain unchanged for multiple years. It tends to adsorb and is distributed in living organisms and gets modified in their chemical structure.

7.3 TYPES OF PHARMACEUTICAL CONTAMINANTS

Pharmaceutical drugs that are used in human and veterinary medicines are emerging as pharmaceutical contaminants in the environment. Different pharmaceuticals have been classified as

- **Analgesics and anti-inflammatories:** Analgesics and anti-inflammatory drugs are chemically unrelated and most commonly used as antipyretic drugs and also to relieve pain. Analgesics include nonsteroidal anti-inflammatory drugs (NSAIDs), acetaminophen (paracetamol), ibuprofen, acetylsalicylic acid, naproxen, ketoprofen, diclofenac, fenoprofen and indomethacin, and opioid or narcotic analgesics (codeine and morphine) are examples of this therapeutic class.
- **Antibiotics:** Antibiotics are widely used drugs worldwide against pathogens of humans and animals and include substances from penicillin, tetracycline, macrolide, quinolone, sulfonamide, aminoglycosides, and cephalosporins classes (Monteiro and Boxall, 2010). However, indiscriminate use has resulted in the emergence of resistant organisms. Antibiotics such as tetracyclines (Dang et al., 2007), aminoglycosides (Shakil et al., 2008), macrolides, beta-lactams, and vancomycin (Akinbowale et al., 2007) have been reported to be present in water. Antibiotic residues found in water have resulted in the formation of resistant organisms such as *Aeromonas, Salmonella, Escherichia, Pseudomonas*, and *Staphylococcus* (Baquero et al., 2008). Other types of antibiotics used include antiviral (acyclovir), antiprotozoal (pyrimethamine), antinematodal (levamisole), and antifungal drugs (miconazole).
- **Beta-blockers:** β-Blockers or β-adrenergic receptor antagonists are used for the treatment of hypertension, ischemic heart disease, congestive heart failure, certain arrhythmias and for the prevention

of migraine headaches. The common beta blockers are propranolol, atenolol, betaxolol, bisoprolol, carazolol, and celiprolol.

- **Hormones and steroids:** Most hormones used belong to proteins and peptides, steroids or amino acid tyrosine derivative and include adrenaline, noradrenaline and dopamine, insulin, norethindrone, progesterone, ethinylo estradiol, oestradiol, and mestranol. The hormones, norethindrone, progesterone, and ethinyloestradiol are used for contraception, whereas insulin is used for treating diabetes.
- **Lipid regulators (LR):** LR are hydrophobic compounds used for lowering levels of triglycerides and low-density lipoproteins (LDL), and increasing levels of high-density lipoproteins (HDL) in the blood. Three kinds of lipid regulators used include fibrates (bezafibrate, gemfibrozil, and fenofibrate), statins (or 3-hydroxy-3-methylglutarylcoenzyme A (HMG-CoA) reductase inhibitors—atorvastatin, simvastatin, and lovastatin), and niacin (or nicotinic acid).
- **Selective serotonin reuptake inhibitors (SSRIs)**: SSRIs include fluoxetine, citalopram, and paroxetine are hydrophobic and generally basic compounds are used in psychopharmacology (Baldessarini, 2001) for relieving depression.
- **Other pharmaceuticals:** Antineoplastic agents or cytotoxic (cyclophosphamide, ifosfamide, methotrexate, and tamoxifen) drugs are used in cancer treatment and include classes like alkylating agents and antimetabolites (Chabner et al., 2001). Antiepileptics or antiseizure (carbamazepine) drugs are used in the treatment of epilepsy. β2-Sympathomimetics or β2-selective adrenergic agonists (albuterol, terbutaline, and fenoterol) are used as bronchodilators in the treatment of asthma (Hoffman, 2001). Iodinated X-ray contrast media are used to intensify contrast of structures during imaging in hospitals (Contrast medium, 2002). The chemicals used in this class include iopromide, iomeprol, diatrizoate, and iopadimol.

7.4 SOURCES OF PHARMACEUTICAL CONTAMINANTS

Pharmaceutical compounds along with their metabolites are released constantly to the environment from both point and nonpoint sources. Various sources of pharmaceuticals are provided in Figure 7.1. Pharmaceutical manufacturing areas are major point sources whereas unregulated domestic effluents, agricultural outputs to fields, and aquatic systems are nonpoint sources.

- **Pharmaceutical manufacturing industries:** The production and manufacturing units act as a major point source of contamination and are of great concern due to high efflux concentrations. In addition, it is a more serious problem in developing countries due to the lack of presence of proper wastewater effluent treatment (Rehman et al., 2015). For example, high concentration (31 mg/L) of ciprofloxacin in India (Larsson et al., 2007) and 28 mg/L of trimethoprim in Croatia (Dolar et al., 2012) were reported. In Pakistan, high concentrations of ciprofloxacin, doxycycline, ofloxacin, levofloxacin, and oxytetracycline were reported pharmaceutical in wastewater (Hussain et al., 2016). The concentrations of pharmaceuticals from manufacturing industries' wastewater are 10–1000 times higher than other wastewaters indicating it as biggest point sources (Patel et al., 2019). In addition, handling and storage of chemicals may contribute to the sources.
- **Human medical compounds:** The major kinds of pharmaceutical compounds are used in hospitals. The contaminants result from the mixture of one or activities like preparation, general compounding, breakages, needles, intravenous sets, partially used ampoules, unused preparations, and outdated pharmaceuticals. As a result, hospital wastewater contains more amounts of pharmaceuticals.
- **Household pharmaceuticals:** Unused and expired pharmaceuticals used in households enter the environment due to flushing into basins, drains, toilets, and or disposal with household wastes. The household wastes reach landfill sites and may leach and enter groundwater when landfill sites are not properly sealed.
- **Agriculture and animal husbandry**: A wide range of compounds are being used in agricultural sector. Antibiotics, anesthetics, antifungal, antiinflammatory, and antiparasitic drugs, hormones, and sedatives are widely used in animals for health care and to improve production. They reach the environment through livestock feed or urine when they are not properly disposed. Pharmaceuticals used in animals are major sources to the terrestrial environment and are major problematic because animal wastes are not treated.

Pharmaceuticals used in aquaculture can enter the aquatic ecosystem (Patel et al., 2019).

7.5 PATHWAY OF PHARMACEUTICAL POLLUTIONS IN GROUND WATER

Pharmaceutical contaminants enter the environment through two main routes (Shabaan et al., 2018). Firstly, pharmaceuticals after administration in humans or animals are absorbed, metabolized, and excreted via urine or faces. Second, expired or unused pharmaceuticals are discarded in toilet or sink. In both cases, pharmaceutical contaminants end up in sewage treatment plants. The pharmaceutical contaminants in the sewage treatment plant may be removed through microbial degradation or sorption into solids that are removed as sludge (Carballa et al., 2004). The more persistent environmental contaminants are not removed during sewage treatment and are released into the environment where they are transported and distributed in various compartments including surface waters, soil, and sediment.

Many pharmaceutical compounds including anti-inflammatories and analgesics, antibiotics, antiepileptics, antidepressants, lipid-lowering agents, antihistamines, β-blockers, and other substances like barbiturates, narcotics, antiseptics, and contrast media have been detected in water (Rivera-Utrilla et al., 2013). Among these compounds, antibiotics are the most commonly detected followed by analgesics. However, the results vary depending upon country, area, consumption pattern, season, and locations of manufacturing industries (Beek et al., 2016). Analgesics are the most frequently found pharmaceuticals globally, whereas antibiotics are the most frequently found pharmaceuticals in Asia (Hughes et al., 2013). Most of the analytical methods detected pharmaceutical contaminants in hospital effluents, effluents, and influents of sewage treatment plants, groundwater, surface water, and drinking water (Hughes et al., 2013; Česen et al., 2018). Based on available literature, pharmaceutical concentration follows the general order: industrial effluents > hospital effluents > wastewater treatment plant effluents > surface water > groundwater > drinking water.

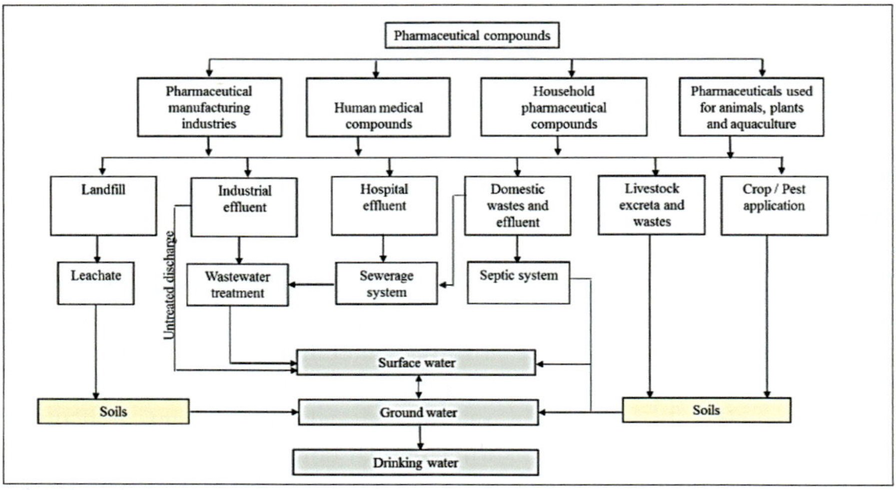

FIGURE 7.1 Sources and pathways for pharmaceuticals entering groundwater.
Source: Adapted from Patel et al., 2019

7.6 IMPACT OF PHARMACEUTICALS POLLUTIONS ON ENVIRONMENT

The multiple classes of pharmaceuticals, antibiotics, nonsteroidal anti-inflammatory drugs, steroids, hormones, antihelmintics, antidepressants, antihypertensives, β-blockers, contraceptive hormones, and anticancerous drugs as original compound or a metabolized form may enter the environment from manufacturing units, hospital effluents, and land applications of bio-solids, manures, and treated sewage water (Daughton and Ternes, 1999;Schwarzenbach et al., 2006; Ternes, 2007; Carter et al., 2016; Shaaban et al., 2018). However, the sewage treatments do not completely remove the active chemicals from wastewater and pollute an aquatic environment and enter the food chain (Patneedi and Prasadu, 2015). Even pharmaceuticals at trace level have potential estrogenic and carcinogenic activity and will possibly affect the quality of ecological and human life (Leal et al., 2010). It also has a negative impact on soil microbes, fish, daphnids, algae, bacteria, earthworms, plants, and dung invertebrates. The antihelmintics >98% of ivermectin excreted through faeces, the environmental contamination with the faeces is usually higher than drug manufacturing unit (Beynon, 2012). It has been reported in aerobic water sediments and is toxic to aquatic invertebrates (Prasse et al., 2009). Ivermectin and abamectin residue in cattle faeces affect the dung-associated insects and also reduce the feeding behavior and growth rate (Halley et al., 1993; Boxall, 2004), levamisole and fenbendazole in the soil can reduce the dung decomposition (Sommer and Bibby, 2002) and ivermectin residue in soil also affects the germination of spores of soil fungus *Fusarium oxysporum* (Prasse et al., 2009), whereas spore production was doubled in *Phanerochaete chrysosporium* and *Mucor racemosus* (Kollmann et al., 2003).

7.6.1 SOIL MICROORGANISM

Soil microorganisms play an important role in the maintenance of soil health, organic matter turnover, nutrients release, stabilization of the soil structure, and fertility (Varma and Buscot, 2005). The antibiotics in the soil affect the soil microorganisms by several ways; changing their enzyme activity, ability to metabolize different carbon sources, inhibition of growth of soil microbe, alteration of soil microbial community, and ecological function of soil (Cycon et al., 2019). The antibiotic residues in soil affect the soil respiration, nitrification, denitrification, sulfate and iron reduction, methanogenesis, and the enzymes responsible for carbon (C), nitrogen (N), and phosphorus (P) turnover process (Ma et al., 2016). The influence of antibiotics on soil nitrification and denitrification process depends on the type and duration of exposure. The sulfonamides (sulfamethoxazole, sulfamethazine, sulfadiazine, and trimethoprim) residue in soil affect the soil respiration (Kotzerke et al., 2008; Liu et al., 2009), the high concentration of oxytetracycline (30 mg/kg) and sulfadiazine (100 mg/kg) decreases the nitrification rate (Ma et al., 2016), whereas ciprofloxacin at a low dose (1 mg/kg soil) stimulates the nitrification rate of soil microbes (Yang et al., 2012; Cui et al., 2014). The low dose of sulfamethoxazole, sulfadiazine, narasin, or gentamicin (0.5 mg/kg soil) inhibits the denitrification rate of soil bacteria. Antibiotics such as sulfamethoxazole and oxytetracycline (>10 mg/kg) affect the turnover rate of iron (Toth et al., 2011) and strongly inhibit the Fe (III) reduction in soil (Molaei et al., 2017; Cycon et al., 2019).

7.6.2 AQUATIC WILDLIFE AND PREDATORS

Algae are important for the maintenance of aquatic ecosystems. The residues of sulfamethoxazole and trimethoprim in the aquatic environment inhibit the growth of green algae (Eguchi et al., 2004). The antibiotic residues in fish were predominantly reported in China. The highest concentrations of antibiotics, erythromycin, lincomycin, and sulfamethazine were observed in aquatic fish plasma. But antidepressants, antihistamines, and calcium channel blockers were mostly reported in fish species of the United States (Aus der Beek et al., 2016). Antidepressants in the aquatic ecosystem have certain implications on fish species (Brodin et al., 2014). The most commonly used antidepressants in humans are SSRIs and serotonin-norepinephrine reuptake inhibitors (SNRIs) (Vaswani et al., 2003). The SNRI in aquatic environment induced aggression and altered reproductive behavior (Kreke and Dietrich, 2008),whereas SSRI and diphenhydramine changed the feeding behavior in fish and fathead minnows (Halford et al., 2005; Berninger et al., 2011). Fluoxetine, TCAs amitriptyline, and mianserin can significantly affect gene transcription in zebrafish (Wu et al., 2017;Melchor-Martinez et al., 2021). Fluoxetine also has an impact on the antipredator behavior of mosquitofish and increases sociality in Arabian killifish (Valenti et al., 2012; Martin et al., 2017; Melchor-Martinez et al., 2021), whereas sertraline increases boldness in fathead minnows (Valenti et al., 2012). Antihistamines also affect wildlife behavior through influence on serotonin level. Similarly, antiepileptic drug carbamazepine, nonsteroidal anti-inflammatory drug diclofenac reduces the feeding activity in Japanese medaka fish (Nassef et al., 2010; Brodin et al., 2014). The sex steroids estrogens, 17-α- ethinylestradiol (EE2) are exclusively used in humans as contraceptive pill, which was identified in the aquatic environment affecting the reproduction in fish even at very low level (ng/L). It alters the female-to-male sex ratio in favor of the females due to the lack of sexual differentiation and feminization in male fish (Shaaban et al., 2018; Lange et al., 2001; Kidd et al., 2007; Kummerer, 2010). The chronic exposure of fish (fathead minnow) to EE2 in the aquatic environment leads to feminization of males through the production of vitellogenin mRNA and protein, which affect the gonadal development evidenced as intersex in males (Kidd et al., 2007). Intersex fish (feminized males) were found for the first time in British rivers near wastewater treatment plants outfalls in 1976 and then they have been reported in many other countries (Jobling et al.,2006). The impact of anticancer drugs on aquatic organisms was unknown, but it can cause chronic toxicity and detrimental impact on their genome on chronic exposure (Jureczko and Kalka, 2020).

The predators are exposed to most of the pharmaceuticals through prey consumption; about 46% of pharmaceuticals are accumulated in predators through bioconcentration (Brodin et al., 2014). Adult *Trichoptera* is an important food source for riparian predators (potential trophic biomagnifications (Kraus et al., 2014). It indicates predators (spiders, birds, and bats) are exposed to various pharmaceuticals and endocrine disruptors in the environment. Earthworms are the key food source for many predator species such as birds; it is the base for many food chains and thus if chemicals are taken up into the earthworms they can facilitate the movement of chemicals into the food web via bioaccumulation and biomagnifications processes (Shore et al., 2014; Carter et al., 2016). Dramatic decline of three vulture species in South East Asia is due to acute kidney failure following ingestion of veterinary diclofenac residues when feeding on carcasses of livestock that had been treated with the drug shortly before their deaths (Dohleet al., 2013). Poisonings of avian scavengers following ingestion of euthanized animal carcass that contained sodium pentobarbital lead to mortality in wild birds (Russell and Franson, 2014).

7.6.3 IMPACT ON HUMAN, ANIMALS, AND PLANTS

Antibiotics are one of the most commonly used pharmaceuticals, which enter the environment in many ways like hospital effluents, WWTPs, farms, aquaculture, urban WWTPs, faeces, and urine from humans and animals (Kummerer, 2010; Karkman et al., 2019). In European countries, treated wastewater mostly used for irrigation of agricultural land, the antibiotics from these enter the ground water through filtration or leaching from soil (Felis et al., 2020). Antibiotics like aminoglycosides streptomycin in marine water and gentamicin in ground water, β-Lactams amoxicillin and cefuroxime in hospital effluent and untreated waste water (Gros et al., 2013), macrolides erythromycin and tylosin in drinking water, fluoroquinolones (Norfloxacin and Ciprofloxacin) in hospital effluent and drinking water (Verlicchi et al., 2012; Reis et al., 2019), sulfonamides (most commonly sulfamethoxazole) in ground and drinking water were reported (Hanna et al., 2018; Felis et al., 2020). The persistence of antibiotics in the soil/environment poses a strong pressure on human, animal and soil microbiome leads to development of antimicrobial resistance genes (ARGs) and antibiotic-resistant bacteria (ARB). The bacteria may acquire resistance through horizontal gene transfer (HGT) such as transformation, conjugation, and transduction mechanism and also transfer resistance genes within and between bacterial species. Emergence of antibiotic resistance leads to poor clinical recovery and poses a serious impact on animal and human health worldwide (Jayalakshmi et al., 2021).

Recently, the chemicals such as phthalates, bisphenol A, alkylphenols, steroid hormones, and plant sterols in the environment have been reported and shown to influence the reproductive capability of humans and animals. These chemicals are called endocrine disruptors and impair fertility in various species via binding with estrogen receptor (Stahlschmidt-Allner et al., 1997). The sex steroids either natural or synthetic estrogens (estrone, estradiol, and estriol) have been reported in soil, sewage waste water treatment plants, animal manure, ground water, and surface water. The natural estrogens in animal and human waste pose a serious risk to the environment (Shrestha et al., 2012). The livestock manure application on agricultural land area is a major source of estrogens in the environment (Arnon et al., 2008). Steroidal estrogens in food and water can induce premature menopause, virilization in young women, breast cancer in women, and low sperm count, feminization, and prostate cancer in men (Bolong et al., 2009; Sumpter and Jobling, 2013; Adeel et al., 2017). Estrogen alone or in combination with progesterone reduces the intraocular pressure (IOP) after menopause in humans and also increases the risk of developing glaucoma. Some plants containing phytoestrogens affect reproduction, immune system, and metabolism in humans (Alexander, 2014) and permanent infertility in sheep after grazing of clover plant. The estrogen can also impair vision and IOP in animals and progesterone also increases the IOP in lions (Hotchkiss et al., 2008; Shemesh and Shore, 2012). The exposure of plants to steroid estrogen hormones or their precursors can affect root and shoot development, flowering, and germination (Adeel et al., 2017).

7.7 ANALYSIS OF PHARMACEUTICALS IN THE ENVIRONMENTAL SAMPLES

7.7.1 SAMPLE COLLECTION

Environmental samples such as wastewater influent, effluent, river water, and pond water (500–2000 mL) are collected in a glass bottle and filtered with glass fiber filters (0.2–1.2 μm) to remove suspended particulate matter and stored at 4°C in dark condition (Guedes-Alonso et al., 2014) and pH adjusted to 2.0–3.0 using hydrochloric acid, sulfuric acid, and formic acid to arrest bacterial activity (Guedes-Alonso et al., 2014). Solid sediments, sludge, and aquatic biota (5 g) were freeze dried and stored at −18 or −20°C and sodium azide used as preservatives (Liu et al., 2011). The soil samples were collected at a depth of 10 cm and stored at −23°C. Prior to analysis soil samples were dried at 35°C in an oven and passed through a 3-mm sieve to remove coarse particles, subsequently ground and sieve (0.63 mm) to obtain a homogeneous sample (Salvia et al., 2012).

7.7.2 SAMPLE PREPARATION AND EXTRACTION TECHNIQUES

Analysis of pharmaceuticals in influent and effluent wastewater, river water, pond water, soil, and biosolids is carried out by various extraction and cleanup techniques, followed by a final analysis typically with chromatographic measurements as shown in Table 7.1 (Guedes-Alonso et al., 2014; Paramasivam et al., 2015; Semreen et al., 2019). Prior to target analysis, sample preparation is the most important step to minimize matrix interference, which may affect the quantification. A number of sample preparation techniques were used for the analysis of various pharmaceuticals in the environmental contaminants. For liquid samples, solid-phase extraction (SPE), liquid–liquid extraction (LLE), and solid-phase microextraction (SPME) are the most commonly used extraction techniques, while microwave-assisted extraction (MAE), accelerated solvent extraction (ASE), Soxhlet extraction, and QuEChERS method are the most commonly used for extraction of analytes from solid samples (Lucchetti et al., 2005; Salvia et al., 2012; Jakimska et al., 2013; Paramasivam et al., 2015).

7.7.3 SOLID-PHASE EXTRACTION (SPE)

SPE is one of the most commonly used extraction techniques for the isolation and concentration of multi-class pharmaceuticals in the aqueous sample (Melchor-Martinez et al., 2021). SPE can be used as off-line (without chromatographic analysis) or online (directly connected to the chromatographic system either LC or GC). The online SPE coupled with liquid chromatography tandem mass spectrometry (SPE-LC-MS/MS) is most commonly used than gas chromatography (Sadutto and Pico, 2020). In the SPE method the sample containing target analytes is passed through a column filled with the appropriate sorbents (which was previously conditioned by appropriate solvent or solvent mixtures). The sample was adsorbed on the sorbent, which is then washed with a small quantity of organic solvents (methanol, ethyl acetate, dichloromethane, acetonitrile, hexane, and acetone). The choice of SPE cartridge depends on the mechanism of interactions between the sorbent and the analytes. Sorbents may be of nonpolar, polar, and ion exchange; their activity is dependent on the properties of the bonded phase. The most common reverse phase sorbents are used in the SPE method that retains a wide range of polar and nonpolar compounds. The polymeric reverse phase sorbent contained vinyl pyrrolidone (Oasis® HLB, Strata®X and Cleanert® PEP), polystyrene-divinylbenzene (Isolute® ENV+, Chromabond® HR-X and Bond ElutTM ENV) was used for polar analytes and octadecyl endcapped

TABLE 7.1 Extraction Techniques and Analytical Methods for Determination of Pharmaceuticals in Environmental Matrices

Pharmaceuticals	Matrix	Extraction method	Detection method	Recovery (%)	Limit of detection (LOD)	Limit of quantification (LOQ)	References
Antibiotics							
Erythromycin Oleandomycin Roxithromycin Salinomycin	Soil	PLE –SPE	LC-MS/MS	38–94	0.2–1.6 μg/kg	0.6–5.3 μg/kg	Schlusener et al. (2006)
Tiamulin	Soil	PLE	LC-MS/MS	118		0.6 μg/kg	Schlusener et al. (2003)
Fluoroquinolones, Tetracycline Sulfonamides	Soil	SPE	LC-MS/MS	56–89 69–97	–	–	Hu et al. (2012)
Sulphamethoxazole Amoxicillin Trimethoprim	Radish	PLE	LC-MS/MS	–	–	–	Al-Farsi et al. (2018)
Coccidiostats 4,40-Dinitrocarbanilide Amprolium, Diclazuril Halofuginone, Lasalocid Maduramicin, Monensin Narasin, Robenidine Salinomycin, Semduramicin Toltrazuril	Fresh pond water	SPE	UHPLC-MS/MS	86.6–115.5	0.7–12.8 ng/L	2.5–25 ng/L	Goessens et al. (2020)
Anthelmintics Flubendazole Ivermectin Levamisole	Fresh pond water	SPE	UHPLC-MS/MS	90.3–112.7	1.0–40 ng/L	2.5–250 ng/L	Goessens et al. (2020)

TABLE 7.1 *(Continued)*

Pharmaceuticals	Matrix	Extraction method	Detection method	Recovery (%)	Limit of detection (LOD)	Limit of quantification (LOQ)	References
Steroid hormone 17α-estradiol,17β-estradiol 17α-ethinylestradiol, 17α-dihydroequilin Progesterone and Norgestrol	Deionized water, wastewater influent, effluent and Surface water	LLE	GC-MS				Andaluri et al. (2017)
Veterinary antibiotics and hormone	Soil	Modified QuEChERS	LC-MS/MS				Salvia et al. (2014)
NSAIDs Tolmetin, Naproxen Fenoprofen, Diflunisal Diclofenac and Ibuprofen	Water sample	SPE	LC-DAD-MS	80–103	0.005–0.095 μg/L		Debska et al. (2005)
Anticancer drugs Methotrexate, Azathioprine, Doxorubicin,Doxorubicinol, Vincristine, Ifosfamide, Cyclophosphamide, Etoposide and Procarbazine	Influent Effluent	SPE	UPLC-ESI-MS/MS	45.3–108.9	0.6–7.0 ng/L and 0.5–3.5 ng/L	1.5–20 ng/L, and 1.5–10 ng/L	Yin et al. (2010)
Antidepressants O-desmethylvenlafaxine Desmethylfluvoxamin Venlafaxine, Mirtazapine, Citalopram, Bupropion-d9, Fluvoxamine, Paroxetine, Norfluoxetine, Nortriptyline, Desmethylsertraline Fluoxetine, Carbamazepine Amitriptyline, Sertraline Dihydrocarbamazepin Desmethylmirtazepine	Effluent, influent, treated biosolids	SPE	LC-MS/MS	72–102, 73–98, 44–101	0.01–0.2 ng/L, 0.03–0.4 ng/L 0.04–0.5 ng/L	0.04–0.7 ng/L, 0.1–1.4 ng/L, 0.1–1.7 ng/L	Lajeunesse et al. (2012)

TABLE 7.1 *(Continued)*

Pharmaceuticals	**Matrix**	**Extraction method**	**Detection method**	**Recovery (%)**	**Limit of detection (LOD)**	**Limit of quantification (LOQ)**	**References**
27 Pharmaceuticals Atenolol, Caffeine Trimethoprim, Metoprolol Propranolol, Carbamezipine Clotrimazole, Salicylic acid Paracetamol, Clofibric acid Ketoprofen ,Naproxen, Warfarin, Flurbiprofen, Diclofenac, Indomethacin, Meclofenamic acid, Gemfibrozil Nimesulide, Sulfamethoxazole and Furosemide	Sludge, Sludge enriched soil	PLE+SPE	LC-MS	>60	<20 ng/g		Barron et al. (2008)
119 pharmaceuticals and 17 hormonal steroids	Sewage sludge	QuEChERS	LC-TOF-MS		1–2500 ng/g		Peysson and Vulliet, 2013
31 pharmaceuticals	Sewage sludge	PLE	LC-MS/MS				Vazquez-Roig et al., 2010
31 pharmaceuticals	Soil	QuEChERS	LC-MS/MS	40–110			Salvia et al.,2012
Determination of pharmaceuticals by GC-MS							
Phenolic endocrine disrupting compounds 4-octylphenol, 4-n-nonylphenol Bisphenol-A (BPA), Estrone, Estradiol , Diethylstilbestrol Triclosan **Acid pharmaceuticals** Clofibric acid, Ibuprofen Gemfibrozil, Fenoprofen Naproxen, Ketoprofen Mefenamic acid, Tolfenamic acid Diclofenac Meclofenamic acid and Indomethacin	Surface water	SPE	GC-MS	–	–	–	Zhao et al. (2009)

TABLE 7.1 *(Continued)*

Pharmaceuticals	Matrix	Extraction method	Detection method	Recovery (%)	Limit of detection (LOD)	Limit of quantification (LOQ)	References
Acidic and endocrine disrupting compounds Ibuprofen, salicylic acid, gemfibrozil, naproxen, triclosan, propranolol, diclofenac, carbamazepine, 4-octylphenol (OP), 4-n-nonylphenol, -monoethoxylate, nonylphenoxyacetic acid estrone and 17ά-ethinyloestradiol	Surface water and tap water	SPE	GC-MS	47–130, 60–109	<5 ng/L <10 ng/L	–	Yu et al. (2007)
Nicotine, Ibuprofen, Acetaminophen, Caffeine, 1,7-Dimethylxanthine DCDD 25, Triclosan, Diclofenac, Bisphenol A and Carbamazepine	Waste water	SPE	GC-IT (EI) MS/MS	66–112	0.2–16 ng/L	1 ng/L and 83215 μg/L.	Gomez et al. (2007)
Clofibric acid, Ibuprofen Salicylic acid, Gemfibrozil Naproxen, Ketoprofen Diclofenac, Carbamazepine 4-Nonylphenols, Triclosan Bisphenol-A, Di-*n*-butylphthalate Butylbenzylphthalate (BuBeP) *Bis*-2 ethylhexylphthalate (DEHP)	Soil	ASE	GC-MS/MS	62–102 75–95	–	–	Duran-Alvarez et al. (2009)
Endocrine disrupting compounds Estrone, 17β-Estradiol 17ά Ethinylestradiol (EE2)	Soil	ASE	GC-MS/MS	54–109	–	–	Duran-Alvarez et al. (2009)

TABLE 7.1 *(Continued)*

Pharmaceuticals	Matrix	Extraction method	Detection method	Recovery (%)	Limit of detection (LOD)	Limit of quantification (LOQ)	References
22 neutral and weakly basic drugs **Neutral drugs** Clofibrate, Phenazone Dimethylaminophenazone Ifosfamide , Cyclophosphamide Carbamazepine, Pentoxifylline Diazepam, Fenofibrate Etofibrate and Dihydrocarbamazepine **Weakly basic drugs** Terbutaline, Clenbuterol, Salbutamol, Metoprolol Timolol, Propranolol Nadolol, Bisoprolol Betaxolol, Fenoterol and Carazolol	River, drinking and wastewater	SPE	GC-MS	>70	5 ng/L	–	Ternes et al. (1998)

silica RP (Supelclean. LC-18 SPE) that was used for nonpolar to moderately polar analytes from aqueous samples (Sadutto and Pico, 2020).

Reversed phase columns are mostly used for the analysis of all families of hormones in environmental waters and phenyl-bonded silica columns for estrogen separation in the river and tap water (Chang et al., 2010; Lucci et al., 2011) and alkyl-amide columns for the separation of glucocorticoids, androgens, and progesterone in river water (Tolgyesi et al., 2010). The solvents acetonitrile is generally used as mobile phase in the analysis of estrogens, while methanol is widely used to achieve the separation of androgens, glucocorticoids, progestogens, and sometimes estrogens (Guedes-Alonso et al., 2014). Oasis HLB (Hydrophillic lipophillic balance) and Oasis WAX cartridges are used for the separation of anticancer drugs (methotrexate, azathioprine, doxorubicin, doxorubicinol, vincristine, ifosfamide, cyclophosphamide, etoposide, and procarbazine) and analyzed by using UHPLC–MS/MS in water samples from sewage treatment plants (Yin et al., 2010). The combination of HLB (to retain antibacterial agents) and SAX cartridges (to remove negatively charged humic material) was used for the separation of antibiotics fluoroquinolones, tetracyclines, and sulfonamides) from soil and analyzed by LC-MS/MS (Vazquez-Roiget al., 2010; Hu et al., 2012). The sorbents like Strata-X, Narc-2, SDB-1, C18 Polar Plus, Quaternary amine, Calixarene, LiChrolut RP-18, and Discovery DSC-18 are used for the extraction of anti-inflammatory drugs (tolmetin, naproxen, fenoprofen, diflunisal, diclofenac, and ibuprofen) in water sample; among these RP-18 was found to be most efficient than others at pH 2.5 (Debska et al., 2005). Goessens et al. (2020) extracted 12 coccidiostat (4,40-dinitrocarbanilide, amprolium, diclazuril, halofuginone, lasalocid, maduramicin, monensin, narasin , robenidine salinomycin, semduramicin, and toltrazuril) and 3 anthelmintics (flubendazole, ivermectin, and levamisole) in the pond water by SPE and confirmed by UHPLC-MS/MS. They reported out of 18 samples, amprolium was found in one and levamisole in all samples. It may be due to contamination of pond with run-off/leaching/drift after liquid manure application in the agricultural land.

7.7.4 SOLID-PHASE MICROEXTRACTION (SPME) AND LIQUID-PHASE MICROEXTRACTION (LPME)

SPME and LPME extraction techniques were mostly used for the analysis of NSAIDs in aquatic and environmental matrices. These techniques required a lesser quantity of organic solvents (μl), eco-friendly, high sensitivity, reproducibility, and reduces matrix interference than the traditional methods (Izadi et al., 2020). Hollow fiber-supported liquid phase microextraction (HFLPME or LPME) was used for the extraction of antidepressants (citalopram, sertraline, paroxetine, fluvoxamine, and fluoxetine) and their metabolites (desmethylcitalopram, didesmethylcitalopram, desmethyl sertraline, and norfluoxetine) in sewage sludge (Lajeunesse et al., 2012), wastewater and analyzed by LC-MS/MS (Melchor-Martinez et al., 2021).

7.7.5 LIQUID–LIQUID EXTRACTION (LLE)

Conventionally, LLE is employed for the manipulation of aqueous pH to extract drugs into an organic solvent (Juhascik and Jenkins, 2009); commonly, solvents of higher affinity for target analytes are used for the extraction of the compound from the solid phase (Melchor-Martinez et al., 2021). LLE coupled with LC-MS/MS analysis is used for the identification and quantification of 14 cytostatic anticancer drugs (Bicalutamide, capecitabine, cyclophosphamide, ifosfamide and mycophenolic acid, doxorubicin, etoposide, megestrol, and prednisone) in wastewater and reported that mycophenolic acid poses a high risk to aquatic organism (Teresa et al., 2020).

7.7.6 MICROWAVE-ASSISTED EXTRACTION (MAE-SPE)

MAE is a process of application of microwave energy to heat solvents to extract target analytes from sample matrix (Eskilsson and Bjorklund, 2000). This method is used to improve the recovery rate of target compound from complex matrices (Evans et al., 2015). Polar solvents are commonly used to absorb microwaves (Ridgway et al., 2007). MAE extract is deposited on the SPE (Qasis Max cartridge for solid matrix). This method is used for the detection of amphetamine, methamphetamine, *3,4-methylenedioxy-methamphetamine* (MDMA), methylenedioxyamphetamine (MDA), ephedrine, pseudoephedrine, and norephedrine from a solid sample. In liquid matrices alprenolol, atenolol, citalopram, desmethylcitalopram, desmethylvenlafaxine, fluoxetine, metoprolol, mirtazapine, propranolol, sotalol, tramadol, and venlafaxine were detected through the passing of extract in Oasis HLB cartridges and analyzed with LC-MS (Evans et al., 2015). Ultrasonic extraction followed by preconcentration by the SPE technique is followed for the extraction of anti-inflammatory drugs (naproxen, fenoprofen, diclofenac, and ibuprofen) and antiepileptic drugs (Carbamazepine) in soil and sediment and the analytes were quantified using HPLC-PDA (Hlengwa and Mahlambi, 2020).

7.7.7 ACCELERATED SOLVENT EXTRACTION (ASE)/PRESSURIZED LIQUID EXTRACTION (PLE)

In the ASE system, conventional solvents are heated at elevated temperatures (up to 200°C) and pressure (up to 3000 psi) for short time periods, resulting in improved extraction of analytes from solid and semisolid samples (Shen et al., 2011). For analysis of veterinary antibiotics and steroid hormone relatively low temperature (Temp 80°C), low pressure (100-140 bars), and 10-min static time with 2 cycles are often reported (Salvia et al., 2014). Nieto et al. (2007) used PLE for the separation of acetaminophen, caffeine, metoprolol, propanolol, carbamazepine, salicylic acid, bezafibrate, naproxen, clofibric acid, diclofenac, and ibuprofen from sewage sludge and it is analyzed by HPLC–MS with ESI (HPLC-(ESI)MS) and their recoveries of pharmaceuticals were more than 68% except for salicylic acid. Radjenovic et al. (2009) employed PLE for the separation of 31 pharmaceuticals from sewage sludge and it was confirmed by LC-MS/MS and for the isolation of acetaminophen, carbamazepine, ciprofloxacin, clofibric acid, codeine, diazepam, fenofibrate, metropolol, ofloxacin, and propanolol in soil and sediments (Vazquez-Roiget al., 2010). The combination of PLE and SPE methods is used for the extraction of macrolide antibiotics, ionophore, and tiamulin in soil and analysed by LC-MS/MS (Schlusener et al., 2006). The extraction of pharmaceuticals in soils and crops (radish) irrigated with treated wastewater was performed by Al-Farsi et al. (2018) and they analyzed the plant uptake of commonly used antibiotics (amoxicillin, sulfamethoxazole, trimethoprim) and NSAIDs (Ibuprofen) by LC-MS/MS, and they reported that the translocation and accumulation rate of sulfamethoxazole was higher than amoxicillin, because sulfamethaoxazole has a high biological half-life in the environment.

7.7.8 QUICK, EASY, CHEAP, EFFECTIVE, RUGGED, AND SAFE (QUECHERS) METHOD

This method is based on a salting-out extraction with a solvent (mainly acetonitrile) followed by a dispersive SPE (d-SPE). It is simple, rapid and does not require a chlorine-based solvent. It was first introduced to extract pesticides from food matrices such as fruits and vegetables (Anastassiades et al., 2003); this method has been recently used for the extraction of various organic compounds such as veterinary drugs, pesticides, phenols and hormones in soil (Salvia et al., 2012; Peysson and Vulliet, 2013; Paramasivam et al., 2015), estrogens and progestogens from sediments, fish and clam tissues (Chen et al., 2012; Jakimska et al., 2013). A total of 31 pharmaceutical substances (14 veterinary products, 11 steroid hormones, and 6 other well-known human contaminant compounds) were extracted in soil using the QuEChERS method and determined by LC-MS/MS (Salvia et al., 2012). They reported that the sulfonamide group of compounds is mainly detected in the soil already treated by manure with a concentration of 0.02 to 0.12 ng/g and also human drugs (carbamazepine and ibuprofen) were detected in soils treated with domestic sludge. Peysson and Vulliet (2013) validated the QuEChERS method for simultaneous determination of 136 pharmaceuticals and hormones in sewage sludge using LC–TOF-MS. They observed antiemetic domperidone, antiepileptic lamotrigine, analgesics tramadol and hydrocodone, the antiarrhythmics amiodarone and lidocaine, antipsychotic pipamperone, antiprotozoal quinacrine, calcium channel blocker verapamil, and the hypnotic zolpidem in sewage sludge for the first time. However, this method is not suitable for the detection of strongly complex molecules like tetracyclines and fluoroquinolones. Currently, TOF-MS has been applied for environmental analysis of organic contaminants for identification and confirmation and for investigation of metabolites or degradation products or in the elucidation of unknown compounds from water, vegetables, and sediments and also for the detection of antimicrobial resistance in the environment rapidly (Jayalakshmi et al., 2021).

7.7.9 DETERMINATION OF PHARMACEUTICALS BY GC-MS

Generally, most of the pharmaceuticals are polar, relatively non-volatile, heat-labile and determined mostly by using liquid chromatography (LC) or liquid chromatography-mass spectrometry (LC-MS) (both conventional LC and ultra high performance LC) than Gas chromatography (GC) because it requires derivatization steps such as silylation, alkylation, and acylation to modify chemical structure before analysis of multi-family pharmaceuticals simultaneously (Reemtsma and Quintana, 2006; Diaz-Cruz et al., 2009). Derivatization was done with pentafluorobenzoyl chloride and pentafluorobenzoyl bromide and acetic anhydride for the analysis of phenolic compound and acidic drugs by GC (Zhao et al., 2009; Kotowska et al., 2014). Most commonly, N-methyl-N-(tert-butyl-dimethysilyl) trifluoroacetamide (MtBSTFA) and N,O-bis (trimethylsilyl) trifluoroacetamide

(BSTFA) are used as derivatization agent for analysis of acidic and neutral pharmaceuticals and endocrine-disrupting substances in GC-MS (Yu et al., 2007; Gibson et al., 2007) and determined the drugs ibuprofen, salicylic acid, gemfibrozil, naproxen, triclosan, propanolol, diclofenac, carbamazepine, estrone, 17ά- ethinylo estrodial in surface and tap water. Gomez et al. (2007) validated GC-IT (EI) MS/MS without derivatization for the analysis of anti-inflammatory drugs (ibuprofen, acetaminophen, and diclofenac); an antiepileptic agent (carbamazepine); stimulants (caffeine and nicotine); an antiseptic (triclosan); plasticizer (bisphenol A) and two of their more relevant metabolites (2,8-dichlorodibenzo-*p*-dioxin and 1,7-dimethylxanthine) in wastewater with SPE technique. Duran-Alvarezet al. (2009) determined the acidic pharmaceuticals, carbamazepine, and potential endocrine disrupting compounds in soil by GC-MS. Ternes et al. (1998) developed the GC-MS method with derivatization silylation for the determination of 22 different neutral and weakly basic drugs belonging to several different medicinal classes like antiphlogistics, beta blockers, β2-sympathomimetics, lipid regulators, antiepileptic agents, psychiatric drugs, and vasodilators in the river, drinking, and wastewater.

7.8 MITIGATION OF PHARMACEUTICALS

Pharmaceutically active compounds can enter the environment in multiple ways viz., normal use (excretion or washed off from dermal application on humans and animals), disposal of unused pharmaceuticals in the garbage, toilet, or drains, hospital effluents, leaching from landfills, runoff from agricultural or animal excreta and industrial effluents (Blair, 2016). This can persist in ground, surface, and drinking waters (Bean et al., 2016) affecting human, animal, and environment health. The pharmaceutical pollutions in the environment may be minimized by the implementation of programmes such as take-back strategies, Pyrolysis-gasification waste treatment technology (PGWTS) (Bean et al., 2016), manure composting (Pruden et al., 2013), site-specific disposal/ destruction of waste from hospital, nursing homes, and hospice care can reduce the substantial amounts of pharmaceutical pollutions in the environment. The active pharmaceuticals from industries and hospital wastewater are 4–150 times higher than urban wastewaters (Verlicchi et al., 2010), policy regulations (banning of substance and authorization restriction), implementation of green pharmacy practice for redesigning of existing pharmaceuticals for increased metabolism in the body, and development of new substance that is more efficiently biodegradable without altering pharmacological activities (Blair, 2016; Toma and Crisan, 2018). Considering the environment, pharmacists should educate patients on the proper disposal of medicines and organize collection programs for unused and expired medicines, in order to develop a greener pharmacy for the future.

7.8.1 TAKE-BACK STRATEGIES

This programme currently exists in some countries to reduce pharmaceutical contamination in the environment (Blair, 2016). In this programme, unused medicines return back to the pharmacy and are destroyed in high-temperature incinerators (Seehusen and Edwards, 2006; Bean et al., 2016). These take-back strategies were followed in Sweden (43%), USA (1.4-5.9%), and UK (17%) (Tong et al., 2011). However, it was costly and transported to long distances for incinerator facilities. In PGWTS, materials other than glass and metals were used in this technology to reduce the volume of waste to <1%. A waste bag or clinical waste bin can be loaded at one end of the unit into a sealed bin in the PGWTS unit, and then heated at temperature 550–700°C with electrical elements (Bean et al., 2016). It has the potential to reduce significant pharmaceutical contaminations in the environment.

7.8.2 MANURE COMPOSITING

Livestock manure application on agricultural land is one of the most important pharmaceutical contaminants in the environment. Composting is an eco-friendly technique for the management of antibiotic residues, ARGs, and pathogenic bacteria in manure (Esperon et al., 2020), which can eliminate 50–70% of antibiotics (Sharma et al., 2009). The degradation of antibiotics most often occurs during thermophilic phase of compositing of manure (Prudent et al., 2013), watering, aerating, and turning off compost can accelerate the decay of chlortetracycline ((>99%), monensin, and tylosin (54-76%) in turkey manure after 35 days of thermophilic phase (Storteboom et al., 2007; Dolliver et al., 2008). The sulfonamides and trimethoprim in livestock manure were reduced during 5 weeks of the fermentation process (Mohring et al., 2009), whereas sulfamethoxazole and oxytetracycline residues in dairy lagoon water were effectively reduced under aerobic than anaerobic incubation (Pei et al., 2007). A100-fold reduction of tetracycline residues and tetracycline-resistant bacteria in manure were observed in long-term storage conditions (Storteboom et al., 2007; Chee-Sanford et al., 2009; Prudent et al., 2013), 90% reduction of ciprofloxacin and doxycycline after 3 weeks, and 73.7-99.9% reduction of ARGs and pathogenic bacteria (*Campylobacter coli* and *E.coli*) in poultry manure were reported after 10 weeks of composting (Esperon et al., 2020). The prevalence of doxycycline residue in pig manure was reduced after 170 days of storage (Widyasari-Mehta et al., 2016). In an experimental study, salinomycin degraded faster (6 days) than erythromycin (41 days), roxithromycin (130 days) and tiamulin remained unchanged even after 180 days in liquid manure (Schlusener et al., 2006). Long-term manure storage and composting can reduce the development of ARGs and ARB as well as minimize the antibiotic pollutions in the environment.

7.8.3 GREEN PHARMACY PRACTICE

The International Pharmaceutical Federation (FIP) designed a document called "Green Pharmacy Practice: taking responsibility for the environmental impact of medicines" (Toma and Crisan, 2018). The pharmaceutical manufacturer can develop a compound that has increased uptake of the body, or has lower doses, maximizes their susceptibility to environmental biodegradation, sorption, photolysis, and is less harmful to the environment (Blair, 2016). Example beta blocker propranolol was redesigned to a more bioavailable form by small molecular changes in the compound and has shown improved degradation (Masoner et al., 2016), the anticancer drug glucofosfamide a green alternative to ifosfamide, which is 70% biodegradable and enhanced absorption in the human gut (Kummerer, 2007).

7.8.4 POLICY REGULATIONS

The policy frameworks such as source-directed approaches, use-oriented approaches, and end- of-pipe measures are directed to reduce environmental contamination with pharmaceuticals. The source-directed approaches are mainly focused on pharmaceutical companies/manufacturers to identify potential environmental risk of existing and new pharmaceuticals, marketing authorization of pharmaceuticals based on predicted risk, and development of drinking water safety plans. Use-oriented approaches include policy instrument for the reduction of excessive consumption of antibiotics, reduction of disease incidence through maintenance of hygiene and sanitations, some regulations such as restrictions on over sale of environmentally harmful pharmaceuticals, and restriction of drugs used for growth promotion in livestock. The end-of-pipe measures involve improved wastewater treatment, engagement of health professionals, veterinarians, and pharmacists to educate the patients, farmers, and consumers about risk associated with inappropriate disposal of unused medicines and bio-solids in the environment (Kummerer, 2010; OECD, 2019).

7.9 CONCLUSIONS

The pharmaceuticals such as antibiotics, endocrine disrupting compounds, antidepressants, and anti-inflammatory drugs are potential contaminants in the environment. Among these antidepressants and endocrine disrupting compounds have more impact on aquatic animals and humans. The implementation of monitoring programme for the analysis of existing and new active pharmaceuticals in the environment, ecological and human health risk assessment, banning of chemicals with high health risk, and determination of safe threshold limit of active pharmaceuticals in the drinking water are essential to minimize the negative impact of chemical drugs on humans, animals, and aquatic wildlife.

KEYWORDS

- **antibiotics**
- **antidepressants**
- **GC-MS**
- **LC-MS**
- **pharmaceuticals**

REFERENCES

Adeel, M.; Song, X.; Wang, Y.; Francis, D.; Yang, Y. Environmental Impact of Estrogens on Human, Animal and Plant Life; A Critical Review. *Environ. Inter.* **2017,** *99*, 107–119.

Akinbowale, O.; Peng, H.; Barton, M. P514 Class 1 Integron Mediates Antibiotic Resistance in *Aeromonas spp*. from Rainbow Trout Farms in Australia. *Int. J. Antimicrob. Agents* **2007,** *29*, S113.

Alexander, V. Phytoestrogens and Their Effects. *Eur. J. Pharmacol.* **2014,** *741*, 230.

Al Farsi, R.; Ahmed, M.; Al Busaidi, A.; Choudri, B. S. Assessing the Presence of Pharmaceuticals in Soil and Plants Irrigated with Treated Wastewater in Oman. *Int. J. Recycl. Oorg. Waste Agric.* **2018,** *7*, 165–172.

Anastassiades, M.; Mastovska, K.; Lehotay, S. J. Fast and Easy Multiresidue Method Employing Acetonitrile Extraction/Partitioning and Dispersive Solid Phase Extraction for the Determination of Pesticide Residues in Produce. *J. Chromatogr. A* **2003,** *1015*, 412–431.

Andaluri, G.; Rominder, P.; Suri, S.; Graham, K. Steroid Hormones in Environmental Matrices: Extraction Method Comparison. *Environ. Monit. Assess.* **2017,** *189*, 626.

Arnon, S.; Dahan, O.; Elhanany, S.; Cohen, K.; Pankratov, I.; Gross, A.; Ronen, Z.; Baram, S.; Shore, L. S. Transport of Testosterone and Estrogen from Dairy-Farm Waste Lagoons to Groundwater. *Environ. Sci. Technol.* **2008,** *42*, 5521–5526.

Aus der Beek, T.; Weber, F. A.; Bergmann, A.; Hickmann, S.; Ebert, I.; Hein, A.; Kuster, A. Pharmaceuticals in the Environment- Global Occurrences and Perspectives. *Environ. Toxicol. Chem.* **2016,** *35* (4), e823–835.

Baldessarini, R. J. Drugs and the treatment of psychiatric disorders: depression and anxiety disorders, In: *The Pharmacological Basis of Therapeutics*; Hardman, J. G., Limbird, L. E., 10th ed.; Eds.; Goodman & Gilman's, McGraw-Hill: New York, 2001; pp 447–484.

Baquero, F.; Martinez, J.; Cantón, R. Antibiotics and Antibiotic Resistance in Water Environments. *Curr. Opin. Biotechnol.* **2008,** *19* (3), 260–265.

Barron, L.; Tobinb, J.; Paull, B. Multi-Residue Determination of Pharmaceuticals in Sludge and Sludge Enriched Soils Using Pressurized Liquid Extraction, Solid Phase Extraction and Liquid Chromatography with Tandem Mass Spectrometry. *J. Environ. Monit.* **2008,** *10* (3), 353–361.

Bean, T. G.; Bergstrom, E.; Thomas-Oates, J.; Wolff, A.; Bart, P.; Eaton, B.; Alistair, B. A.; Boxall, A. B. A. Evaluation of a Novel Approach for Reducing Emissions of Pharmaceuticals to the Environment. *Environ. Manage.* **2016,** *58*, 707–720.

Beek, T.; Weber, F. A.; Bergmann, A.; Hickmann, S.; Ebert, I.; Hein, A.; Küster, A. pharmaceuticals in the Environment Global Occurrences and Perspectives. *Environ. Toxicol. Chem.* **2016,** *35*, 823–835.

Berninger, J. P.; Du, B.; Connors, K. A.; Eytcheson, S. A.; Kolkmeier, M. A.; Prosser, K. N.; Valenti, T. W.; Chambliss, C. K.; Brooks, B. W. Effects of the Antihistamine Diphenhydramine on Selected Aquatic Organisms. *Environ. Toxicol. Chem.* **2011,** *30*, 2065–2072.

Beynon, S. A. Potential Environmental Consequences of Administration of Anthelmintics to Sheep. *Vet. Parasitol.* **2012,** *189*, 113–124.

Blair, B. D. Potential Upstream Strategies for the Mitigation of Pharmaceuticals in the Aquatic Environment: A Brief Review. *Curr. Environ. Health Rep.* **2016,** *3*,153–160.

Seehusen, D. A.; Edwards, J. Patient Practices and Beliefs Concerning Disposal of Medications. *J. Am. Board Fam. Med.* **2006,** *19* (6), 542–547.

Bolong, N.; Ismail, A.; Salim, M. R.; Matsuura, T. A Review of the Effects of Emerging Contaminants in Wastewater and Options for Their Removal. *Desalination* **2009,** *239*, 229–246.

Boxall, A. B.; Johnson, P.; Smith, E. J.; Sinclair, C. J.; Stutt, E.; Levy, L. S. Uptake of Veterinary Medicines from Soils Into Plants. *J. Agric. Food Chem.* **2006,** *54*, 2288–2297.

Boxall, A. B. A. The Environmental Side Effects of Medication. *EMBO Rep.* **2004,** *5* (12), 1110–1116.

Brodin, T.; Piovano, S.; Fick, J.; Klaminder, J.; Heynen, M.; Jonsson, M. Ecological Effects of Pharmaceuticals in Aquatic Systems-Impacts Through Behavioural Alterations. *Phil. Trans. R. Soc. Lond. B Biol. Sci.* **2014,** *369* (1656), 20130580.

Carballa, M.; Omil, F.; Lema, J. M.; Llompart, M.; García-Jares, C.; Rodríguez, I.; Gomez, M.; Ternes, T. Behavior of Pharmaceuticals, Cosmetics and Hormones in a Sewage Treatment Plant. *Water Res*. **2004,** *38*, 2918–2926.

Carter, L. J.; Ryan, J. J.; Alistair, B.; Boxall, A. Effects of Soil Properties on the Uptake of Pharmaceuticals into Earthworms. *Eniron. Pollut.* **2016,***213*, 922e931.

Česen, M.; Heath, D.; Krivec, M.; Kosmrlj, J.; Kosjek, T.; ˇ Heath, E. Seasonal and Spatial Variations in the Occurrence, Mass Loadings and Removal of Compounds of Emerging Concern in the Slovene Aqueous Environment and Environmental Risk Assessment. *Environ. Pollut.* **2018,** *242*, 143–154.

Chabner, B. A.; Ryan, D. P.; Paz-Ares, L.; Garcia-Carbonero, R.; Calabresi, P. Antineoplastic Agents. In: *The Pharmacological Basis of Therapeutics*; Hardman, J. G., Limbird, L. E., Eds., 10th ed.; Goodman & Gilman's, McGraw-Hill: New York, 2001; pp 1389–1460.

Chang, C. C.; Huang, S. D. Determination of the Steroid Hormone Levels in Water Samples by Dispersive Liquid-Liquid Microextraction with Solidification of a Floating Organic Drop Followed by High-Performance Liquid Chromatography. *Anal. Chim. Acta* **2010,** *662*, 39–43.

Chee-Sanford, J. C.; Mackie R. I.; Koike, S.; Krapac, I. G.; Lin, Y. F.; Yannarell, A. C.; Maxwell, S.; Aminov, R. I. Fate and Transport of Antibiotic Residues and Antibiotic Resistance Genes Following Land Application of Manure Waste. *J. Environ. Qual.* **2009,** *38*, 1086–1108.

Chen, W. L.; Wang, G. S.; Gwo, J. C.; Chen, C. Y. Ultra-High Performance Liquid Chromatography/Tandem Mass Spectrometry Determination of Feminizing Chemicals in River Water, Sediment and Tissue Pretreated Using Disk-Type Solid-Phase Extraction and Matrix Solid-Phase Dispersion. *Talanta* **2012**, *89*, 237–245.

Contrast Medium. In: Concise Medical Dictionary. Oxford University Press, 2002; Oxford Reference Online. Oxford University Press: York University. http://www. Oxfordreference. com/views/ENTRY.html?subview=Main&entry= t60.e2140

Cui, H.; Wang, S. P.; Fu, J.; Zhou, Z. Q.; Zhang, N.; Guo, L. Influence of Ciprofloxacin on Microbial Community Structure and Function in Soils. *Biol. Fertil. Soils* **2014,** *50*, 939–947.

Cycon, M.; Mrozik, A.; Piotrowska-Seget, Z. Antibiotics in the Soil Environment-Degradation and Their Impact on Microbial Activity and Diversity. *Front. Microbiol.* **2019,** *10*, 338.

Dang, H.; Zhang, X.; Song, L.; Chang, Y.; Yang, G. Molecular Determination of Oxytetracycline-Resistant Bacteria and Their Resistance Genes from Mariculture Environments of China. *J. Appl. Microbiol*. **2007,** *103* (6), 2580–2592.

Daughton, C. G.; Ternes, T. A. Pharmaceuticals and Personal Care Products in the Environment; Agents of Subtle Change. *Environ. Health Perspect*. **1999,** *107*, 907–938.

Daughton, C. G. Non-Regulated Water Contaminants: Emerging Research. *Environ. Impact Assess. Rev*. **2004,** *24*, 711–732.

Debska, J.; Kot-Wasik. A.; Namiesnik, J. Determination of Nonsteroidal Anti-Inflammatory Drugs in Water Samples Using Liquid Chromatography Coupled with Diode-Array Detector and Mass Spectrometry. *J. Sep. Sci.* **2005,** *28*, 2419–2426.

Diaz-Cruz, M. S.; Garcia-Galan, M. J.; Guerra, P.; Jelic, A.; Postigo, C.; Elarrat, E.; Farre, M., Lopez de Alda, M. J.; L. D.; Petrovic, M.; Barcelo, D. Analysis of Selected Emerging Contaminants in Sewage Sludge. *Trends Analyt. Chem.* **2009,** *28* (11), 1263–1275.

Dohle, S.; Campbell, V. E. A.; Arvai, J. L. Consumer-Perceived Risks and Choices About Pharmaceuticals in the Environment: A Cross-Sectional Study. *Environ. Health.* **2013,** *12,* 45.

Dolar, D.; Vukovic, A.; Asperger, D.; Kosutic, K. Efficiency of´ RO/NF Membranes at the Removal of Veterinary Antibiotics. *Water Sci. Technol.* **2012,** *65*, 317–323.

Dolliver, H.; Gupta, S.; Noll, S. Antibiotic Degradation During Manure Composting. *Int. J. Environ. Qual.* **2008,** *37*, 1245–1253.

Dolliver, H.; Kumar, K.; Gupta, S. Sulfamethazine Uptake by Plants from Manure-Amended Soil. *J. Environ. Qual.* **2007,** *36*, 1224–1230.

Duran-Alvareza, J. C.; Becerril-Bravoa, E.; Castroa, V. S.; Jimeneza, B.; Gibson, R. The Analysis of a Group of Acidic Pharmaceuticals, Carbamazepine, and Potential Endocrine Disrupting Compounds in Wastewater Irrigated Soils by Gas Chromatography–Mass Spectrometry. *Talanta* **2009,** *78*, 1159–1166.

Eguchi, K.; Nagase, H.; Ozawa, M.; Endoh, Y. S.; Goto, K.; Hirta, K.; Miyamoto, K.; Yoshimura, H. Evaluation of Antimicrobial Agents for Veterinary Use in the Ecotoxicity Test Using Microalgae. *Chemosphere* **2004,** *57* (11), 1733–1738.

Eskilsson,C. S.; Bjorklund, E. Analytical-Scale Microwave- Assisted Extraction. *J. Chromatogr. A* **2000,** *902* (1), 227–250.

Esperon, F.; Albero, B.; Ugarte-Ruiz, M.; Domínguez, L.; Carballo, M.; Tadeo, J. L.; Delgado, M. D. M.; Moreno, M. A.; Torre, A. D. L. Assessing the Benefits of Composting Poultry Manure in Reducing Antimicrobial Residues, Pathogenic Bacteria, and Antimicrobial Resistance Genes: A Field-Scale Study. *Environ. Sci.Poll. Res.* **2020,** *27*, 27738–27749.

Evans, S. E.; Davies, P.; Lubben, A.; Kasprzyk-Hordern, B. Determination of Chiral Pharmaceuticals and Illicit Drugs in Wastewater and Sludge Using Microwave Assisted Extraction, Solid-Phase Extraction and Chiral Liquid Chromatography Coupled with Tandem Mass Spectrometry. *Anal. Chim. Acta* **2015,** *882*, 112–126.

Felis, E.; Kalkaa, J.; Sochacki, A.; Kowalska, K.; Bajkacz, S.; Harniszc, M.; Korzeniewska, E. Antimicrobial Pharmaceuticals in the Aquatic Environment—Occurrence and Environmental Implications. *Eur. J. Pharmacol.* **2020,** 866.

Fick, J.; Söderström, H.; Lindberg, R. H.; Phan, C.; Tysklind, M.; Larsson, D. Contamination of Surface, Ground, and Drinking Water from Pharmaceutical Production. *Environ. Toxicol. Chem.* **2009,** *28*, 2522−2527.

Gibson, R.; Becerril-Bravo, E.; Silva-Castro, V.; Jimenez, B. Determination of Acidic Pharmaceuticals and Potential Endocrine Disrupting Compounds in Wastewaters and Spring Waters by Selective Elution and Analysis by Gas Chromatography-Mass Spectrometry. *J. Chromatogr. A* **2007,** 1169, 31–39.

Goessens, T.; Baere, S. D.; Troyer, N. D.; Deknock, A.; Goethals, P.; Lens, L.; Pasmansd, F.; Croubels, S. Highly Sensitive Multi-Residue Analysis of Veterinary Drugs Including Coccidiostats and Anthelmintics in Pond Water Using UHPLC-MS/MS: Application to Freshwater Ponds in Flanders, Belgium. *Environ. Sci. Processes Impacts* **2020,** *22*, 2117.

Gomez, M. J.; Aguera, A.; Mezcua, M.; Hurtado, J.; Mochol, F.; Fernandez-Alba, A. R. Simultaneous Analysis of Neutral and Acidic Pharmaceuticals as Well as Related Compounds by Gas Chromatography–Tandem Mass Spectrometry in Wastewater. *Talanta.* **2007,** *73*, 314–320.

Gros, M.; Rodriguez-Mozaz, S.; Barcelo, D. Rapid Analysis of Multiclass Antibiotic Residues and Some of Their Metabolites in Hospital, Urban Wastewater and River Water by Ultra-High-Performance Liquid Chromatography Coupled to Quadrupole-Linear Ion Trap Tandem Mass Spectrometry. *J. Chromatogr. A* **2013,** *1292*, 173–188.

Guedes-Alonso, R.; Montesdeoca-Esponda, S.; Sosa-Ferrera, Z.; Santana-Rodríguez, J. J. Liquid Chromatography Methodologies for the Determination of Steroid Hormones in Aquatic Environmental Systems. *Trends Environ. Anal. Chem.* **2014,** *3*, 14–27.

Halford, J. C. G.; Harrold, J. A.; Lawton, C. L.; Blundell, J. E. Serotonin (5-HT) Drugs: Effects on Appetite Expression and Use for the Treatment of Obesity. *Curr. Drug Targets* **2005,** *6*, 201–213.

Halley, B. A.; William J. A.; Heuvel , V.; Wislocki, P. G. Environmental Effects of the Usage of Avermectins in Livestock.*Vet. Parasitol.* **1993,** *48,* 109–125.

Hanna, N.; Sun, P.; Sun, Q.; Li, X.; Yang, X.; Ji, X.; Zoub, H.; Ottosond, J.; Nilssone, L. E.; Berglunde, B;, Dyara, O. J.; Tamhankar, A. J.; Stalsby Lundborg, C. Presence of Antibiotic Residues in Various Environmental Compartments of Shandong Province in Eastern China: Its Potential for Resistance Development and Ecological and Human Risk. *Environ. Int.* **2018,** *114*, 131–142.

Hlengwa, N. B.; Mahlambi, P. N. Ultrasonic Followed by Solid Phase Extraction and Liquid Chromatography Photodiode Array for Determination of Pharmaceutical Compounds in Sediment and Soil. *Bull. Environ. Contam. Toxicol.* **2020,** *104*, 464–470.

Hoffman, B. B. Catecholamines, Sympathomimetic Drugs and Adrenergenic Receptor Antagonists. In: *The Pharmacological Basis of Therapeutics*; Hardman, J. G., Limbird, L. E., Eds., 10th ed.; Goodman & Gilman's, McGraw-Hill: New York, 2001; pp 215–268.

Homem, V.; Santos, L. Degradation and Removal Methods of Antibiotics from Aqueous Matrices—A Review. *J. Environ. Manage.* **2011,** *92*, 2304–2347.

Hotchkiss, A. K.; Rider, C. V.; Blystone, C. R.; Wilson, V. S.; Hartig, P. C.; Ankley, G. T.; Foster, P. M.; Gray, C. L.; Gray, L. E. Fifteen Years After "Wingspread-Environmental Endocrine Disrupters and Human and Wildlife Health: Where We Are Today and Where We Need To Go. *Toxicol. Sci.* **2008,** *105*, 235–259.

Hu, W.; Ma, L.; Guo, C.; Sha, J.; Zhu, X. Wang, Y. Simultaneous Extraction and Determination of Fluoroquinolones, Tetracyclines and Sulfonamides Antibiotics in Soils Using Optimised Solid Phase Extraction Chromatography—Tandem Mass Spectrometry. *Intern. J. Environ. Anal. Chem.* **2012,** *92* (6), 698–713.

Hughes, S. R.; Kay, P.; Brown, L. E. Global Synthesis and Critical Evaluation of Pharmaceutical Data Sets Collected from River Systems. *Environ. Sci. Technol.* **2013,** *47*, 661–677.

Hussain, S.; Naeem, M.; Chaudhry, M. N. Estimation of Residual Antibiotics in Pharmaceutical Effluents and Their Fate in Affected Areas. Pol. *J. Environ. Stud.* **2016,** *25*, 607–614.

Izadi, P.; Izadi, P.; Salem, R.; Papry, S. A.; Magdouli , S.; Pulicharla, R.; Brar, S. K. Non-Steroidal Anti-Inflammatory Drugs in the Environment: Where Were We and How Far We Have Come? *Environ. Pollut.* **2020,** *267*, 115370.

Jakimska, A.; Huerta, B.; A.; Barganska, Z.; Kot-Wasik, A.; Rodriguez-Mozaz, S.; Barcelo, D. Development of a Liquid Chromatography—Tandem Mass Spectrometry Procedure for Determination of Endocrine Disrupting Compounds in Fish from Mediterranean Rivers. *J. Chromatogr. A* **2013,** *1306*, 44–58.

Jayalakshmi, K.; Veeraselvam, M.; Ramkumar, P. K.; Venkatesan, M.; Yogeshpriya, S.; Premalatha, N. A Review on Antimicrobial Resistance, Diagnosis and Alternative Approach. *J. Ento. Zool. Studies* **2021,** *9* (1), 1058–1071.

Jobling, S.; Williams, R.; Johnson, A.; Taylor, A.; Gross-Sorokin, M.; Nolan, M.; Tyler, C. R.; Aerle, R. V.; Santos, E.; Brighty, G. Predicted Exposures to Steroid Estrogens in UK Rivers

Correlate with Widespread Sexual Disruption in Wild Fish Populations. *Environ. Health Persp.* **2006,** *114* (1), 32–39.

Juhascik, M. P.; Jenkins, A. J. Comparison of Liquid/Liquid and Solid Phase Extraction for Alkaline Drugs. *J. Chromatogr. Sci.* **2009,** *47*, 553–557.

Jureczko, M.; Joanna Kalka, J. Cytostatic Pharmaceuticals as Water Contaminants. *Eur. J. Pharmacol*.**2020,** *866*, 172816.

Kallenborn, R.; Fick, J.; Lindberg, R.; Moe, M.; Nielsen, K.; Tysklind, M.; Vasskog, T. Pharmaceutical Residues in Northern European Environments: Consequences and Perspectives. In: *Pharmaceuticals in the Environment*; Kümmerer, K., Ed.; Springer: Berlin, Heidelberg, 2008; pp 61–74.

Kapoor, K. Impact of Pharmaceutical Industries on Environment, Health and safety. *J. Crit. Reviews* **2015,** *2* (4), 25–30.

Karkman, A., Parnanen, K.; Larsson, D. G. J. Fecal Pollution Can Explain Antibiotic Resistance Gene Abundances in Anthropogenically Impacted Environments. *Nat. Commun.* **2019,** *10*, 80.

Kidd, K. A.; Blanchfield, P. J.; Mills, K. H.; Palace, V. P.; Evans, R. E.; Lazorchak, J. M; Flick, R. W. Collapse of a Fish Population After Exposure to a Synthetic Estrogen. *Proc. Natl. Acad. Sci. USA.* **2007,** *10*, 8897–901.

Kim, K. R.; Owens, G.; Kwon, S. I.; So, K. H.; Lee, D. B.; Ok, Y. S. Occurrence and Environmental Fate of Veterinary Antibiotics in the Terrestrial Environment. *Water, Air, Soil Pollut.* **2011,** *214*, 163–174.

Kollmann, A.; Touton, I.; Brault, A.; Alvinerie, M.; Galtier, P.; Mougin, C. Effect of the Endectocide Ivermectin on Filamentous Fungi. *Environ. Chem. Lett.* **2003,** *1*, 215–218.

Kotowska, U.; Kapelewska, J.; Sturgulewska, J. Determination of Phenols and Pharmaceuticals in Municipal Wastewaters from Polish Treatment Plants by Ultrasound-Assisted Emulsification—Microextraction Followed by GC-MS. *Environ. Sci. Pollut. Res.* **2014,** *21*, 660–673.

Kotzerke, A., Sharma, S., Schauss, K., Heuer, H., Thiele-Bruhn, S., Smalla, K.; Wilke, B. M.; Schloter, M. Alterations in Soil Microbial Activity and N-Transformation Processes Due to Sulfadiazine Loads in Pig-Manure. *Environ. Pollut.* **2008,** *153*, 315–322.

Kraus, J. M.; Walters, D. M.; Wesner, J. S.; Stricker, C. A.; Schmidt, T. S.; Zuellig, R. E. Metamorphosis Alters Contaminants and Chemical Tracers in Insects: Implications for Food Webs. *Environ. Sci. Technol.* **2014,** *48* (18), 10957–10965.

Kreke, N.; Dietrich, D. R. Physiological Endpoints for Potential SSRI Interactions in Fish. *Crit. Rev. Toxicol.* **2008,** *38*, 215–247.

Kummerer, K. Pharmaceuticals in the Environment. *Annu. Rev. Environ. Resour*. **2010,** *35*, 57–75.

Kummerer, K. Sustainable from the Very Beginning; Rational Design of Molecules by Life Cycle Engineering as an Important Approach for Green Pharmacy and Green Chemistry. *Green Chem*, **2007,** *9* (8), 899.

Lajeunesse , A.; Smyth , S. A.; Barclay, K.; Sauve, S.; Gagnon, C. Distribution of Antidepressant Residues in Wastewater and Biosolids Following Different Treatment Processes by Municipal Wastewater Treatment Plants in Canada. *Water Res.* **2012,** *46*, 5600–5612.

Lange, R.; Hutchinson. T. H.; Croudace, C. P.; Siegmund, F. Effects of the Synthetic Estrogen 17 Alphaethinylestradiol on the Life-Cycle of the Fathead Minnow (*Pimephales promelas*). *Environ. Toxicol. Chem,* **2001,** *20*, 1216–1627.

Lapworth, D.; Baran, N.; Stuart, M.; Ward, R. Emerging Organic Contaminants in Groundwater: A Review of Sources, Fate and Occurrence. *Environ. Pollut.* **2012,** *163*, 287–303.

Larsson, D. J.; de Pedro, C.; Paxeus, N. Effluent from Drug Manufactures Contains Extremely High Levels of Pharmaceuticals. *J. Hazard. Mater.***2007,** *148*, 751–755.

Larsson, D. G. J.; Pollution from Drug Manufacturing: Review and Perspectives. *Phil. Trans. R. Soc. Lond. B Biol. Sci.* **2014,** *369* (1656), 20130571.

Leal, L. H.; Vieno, N.; Temmink, H.; Zeeman, G.; Buisman, C. J. N. Occurrence of Xenobiotics in Gray Water and Removal in Three Biological Treatment Systems. *Environ. Sci. Technol.* **2010,** *44* (17), 6835–6842.

Liu, S.; Ying, G. G.; Zhao, J. L.; Chen, F.; Yang, B.; Zhou, L. J. Trace analysis of 28 Steroids in Surface Water, Wastewater Sludge Samples by Rapid Resolution Liquid Chromatography-Electrospray Ionization Tandem Mass Spectrometry. *J. Chromatogr. A* **2011,** *1218*, 1367–1378.

Liu, F.; Ying, G. G.; Tao, R.; Zhao, J. L.; Yang, J. F.; Zhao, L. F. Effects of Six Selected Antibiotics on Plant Growth and Soil Microbial and Enzymatic Activities. *Environ. Pollut.* **2009,** *157* (5), 1636–1642.

Lucchetti, D.; Fabrizi, L.; Esposito, A.; Guandalini, E.; Pasquale, M. D.; Coni, E. Simple Confirmatory Method for the Determination of Erythromycin Residues in Trout: a Fast Liquid–Liquid Extraction Followed by Liquid Chromatography—Tandem Mass Spectrometry. *J. Agric. Food Chem.* **2005,** *53*, 9689–9694.

Lucci, P.; Nunez, O.; Galceran, M. T. Solid-Phase Extraction Using Molecularly Imprinted Polymer for Selective Extraction of Natural Synthetic Estrogens from Aqueous Samples. *J. Chromatogr. A* **2011,** *1218*, 4828–4833.

Ma, T.; Pan, X.; Chen, L.; Liu, W.; Christie, P.; Luo, Y.; Wu, L. Effects of Different Concentrations and Application Frequencies of Oxytetracycline on Soil Enzyme Activities and Microbial Community Diversity. *Eur. J. Soil Biol.* **2016,** *76*, 53–60.

Martin, J. M.; Saaristo, M.; Bertram, M. G.; Lewis, P. J.; Coggan, T. L.; Clarke, B. O.; Wong, B. B. The Psychoactive Pollutant Fluoxetine Compromises Antipredator Behaviour in Fish. *Environ. Pollut.* **2017,** *222*, 592–599.

Masoner, J. R.; Kolpin, D. W.; Furlong, E. T.; Cozzarelli, I. M.; Gray, J. L. Landfill Leachate as a Mirror of Today's Disposable Society: Pharmaceuticals and Other Contaminants of Emerging Concern in Final Leachate from Landfills in the Conterminous United States. *Environ. Toxicol. Chem.* **2016,** *35* (4), 906–918.

Melchor-Martinez, E. M.; Jimenez-Rodríguez, M. G.; Martínez-Ruiz, M.; Pena-Benavides, S. A.; Iqbal, H. M. N.; Parra-Saldívar, R.; Juan Eduardo Sosa-Hernandez, J. E. Antidepressants Surveillance in Wastewater: Overview Extraction and Detection. *Case Stud. Chem. Environ. Eng.* **2021,** *3*, 100074.

Mohring, S. A. I.; Strzysch, I.; Fernandes, M. R.; Kiffmeyer, T. K.; Tuerk, J.; Hamscher, G. Degradation and Elimination of Various Sulfonamides During Anaerobic Fermentation: A Promising Step on the Way to Sustainable Pharmacy. *Environ. Sci. Technol.***2009,** *43*, 2569–2574.

Molaei, A.; Lakzian, A.; Haghnia, G.; Astaraei, A.; Rasouli-Sadaghiani, M.; Ceccherini, M. T.; Datta, R. Assessment of Some Cultural Experimental Methods to Study the Effects of Antibiotics on Microbial Activities in a Soil: An Incubation Study. *PLoS ONE.* **2017,** *12*, e0180663.

Monteiro, S. C.; Boxall, A. B. Occurrence and Fate of Human Pharmaceuticals in the Environment. *Rev. Environ. Contam. Toxicol.* **2010,** *202*, 53–154.

Nassef, M.; Matsumoto, S.; Seki, M.; Khalil, F.; Kang, I. J.; Shimasaki, Y.; Oshima, Y.; Honjo, T. Acute Effects of Triclosan, Diclofenac and Carbamazepine on Feeding Performance of Japanese Medaka Fish (*Oryzias latipes). Chemosphere* **2010,** *80*, 1095–1100.

Nieto, A.; Borrull, F.; Pocurull, E.; Marc, R. M. Pressurized Liquid Extraction of Pharmaceuticals from Sewage-Sludge. *J. Sep. Sci.* **2007,** *30*, 979–984.

Nikolaou, A.; Meric, S.; Fatta, D. Occurrence Patterns of Pharmaceuticals in Water and Wastewater Environments. *Anal. Bioanal. Chem.* **2007,** *387*, 1225–1234.

OECD. *Pharmaceutical Residues in Freshwater: Hazards and Policy Responses, OECD Studies on Water*; OECD Publishing: Paris, 2019. https://doi.org/10.1787/c936f42d-en.

Paramasivam, M.; Jayalakshmi, K.; Chandrasekaran, S.; Kuttalam, S. Sample Preparation and Extraction Techniques. In: *Handbook of Food Analysis*; Leo, M. L.N., Toldra, F., Ed., 3rd ed.; CRC Taylor & Francis Group, 2015; pp 415–426.

Patel, M.; Kumar, R.; Kishor, K.; Mlsna, T.; Pittman, C. U. Jr.; Mohan, D. Pharmaceuticals of Emerging Concern in Aquatic Systems: Chemistry, Occurrence, Effects, and Removal Methods. *Chem. Rev.* **2019,** *119* (6), 3510–3673.

Patneedi, C. B.; Prasadu, K. D. Impact of Pharmaceutical Waste on Human Life and Environment. *Rasayan. J. Chem.* **2015,** *8* (1), 67–70.

Pei, R.; Cha, J.; Carlson, K. H.; Pruden, A. Response of Antibiotic Resistance Genes (ARG) to Biological Treatment in Dairy Lagoon Water. *Environ. Sci. Technol.* **2007,** *41*, 5108–5113.

Peysson, W.; Vulliet, E. Determination of 136 Pharmaceuticals and Hormones in Sewage Sludge Using Quick, Easy, Cheap, Effective, Rugged and Safe Extraction Followed by Analysis with Liquid Chromatography-Time-of-Flight-Mass Spectrometry. *J. Chromatogr. A* **2013,** *1290*, 46–61.

Prasse, C.; Loffler, D.; Ternes, T. A. Environmental Fate of the Anthelmintic Ivermectin in an Aerobic Sediment/Water System. *Chemosphere* **2009,** *77*, 1321–1325.

Pruden, A.; Larsson, D. G. J.; Amezquita, A.; Collignon, P.; Brandt, K. K.; Graham, D. W.; Lazorchak, J. M.; Suzuki, S.; Silley, P.; Snape, J. R.; Topp, E.; Zhang, T.; Zhu, Y. G. Management Options for Reducing the Release of Antibiotics and Antibiotic Resistance Genes to the Environment. *Environ. Health Perspect.* **2013,** *121*, 878–885.

Radjenovic, J.; Jelic, A.; Petrovic, M.; Barcelo, D. Determination of Pharmaceuticals in Sewage Sludge by Pressurized Liquid Extraction (PLE) Coupled to Liquid Chromatography-Tandem Mass Spectrometry (LC-MS/MS). *Anal. Bioanal. Chem.* **2009,** *393*, 1685–1695.

Reemtsma, T.; Quintana, J. B. Analytical Methods for Polar Pollutants. In: *Organic Pollutants in the Water Cycle: Properties, Occurrence, Analysis and Environmental Relevance of Polar Compounds*; Reemtsma, T., Jekel, M., Ed.; Weinheim: Wiley, **2006,** pp 1–40.

Rehman, M. S. U.; Rashid, N.; Ashfaq, M.; Saif, A.; Ahmad, N.; Han, J. I. Global Risk of Pharmaceutical Contamination from Highly Populated Developing Countries. *Chemosphere* **2015,** *138*, 1045–1055.

Reis, E. O.; Foureaux, A. F. S.; Rodrigues, J. S.; Moreira, V. R.; Lebron, Y. A. R.; Santos, L. V. S.; Amaral, M. C. S.; Lange, L. C. Occurrence, Removal and Seasonal Variation of Pharmaceuticals in Brazilian Drinking Water Treatment Plants. *Environ. Pollut.* **2019,** *250*, 773–781.

Ridgway, K.; Lalljie, S. P. D.; Smith, R. M. 2007. Sample Preparation Techniques for the Determination of Trace Residues and Contaminants in Foods. *J. Chromatogr. A* **2007**, *1153*, 36–53.

Rivera-Utrilla, J.; Sanchez-Polo, M.; Ferro-García, M. A.; PradosJoya, G.; Ocampo-Perez, R. Pharmaceuticals as Emerging Contaminants and Their Removal from Water: A Review. *Chemosphere* **2013,** *93*, 1268–1287.

Russell, R. E.; Franson, J. C. Causes of Mortality in Eagles Submitted to the National Wildlife Health Center: 1975–2013. *Wildl. Soc. B* **2014,** *38* (4), 697–704.

Sadutto, D.; Pico, Y. Sample Preparation to Determine Pharmaceutical and Personal Care Products in an All-Water Matrix: Solid Phase Extraction. *Molecules* **2020,** *25*, 5204.

Salvia, M. V.; Cren-Olive, C.; Wiest, L.; Baudot, R.; Vulliet, E. Comparison of Two Analytical Methods for the Determination of Traces of Veterinary Antibiotics and Steroid Hormones in Soil Based on Pressurised Liquid Extraction (PLE) and Quick, Easy, Cheap, Effective, Rugged, Safe (Modified-QuEChERS) extraCtion. *Pharma. Anal. Acta* **2014,** *5*, 9.

Salvia, M. V.; Vulliet, E.; Wiest, L.; Baudot, R.; Cren-Olive, C. Development of a Multi-Residue Method Using Acetonitrile-Based Extraction Followed by Liquid Chromatography-Tandem Mass Spectrometry for the Analysis of Steroids and Veterinary and Human Drugs at Trace Levels in Soil. *J. Chromatogr. A* **2012,** *1245*, 122–133.

Schlusener, M. P.; Arb, M. A. V.; Bester, K. Elimination of Macrolides, Tiamulin, and Salinomycin During Manure Storage. *Arch. Environ. Contam. Toxicol.* **2006,** *51*, 21–28.

Schlusener, M. P.; Spiteller, M.; Bester, K. Determination of Antibiotics from Soil by Pressurized Liquid Extraction and Liquid Chromatography-Tandem Mass Spectrometry. *J. Chromatogr. A* **2003,** *1003*, 21–28.

Schwarzenbach, R. P.; Escher, B. I.; Fenner, K.; Hofstetter, T. B.; Johnson, A.; Gunten, U. V.; Wehrli, B. The Challenge of Micropollutants in Aquatic Systems. *Science* **2006,** *313* (5790), 1072–1077.

Seifrtova, M.; Pena, A.; Lino, C.; Solich, P. Determination of Fluoroquinolone Antibiotics in Hospital and Municipal Wastewaters in Coimbra by Liquid Chromatography with a Monolithic Column and Fluorescence Detection. *Anal. Bioanal. Chem.* **2008,** *391*, 799–805.

Semreen, M. H.; Shanableh, A.; Semerjian, L.; Alniss, H.; Mousa, M.; Bai, X.; Kumud Acharya, K. Simultaneous Determination of Pharmaceuticals by Solid-Phase Extraction and Liquid Chromatography- Tandem Mass Spectrometry: A Case Study from Sharjah Sewage Treatment Plant. *Molecules* **2019,** *24*, 633.

Shaaban, H.; Alghamdi, H.; Alhamed, N.; Alzaydi, A.; Mostafa, A. Environmental Contamination by Pharmaceutical Waste: Assessing Patterns of Disposing Unwanted Medications and Investigating the Factors Influencing Personal Disposal Choices. *J. Pharmacol. Pharm. Res.* **2018,** *1* (1), 003.

Shakil, S.; Khan, R.; Zarrilli, R.; Khan, A. Aminoglycosides versus Bacteria-A Description of the Action, Resistance Mechanism, and Nosocomial Battleground. *J. Biomed. Sci.* **2008,** *15* (1), 5–14.

Sharma, R.; Larney, F. J.; Chen, J.; Yanke, L. J.; Morrison, M.; Topp, E.; McAllister, T. A.; Yu, Z. Selected Antimicrobial Resistance During Composting of Manure from Cattle Administered Sub-Therapeutic Antimicrobials. *J. Environ. Qual.* **2009,** *38*, 567–575.

Shemesh, M.; Shore, L. Effects of Environmental Estrogens on Reproductive Parameters in Domestic Animals. *Israel. J. Vet. Med.* **2012,** *67* (1), 6–10.

Shen, Z. L.; Yuan, D.; Zhang, H.; Hu, M.; Jian-Hua, Z.; Zhang, X. Q.; Qing-De, S. Matrix Solid Phase Dispersion-Accelerated Solvent Extraction for Determination of OCP Residues in Fish Muscles. *J. Chin. Chem. Soc.* **2011,** *58*, 494–502.

Shenker, M.; Harush, D.; Ben-Ari, J.; Chefetz, B. Uptake of Carbamazepine by Cucumber Plants-A Case Study Related to Irrigation with Reclaimed Wastewater. *Chemosphere* **2011,** *82*, 905- 910.

Shore, R. F.; Taggart, M. A.; Smits, J.; Mateo, R.; Richards, N. L.; Fryday, S. Detection and Drivers of Exposure and Effects of Pharmaceuticals in Higher Vertebrates. *Phil. Trans. R. Soc. B* **2014,** *369* (1656), 20130570.

Shrestha, S. L.; Casey, F. X.; Hakk, H.; Smith, D. J.; Padmanabhan, G. Fate and Transformation of an Estrogen Conjugate and Its Metabolites in Agricultural Soils. *Environ. Sci. Technol.* **2012,** *46*, 11047–11053.

Smith, A.; Balaam, J.; Ward, A. The Development of a Rapid Screening Technique to Measure Antibiotic Activity in Effluents and Surface Water Samples. *Mar. Pollut. Bull.* **2007,** *54*, 1940–1946.

Sommer, C.; Bibby, B. M. The Influence of Veterinary Medicines on the Decomposition of Dung Organic Matter in Soil. *Eur. J. Soil Biol.* **2002,** *38*, 155–159.

Stahlschmidt-Allner, P.; Allner, B.; Rombke, J.; Knacker, T. Endocrine disrupters in the aquatic environment. *Environ. Sci. Pollut. Res.* **1997,** *4* (3), 1550–162.

Storteboom, H. N.; Kim S. C.; Doesken K. C.; Carlson, K. H.; Davis, J. G.; Pruden, A. Response of Antibiotics and Resistance Genes to High-Intensity and Low-Intensity Manure Management. *J. Environ. Qual.* **2007,** *36*, 1695–1703.

Sumpter, J. P.; Jobling, S. The Occurrence, Causes, and Consequences of Estrogens in the Aquatic Environment. *Environ. Toxicol. Chem.* **2013,** *32*, 249–251.

Taylor, D.; Senac, T. Human Pharmaceutical Products in the Environment-the "Problem" in Perspective. *Chemosphere* **2014,** *115*, 95–99.

Teresa, I. A.; Adrian, G.; Silva, M. T.; Ribeiro, A. R.; Alves, A.; Santos, M. S. F. Liquid-Liquid Extraction as a Simple Tool to Quickly Quantify Fourteen Cytostatics in Urban Wastewaters and Access Their Impact in Aquatic Biota. *Sci. Total Environ.* **2020,** *740*, 139995.

Ternes, T. The Occurrence of Micropollutants in the Aquatic Environment: A New Challenge for Water Management. *Water Sci. Technol.* **2007**, *55* (12), 327–332.

Ternes, T. A.; Hirsch, R.; Mueller, J.; Haberer, K. Methods for the Determination of Neutral Drugs as Well as Betablockers and β_2-Sympathomimetics in Aqueous Matrices Using GC/MS and LC/MS/MS. *Fresenius J. Anal. Chem.* **1998,** *362*, 329–340.

Tolgyesi, A.; Verebey, Z.; Sharma, V. K.; Kovacsics, L.; Fekete, J. Simultaneous Determination of Corticosteroids, Androgens, and Progesterone in River Water by Liquid Chromatography-Tandem Mass Spectrometry. *Chemosphere* **2010,** *78*, 972–979.

Toma, A.; Crisan, O. Green Pharmacy—A Narrative Review. *Clujul Med.* **2018,** *91* (4), 391–398.

Tong, A. Y. C.; Peake, B. M.; Braund, R. Disposal Practices for Unused Medications Around the World. *Environ. Int.* **2011,** *37* (1), 292–298.

Toth, J. D.; Feng, Y.; Dou, Z. Veterinary Antibiotics at Environmentally Relevant Concentrations Inhibit Soil Iron Reduction and Nitrification. *Soil Biol. Biochem.* **2011,** *43*, 2470–2472.

Valenti, T. W.; Gould, G. G.; Berninger, J. P.; Connors, K. A.; Keele, N. B.; Prosser, K. N.; Brooks, B. W. Human Therapeutic Plasma Levels of the Selective Serotonin Reuptake Inhibitor (SSRI) Sertraline Decrease Serotonin Reuptake Transporter Binding and Shelter-Seeking Behavior in Adult Male Fathead Minnows. *Environ. Sci. Technol.* **2012,** *46*, 2427–2435.

Varma, A.; Buscot, F. *Microorganisms in Soils: Roles in Genesis and Functions*; Springer-Verlag: Berlin Heidelberg: 2005.

Vaswani, M.; Linda, F. K.; Ramesh, S. Role of Selective Serotonin Reuptake Inhibitors in Psychiatric Disorders: A Comprehensive Review. *Prog. Neuropsychopharmacol. Biol. Psychiatry* **2003,** *27*, 85–102.

Vazquez-Roiga, P.; Segarra, R.; Blasco, C.; Andreu, V.; Pico, Y. Determination of Pharmaceuticals in Soils and Sediments by Pressurized Liquid Extraction and Liquid Chromatography Tandem Mass Spectrometry. *J. Chromatogr. A* **2010,** *1217*, 2471–2483.

Verlicchi, P.; Al Aukidy, M.; Galletti, A.; Petrovic, M.; Barcelo, D. Hospital Effluent: Investigation of the Concentrations and Distribution of Pharmaceuticals and Environmental Risk Assessment. *Sci. Total Environ.* **2012,** *430*, 109–118.

Verlicchi, P.; Galletti, A; Petrovic, M.; Barcelo, D. Hospital Effluents as a Source of Emerging Pollutants: An Overview of Micropollutants and Sustainable Treatment Options. *J. Hydrol.***2010,** *389*, 416–428.

Widyasari-Mehta, A.; Suwito, H. R.; Kreuzig, R. Laboratory Testing on the Removal of the Veterinary Antibiotic Doxycycline During Long-Term Liquid Pig Manure and Digestate Storage. *Chemosphere* **2016,** *149*, 154–160.

Wu, M.; Liu, S.; Hu, L.; Qu, H.; Pan, C.; Lei, P.; Shen, Y.; Yang, M. Global Transcriptomic Analysis of Zebrafish in Response to Embryonic Exposure to Three Antidepressants, Amitriptyline, Fluoxetine and Mianserin. *Aquat. Toxicol.* **2017,** *192*, 274–283.

Yang, J. F.; Ying, G. G.; Liu, S.; Zhou, L. J.; Zhao, J. L.; Tao, R.; Peng, P. A. Biological Degradation and Microbial Function Effect of Norfloxacin in a Soil Under Different Conditions. *J. Environ. Sci. Heal.* **2012,** *47*, 288–295.

Yin, J.; Yang, Y.; Li, K.; Zhang, J.; Shao, B. Analysis of Anticancer Drugs in Sewage Water by Selective SPE and UPLC-ESI-MS-MS. *J. Chromatogr. Sci.* **2010,** *48*, 781–789.

Yu, Z.; Peldszus, S.; Huck, P. M. Optimizing Gas Chromatographic-Mass Spectrometric Analysis of Selected Pharmaceuticals and Endocrine-Disrupting Substances in Water Using Factorial Experimental Design. *J. Chromatogr. A* **2007,** *1148*, 65–77.

Zhao, J. L.; Ying, G. G.; Wang, L.; Yang, J. F.; Yang, X. B.; Yang, L. H.; Li, X. Determination of Phenolic Endocrine Disrupting Chemicals and Acidic Pharmaceuticals in Surface Water of the Pearl Rivers in South China by Gas Chromatography-Negative Chemical Ionization-Mass Spectrometry. *Sci. Total Environ.* **2009,** *407* (2), 962–974.

Ziylan, A.; Ince, N. H. The occurrence and Fate of Antiinflammatory and Analgesic Pharmaceuticals in Sewage and Fresh Water: Treatability by Conventional and Non-Conventional Processes. *J. Hazard. Mater*. **2011,** *187*, 24–36.

Zwiener, C. Occurrence and Analysis of Pharmaceuticals and Their Transformation Products in Drinking Water Treatment. *Anal. Bioanal. Chem*. **2007,** *387*, 1159–1162.

CHAPTER 8

Environmental Pollution Due to Improper Use of Chemical Fertilizers and Their Remediation

BHOLANATH SAHA[1], AYESHA FATIMA[2], SUSHANTA SAHA[3], SANJIB KUMAR SAHOO[4], and PARTHENDU PODDAR[5]

[1]*Department of Soil Science, Dr. Kalam Agricultural College, Bihar Agricultural University, Kishanganj, Bihar, India*

[2]*Department of Agronomy, Dr. Kalam Agricultural College, Bihar Agricultural University, Kishanganj, Bihar, India*

[3]*Directorate of Research, Bidhan Chandra Krishi Viswavidyalaya, Kalyani, Nadia, West Bengal, India*

[4]*Department of Soil Science & Agricultural Chemistry, Odisha University of Agriculture & Technology, Bhubaneswar, Odisha, India*

[5]*Deparrtment of Agronomy, Uttar Banga Krishi Viswavidyalaya, Pundibari, Coochbehar, West Bengal, India*

ABSTRACT

The two primary or major nutrients that cause soil and environmental health problems are nitrogen (N) and phosphorus (P). These two nutrients are together the most widespread cause of water quality impairment in surface water and groundwater pollution. Both point and nonpoint sources of pollution through fertilizer usage have been identified since green revolution as indiscriminate and injudicious application of fertilizers has been practiced globally. The primary means of avoiding the environmental pollution with

Environmental Contaminants: Impact, Assessment, and Remediation.
Pritam Ganguly, Jajati Mandal, M. Paramsivam, & Sandip Patra, (Eds.)

excessive use of N and P fertilizers is by avoiding the excessive application of these high-analysis fertilizers and their accumulation in soil and/or groundwater. Nutrient supply through right manner and with judicious fertilizer application could meet the need of growing plants vis-a-vis could reduce the environmental pollution. In most of the surface water, phosphorus is the limiting nutrient that can cause eutrophication. Nitrogen in the form of ammonia in aquatic ecosystems can induce toxicity to aquatic macrofauna in addition to causing eutrophication. The nitrogen in the form of nitrate is also a major issue in certain states of India, where intensive agriculture is practiced, as nitrate levels above 10.0 mg/L are considered toxic for the drinking water. Several issues have been identified that link excessive fertilizer use to environmental issues. The major fertilizer-related pollution issue is the buildup of nitrates in drinking water, a trend that the general public is concerned about and is thought to be caused by the excessive use of nitrogenous fertilizers and manures derived from animals. Another issue is nutrient enrichment, eutrophication, and deterioration of fresh water quality as a result of nutrients transported through fertilizers being transported via leaching and/or runoff, as well as sediment erosion. Nutrient management impacts the environment most directly in the area of water quality and soil quality. Widespread nutrient pollution through fertilizer usage has highlighted the need for better nutrient management to reduce the contamination of environment.

8.1 INTRODUCTION

The per capita availability of land for agricultural production has decreased substantially in recent decades, not only in India but also around the world, and is predicted to continue to shrink in the near future. The amount of land used in production per person is predicted to be 56% lower by 2025 than it was in 1965. As per the UN report on "The World Population Prospects 2019," the world's population will grow by 2 billion people in the next 30 years, from 7.7 billion presently to 9.7 billion in 2050. Because of these tendencies, market demand for food will continue to rise, and feeding a world population of 9.7 billion people in 2050 will necessitate increased overall food production, which will further necessitate increased nutrient inputs to meet demand. It is also reasonable to expect that, as public control over production grows, the impact of agriculture on the environment will be evaluated more closely. India removes 25 Mt of plant nutrients from soil each year to harvest 200

MT of food grains, although the annual input from external sources such as fertilizers and manures is only 33% of the actual need. The situation will become even worse in the following years, when the production demand will be substantially higher. Though, this has been estimated only for major nutrients, but taking into account the uptake and demand of micronutrients and secondary nutrients for correcting their widespread deficiency, it can be inferred that the future for these nutrients may also be alarming. In Indian agriculture, less application of nutrients may be of common occurrence, but imbalanced and untimely application of nutrients can lead to detrimental consequences including water, soil, and air pollution. Therefore, the thrust should be on balanced and judicious application of nutrients to achieve higher nutrient use efficiency on one hand and lesser environmental impact on the other. Fertilizers and other agro-chemicals are a non-point source of pollution in intensive agriculture in three ways: groundwater contamination due to leaching, surface water contamination due to runoff of excess nutrients or their derivatives, and emission of greenhouse gases that alter the composition of atmospheric gases, ultimately leading to climate warming.

Several issues have been identified that link increased fertilizer use to environmental concerns. The increase of nitrates in the drinking water is mainly due to the excessive use of N fertilizers and animal-based manures, and is considered to be one of the most important pollution issues related to fertilizers. As the reactive nitrogen can swiftly move across different medium such as air, soil, and water, it can lead to increased levels of ozone in the lower atmosphere, thereby causing respiratory ailments and damage to the vegetation. A significant part of fertilizer nitrogen applied to soil is converted into nitrous oxide which results in both the greenhouse effect and ozone depletion from the stratosphere. Nitrogen can be lost to atmosphere also as ammonia gas which can fall to the surface in the form of atmospheric deposition after undergoing oxidation to nitric acid, leading to acidification of water bodies and soil, unintended fertilization of vegetation leading to unnatural growth rates, nutrient imbalances in soil, decreasing or altering natural biodiversity and abrasion of buildings, bridges and other man-made and natural structures. Environmentalists are increasingly concerned about heavy metal contamination of soils from fertilizers, notably as cadmium from phosphatic fertilizers. The application of fertilizer to cultivated soil as part of intensive agriculture is also to blame. In the developed countries, fertilizer use has been intensive for the past 3–4 decades. If the detrimental effects of fertilizers are being perceived there now, then similar problems in the developing countries like India can be witnessed in the near future.

8.2 SOURCES OF ENVIRONMENTAL POLLUTION DUE TO FERTILIZER USE

Several issues have been identified that link excessive fertilizer use to the environmental issues. One of the major fertilizer-related pollution issues is the growth of nitrates in the drinking water, while some other related issues are nutrient enrichment, eutrophication, and worsening of surface water quality as a result of nutrients provided through fertilizers being transported via leaching and/or runoff, as well as sediment erosion. The reactive nitrogen can flow quickly across multiple media such as air, soil, and water. It may contribute to higher ozone concentration in the lower atmosphere, which causes respiratory problems and wreaking havoc on plants. A portion of nitrogenous fertilizer when added to the soil is transformed into nitrous oxide, which increases the greenhouse effect and causes ozone depletion in the stratospheric layer. Also, nitrogen can be lost to atmosphere as ammonia gas which can fall to the land surface as atmospheric deposition after undergoing oxidation to nitric acid, resulting in soil and water body acidification, as well as unintentional fertilization of grasslands and forests, nutritional imbalances, and decreased or altered biodiversity and corrosion of buildings, bridges, and other man-made and natural structures. Environmentalists are increasingly concerned about heavy metal contamination in soils caused by fertilizers like cadmium from phosphatic fertilizers. Fertilizer application to cultivated soil as part of intensive agriculture is also a contributing factor. Fertilizer has been extensively used in developed countries for the past three to four decades. If the damaging impacts of fertilizers are being detected now, comparable issues in developing countries such as India might be expected in the not-too-distant future.

8.3 EXCESSIVE USE OF NITRATES (NO_3^-) IN SOIL

Nitrate may be assimilated by agricultural plants, leached away from the rooting zone or denitrified into Nitrogen and Nitrous oxide gases. Leaching of nitrate below the crop root zone poses a serious pollution concern to surface and groundwater resources. The formation of N_2O via the nitrification-denitrification reaction poses a possible threat to the stratospheric ozone as well as to the greenhouse effect. Plants, particularly fodders and vegetables, can absorb excessive levels of nitrate, which can induce nitrate toxicity in consumers. Because nitrate uptake by roots surpasses reduction and assimilation, particularly in the plants of *Chenopodiaceae* family, that thrive in soils with high nitrate levels tends to accumulate nitrates. As a result, the majority of environmental concerns about nitrogen use in agriculture revolve around soil transformations that result in nitrate production. The extent of nitrate accumulation in plants is controlled by environmental conditions, fertilizer N management, and crop production strategies, in addition to the plant's genetic potential. The impacts of nitrogenous fertilizers application on nitrate concentration in plants are governed by the amount, source, time, and manner of application. In general, the longer a plant is exposed to nitrate, the higher its proclivity to absorb nitrate. As a result, side-dressing of urea or ammonium fertilizer will result in less nitrate deposition in plants than nitrate fertilizer. Plants accumulate less nitrate when organic manures are slowly mineralized.

8.4 NITRATE POLLUTION OF GROUND WATER

The nitrate concentration in the soil profile during leaching and the volume of water entering through the soil profile are two important parameters controlling nitrate leaching losses. Soils with high nitrate levels and substantial downward movement of water to transfer nitrates below the rooting depth are common in humid and sub-humid zone soils, to a lesser extent in semiarid zone soils, and very seldom in arid zone soils (Bijay-Singh and Yadvinder-Singh, 1993). During rainfall, when the soils are at or near field capacity, significant leaching losses occur. The rate and extent of leaching are affected by the amount and intensity of rainfall or irrigation. Leaching losses of nitrate can be significant when high fertilizer rates are paired with intensive irrigation regimes on coarse-textured soils. Leaching of nitrogen can occur to a significant level when summer fallow is followed by monsoon-type heavy rainfall (Bijay-Singh and Sekhon, 1979).

8.5 HEAVY METAL CONTAMINANTS IN FERTILIZERS AS SOIL POLLUTANTS

Due to the health risks related to heavy metals infiltrating the food chain via the soil-water-plant continuum, an alarming scenario has arisen in recent days. Fertilizers contain heavy metals as contaminants and undesirable elements in their admixtures. In this regard, rock phosphate, among all other fertilizers, is a major heavy metal source. When rock phosphate or its products are applied to soil, a substantial quantity of lead and cadmium is introduced into the soil. The lead and cadmium content of several commercial fertilizers, a low analysis and straight fertilizers combination may add more amount of lead and cadmium to soil than a combination of high analysis and mixed fertilizers. Heavy metals supplied to soil using various rock phosphate derivatives deposit almost entirely in the soil's surface layer and in forms that are generally readily available to plants. Heavy metals added through fertilizer are more available in soils with coarser textures and acidic reactions than in those with high clay content and alkaline reactions. Although relevant data from Indian soils is unavailable, it is interesting to note that phosphate fertilizer use accounts for less than 6% of annual cadmium deposits in European Economy Community soils, with phosphoric acid manufacturing accounting for another 2%. Solid wastes and excrement, airborne deposition, and the usage of colors and stabilizers account for two-thirds of the total.

8.6 POLLUTION DUE TO MAJOR NUTRIENTS

Developing countries, which use more nitrogen fertilizer than developed ones, account for roughly 76% of the global population. Negative nitrogen imbalances in the soil have been a common feature of crop production systems in developing nations throughout the previous 5 or 6 decades because of imbalanced nitrogen fertilization rates. The threat of nitrate pollution of groundwater in developing nations will be strongly connected to fertilizer-N use efficiency in the future, as fertilizer-N application rates increase. A limited number of studies from developing nations suggest that nitrate pollution of groundwater exists in Asia's irrigated soils and Africa's humid tropics, especially when fertilizer-N is applied inefficiently. Although no attempts have been done to correlate fertilizer-N use with nitrate levels in groundwater in developing countries in the humid tropics, fertilizers are becoming more widely employed. Aside from heavy rainfall, irrigation is becoming more accessible to farmers in the humid tropical regions, and significant nitrogen leaching may develop as a result. (Singh et al.,1995).

Thus, non-point pollution from agricultural fertilizers and pesticides, which is often distributed over large regions, poses a significant threat to fresh groundwater ecosystems. Excessive use of synthetic fertilizers in farmland and indiscriminate discharge of human excreta and animal wastes on the surface of land causes residual nitrate to leach into groundwater, resulting in high nitrate concentrations. In 11 Indian states, nitrate concentrations are higher than the permissible limit of 45 ppm. However, soil texture, fertilizer and pesticide use patterns, degradation products, and total soil organic matter influence the susceptibility of groundwater to fertilizer and pesticide pollution. Nutrient-type water pollution is also caused by atmospheric nitrogen deposition (from nitrogen oxides). Nutrients, in excess, enhance the growth of aquatic plants and algae. Unwarranted development of these organisms clogs local waterways and inhibits light from reaching deeper waters while they are alive. When they die, the organisms take up dissolved oxygen as they disintegrate, resulting in oxygen-depleted waters that support only a small number of marine lives. Nutrient pollution in particular is a serious issue in estuaries and deltas, wherein watershed runoff is finally discharged at the major rivers' mouths (Kumar and Shah, http://www.iwmi.cgiar org/ iwmi-tata/files/pdf/ground-pollute4_FULL_ pdf).

8.7 ENVIRONMENTAL CONSEQUENCES OF IMPROPER FERTILIZER USE

Application of nutrients through fertilizers and manures more than the crop requirements increases the cost of production and the risk of detrimental environmental effects leading to human and animal health problems. If the nitrate-N in the drinking water exceeds the safe limit of 10 mg/L, ingested nitrate is converted into nitrite which is absorbed in blood, causing methemoglobinemia, popularly known as "Blue baby syndrome," and gastrointestinal cancer. In intensively cultivated soils, where fertilizer N consumption is very high, nitrate-N concentration in the groundwater increased by almost 2 mg/L (Aulakh and Bijay-Singh, 1997). Animal wastes are believed to be the major source of high $NO_{3-}N$ in groundwater. Animal wastes are generally discarded near feedlots in the village outskirts; there occurs a substantial increase in the level of $NO_{3-}N$ in soil as well as in groundwater. The level of $NO_{3-}N$ in 367 hand-pump water samples from four districts, as well as 45 water samples collected below feedlots, was quite higher as compared to 236 tubewell water samples of nearby areas, clearly indicating that animal wastes and feedlots act as a point source of nitrates (Bajwa et al., 1993). Use of judicious fertilizer nitrogen in conjunction with higher application frequency of irrigation water, along with split N fertilizer application, can help in minimizing the movement of $NO_{3-}N$ to deeper soil layers.

Excessive residual P accumulation in soil may accelerate downward flow of P, which may subsequently leach down into groundwater. Long-term studies on fertilizer P when applied in different rates, frequencies, and intervals have demonstrated that there P leaching is possible, particularly in coarse-textured soils. Continuous application of different water-soluble as well as partially soluble phosphatic fertilizers after 40 years increased Olsen's available P significantly 5–7 fold over control in soil (Saha et al., 2013). Aulakh et al. (2007) also reported that there was substantial increase in residual P fertilizer accumulated as Olsen-P in deeper layers of soil profile. Interchange between the fertilizer P management practices (alternative *vis-a-vis* cumulative P application, and optimal *vis-a-vis* high rates of P fertilizer in various crop rotations), amount of labile P deposited in soil profile, and soil properties (texture and SOC) largely influenced the downward movement and P leaching in subtropical irrigated soils.

Many fertilizers and amendments contain varying amounts of toxic elements. The rock phosphate is the major source of lead (Pb) and cadmium (Cd). Emission of greenhouse gases such as N_2O through denitrification and nitrification, CH4 from submerged and waterlogged soils could be induced by management practices.

8.7.1 NITROUS OXIDE (N_2O) EMISSIONS FROM SOIL IN RELATION TO CLIMATE CHANGE EFFECTS

Nitrous oxide (N_2O) is a trace gas that is formed by microbial nitrification and denitrification processes in soils. The natural emissions of nitrous and nitric oxide gases may be increased due to improper agricultural practices by a variety of practices and activities, including the use of synthetic chemical and organic fertilizers, nitrogen-fixing crops cultivation, cultivation on high organic content soils, and the application of livestock manure to agricultural lands and pastures. All these practices directly add additional nitrogen to soils, which then gets converted into NO. Water management in fields greatly influences the extent of N_2O emission. Up to 50% of the applied-N can be lost through denitrification when alternating aerobic-anaerobic conditions prevail in the paddy fields. In India's Indo-Gangetic Plains, a vast majority of soils are relatively coarse-textured and experience frequent alternating aerobic anaerobic conditions. Nitrous oxide is one of the greenhouse gases and contributes around 6% to the anthropogenic greenhouse effect (IPCC, 2001) and is also a contributor to stratospheric ozone layer depletion. Following the methodologies of Intergovernmental Panel on Climate Change (IPCC), Garg et al. (2006) estimated direct N_2O emissions from soils (including application of synthetic fertilizers), on-farm burning of agricultural crop residues, manure management, indirect soil emissions, combustion of fossil fuels, industrial activities, namely production of nitric acid and waste management sector in India and found that agricultural sector activities account for greater than 80% of the total N_2O-emission, including 60% from the use of synthetic fertilizer, about 12% each from agriculture residue burning and indirect soil emissions and approximately 3% from manure management. Emission of N_2O occurs by both nitrification and denitrification processes during the storage of animal manures. Emission factor for NO for solid storage of animal manure in India, as worked out by Gupta et al (2007), varies from 3 to 11.7 mg/head/year for different categories of bovine with an estimated annual emission of about 2344 kg NO in India.

8.7.2 AMMONIA EMISSIONS FROM SOIL AND ENVIRONMENTAL DEGRADATION

Application of nitrogen fertilizers to soil and the livestock are the major contributors to ammonia emissions. Out of all the fertilizers, application of urea is responsible for greater than 90% of the total contribution due to fertilizers. Substantial loss of N via ammonia volatilization can be expected when urea is top dressed, particularly on alkaline soils. Deep placement of urea in the soil as such or its application before irrigation results in substantial reduction in volatilization losses from agricultural fields. In the atmosphere, ammonia either gets deposited or transformed into an ammonium aerosol and before deposition, these aerosols contribute fine particulate matter to the atmosphere. Because of the direct and indirect impacts of aerosols on radiative forcing, ammonia plays a crucial role in global climate change (Bijay-Singh et al., 2008). About half of the volatilized ammonia is deposited in downwind ecosystems within a radius of 50 km, while the other half gets deposited over a much broader region.

8.7.3 EUTROPHICATION OF WATER BODIES

The process of nutrient enrichment of surface water is known as eutrophication. The excessive multiplication and deposition of algae and higher aquatic plants caused by nutrient enrichment of water bodies can result in adverse changes in water quality and considerably interfere with man's usage of water resources. For example, decomposition of the algae and macrophyte biomass reduces dissolved oxygen and leads to fish mortality, thus further impairing the water quality. Not only groundwater, but surface water too receives large contribution of nutrients from non-point sources such as agricultural lands. Estimates indicate that higher than 90% of the N entering the surface waters originates from non-point sources and that more than 80% of that portion is from agricultural lands. Agricultural activities are often related to increased concentrations of total N in water bodies. The spreading of fertilizers and manures in the drainage basin enhances the nutrient load of the water body. A considerable portion of the nutrient load released by the agricultural land is related to the natural nutrient-content of the soil. Practices that do not allow for the proper incorporation of fertilizers applied at excessive rates can cause short-term elevated nutrient concentrations in the runoff. N and P both play a crucial role in enhancing eutrophication. In most of the low producing oligotrophic (low in nutrients) surface water bodies, P rather than N is a limiting constituent. In many of the eutrophic lakes, biotic productivity is controlled by N because the N/P ratios of pollutants from many sources are far below the ratios required for the plant growth. Also, in some oligotrophic lakes where N is the limiting nutrient, the input from groundwater, surface runoff, or precipitation may be essential to maintain the biological productivity. The urban and industrial wastes usually account for most of the phosphate loading. However, surface runoff (including soil erosion) from cultivated land can contribute to phosphate loading of surface water. The best management strategies for a certain soil, crop, and area are highly effective in avoiding this risk, while also allowing crops to make the utmost efficient use of fertilizers for optimum food production.

8.7.4 ALGAL BOOMING

An algal bloom is a quick rise in the population of algae in aquatic ecosystems, including in fresh and marine waters, and is referred to as a water bloom or a marine bloom respectively. They flourish where there is an abundant supply of two nutrients, phosphorus and nitrogen, mainly caused due to runoff (fertilizers, animal waste, sewage) of these nutrients from the land. Algal blooms are mainly composed of bio-toxins and are distinguished by a characteristic discoloration of water owing to the presence of a large number of pigmented algal cells. The color of water affected by alagal blooms varies from green, yellow, brown, and red. The two most common types of algal blooms are cyanobacteria (blue-green algae) and red tides (red algal blooms).

The presence of excessive levels of nitrogen and phosphorus in water usually causes an algal bloom. These minerals are carried away from agricultural farms where nitrogenous and phosphatic fertilizers are used frequently.

Rain serves as a dispersing agent by transporting these soluble nutrients from soil into nearby water bodies such as streams and rivers, eventually ending up in large water bodies like lakes and oceans. Drainage systems also discharge these nutrients into waterways. Untreated raw sewage finds its way into water bodies due to poor sewage treatment, and since it is rich in nitrogenous compounds like nitrates and ammonia, it results in algal bloom. Water pollution caused due to the discharge of untreated or improperly treated industrial wastes into waterways results in release of raw toxic pollutants into water systems which consists of nutrients and compounds like nitrogen, phosphorus, and lead; the outcome is the dense algal growth.

8.8 HEALTH PROBLEMS ASSOCIATED WITH NITRATE POLLUTION OF GROUNDWATER

Methaemoglobinaemia (blue baby syndrome) can be caused by consuming water with high amounts of nitrate-N. This is especially true in the case of infants (6 months old). The World Health Organization's threshold for nitrate in drinking water is 10 mg NO_{3-}N/L. Nitrate has similar effects on livestock as it is on humans. The ruminants are particularly susceptible because rumen microorganisms can easily reduce nitrate to nitrite which leads to methaemoglobinaemia. Cattle are more vulnerable to nitrate toxicity than sheep and may even abort if pregnant. Excessive nitrate consumption may also increase cancer risk in humans by forming carcinogenic nitrosamines in the human stomach as a result of ingested amines reacting with nitrite.

8.9 ALTERNATIVES OF INDISCRIMINATE USE OF CHEMICAL FERTILIZERS

The utilization of same chemical fertilizer for a long period in a particular land may accelerate soil degradation, loss of beneficial soil microorganisms, and many other detrimental effects on soil and environment as discussed above. Therefore, it has become necessary to save our environment for sustainable agricultural production by integrated use of different types of nutrient sources such as chemical fertilizers, organic manures, legume-based cropping systems, biofertilizers, and slow release or controlled release fertilizers (Pandiselvi et al., 2017). It has been proven in different research experiments that integrating organic manures with chemical fertilizers leads to a very high positive effect on the enhancement of soil microbial population and ultimately improvement of soil health in comparison with addition of chemical fertilizers alone (Usman et al., 2015). Different new forms of fertilizer can be used to avoid the negative impact on environment caused by different fertilizers, as follows.

8.9.1 USE OF BIOFERTILIZER

Biofertilizer use is one of the best modern tools for sustainable agriculture. Biofertilizer is an integral part of integrated nutrient management in agriculture. It is a low-cost and sustainable source of plant nutrients that can be used to supplement synthetic chemical fertilizers (Santra et al., 2015). Biofertilizer is defined as a substance containing living cells of various microorganisms which, when applied to seed, plant surface, or soil, colonize the rhizosphere or the interior of the plant, and promote the growth of plants by transforming nutritionally important elements such as nitrogen and phosphorus from unavailable to available form through biological processes such as nitrogen fixation and rock mineral solubilization (El-Yazeid et al., 2007). A healthy plant will usually have a healthy rhizosphere that is dominated by beneficial microorganisms. Biofertilizers differ from organic and chemical fertilizers in that they are cultures of specific bacteria and fungus that do not directly deliver nutrients to the crops. Biofertilizers have become a hope for the future world as far as economic and environmental viewpoints are concerned. Biofertilizer production technology is generally simple with a very low installation cost as compared to chemical fertilizer units (Chen, 2006) and its use does not cause environmental pollution.

8.9.2 USE OF NATURALLY OCCURRING OR MANUFACTURED ORGANIC FERTILIZERS

The organic fertilizer registers a vital role in the soil nourishment because it mobilizes the nutrients present in the soil and makes them available to the plants in adequate and balanced proportions. Some of the naturally occurring organic fertilizers include animal manure, slurry waste, guano waste, seaweeds, peat, sewage, and other bio-degradable wastes. This category also includes mined rock phosphate, sulfate, potash, and limestone also. Decomposing crop plants also act as an organic fertilizer that occurs naturally (Bokhtiar et al., 2005).

The fertilizers which are prepared by the degradation of natural waste materials are known as organically manufactured fertilizers. Different types of eco-friendly organic manure like bio-compost, compost, vermin-compost, phospo-compost, bone meal, blood meal, and seaweed extracts which are made from organic crop residue can be effectively used in reducing environmental pollution caused by excessive and imbalanced chemical fertilizer use.

8.9.3 USE OF SLOW-RELEASE FERTILIZERS

The term slow release fertilizers (SRFs) is used as a general description of fertilizers from which nutrient release into the environment occurs in more or less controlled and slow manner. However, the rate, pattern, and timing of release are not well regulated in slow release fertilizers, but these are well known in the case of controlled release fertilizers (Shaviv, 2000).

SRFs can be generally classified into 4 types according to Hauck (1985):

i) water-soluble fertilizers controlled by a physical barrier (e.g. coating, matrix formation);
ii) inorganic materials of low solubility (e.g. metal ammonium phosphates);
iii) low solubility materials that are biologically or chemically degradable (e.g. urea form); and
iv) soluble materials that eventually decompose in the soil.

8.9.4 USE OF NANOFERTILIZERS

About 40–70% of Nitrogen, 80–90% of Phosphorus, and 50–70% of Potassium from the fertilizers applied to the soil are lost to the environment leading to significant environmental pollution (Duhan et al., 2017). A current alternative is to employ nano-coated urea or other nano-encapsulated fertilizers which have the potential to reduce environment pollution by virtue of slow and consistent release of nutrients to the plants. Nanofertilizers are a type of fertilizer that is synthesized or modified from traditional fertilizers, fertilizer bulk materials, or derived from various vegetative or reproductive parts of the plant using various chemical, physical, mechanical, or biological methods using nanotechnology to enhance soil fertility, productivity, and quality of agricultural output. Fully bulk materials can be utilized to synthesize nanoparticles. In comparison to the control, seed treated with nano-TiO2 produced plants with higher dry weight, photosynthetic rate, and chlorophyll-a production (Singh et al., 2017). It is nontoxic and can also be used as fertilizer additives (Duhan et al., 2017). Nanoscale fertilizers, nanoscale additives, and nanoscale coatings are the three types of nanofertilizers (Mikkelsen, 2018).

Nanotechnology has the potential to exert profound effect on some of the most challenging environmental cleanup by providing a cost-effective solution to these problems. The use of nano particles such as nano-scale zero valent iron “nZVI,” nZVI-bentonite, nano-alginite, nano-carbon, bentonite, and dendrimers can effectively eliminate Cd and Pb from polluted soil. Application of “magnetic” bacteria has been found to be useful for metallic ion and heavy metal removal from aqueous solutions (e.g. Ag, Hg, Pb, Cu, Zn, Sb, Mn, Fe, As, Ni, Al, Pt, Pd, and Ru). In the presence of magnetic ions such as iron sulfide, heavy metal precipitates onto bacterial cell walls, making the bacteria sufficiently magnetized for removal from suspension by the magnetic separation procedure (Elrahman and Mostafa, 2015). Many researchers have reported that certain bacteria could produce iron sulfonide, which would effectively act as an adsorbent for several metallic ions.

8.9.5 APPROPRIATE METHOD OF APPLICATION FOR IMPROVING USE EFFICIENCY

For achieving higher use efficiency of applied fertilizers, any fertilizer should be applied at an economic dose rather than optimum dose. In addition, application at right time, amount, place, and source will limit the negative impact on the crop and the environment. Other practices that can enhance use efficiency while reducing environmental pollution are adoption of fertilizer best management practices such as 4R Nutrient Stewardship, site-specific nutrient management (SSNM), split application of N and K fertilizers, especially in coarse-textured soils, foliar fertilization, fertigation, and integrated soil fertility management (IFSM). Use of enhanced-efficiency N fertilizers is one of several possible management practices that can be effective in reducing environmental N losses and thus increasing N-use efficiency. Understanding the factors that influence N loss and the technical characteristics of the different enhanced-efficiency fertilizers is critical for determining their optimal use. In addition, quantification of reductions in environmental N loss that can occur with the use of these enhanced-efficiency

N fertilizers under diverse cropping systems and environmental conditions is to be known.

8.10 CONCLUSIONS AND POLICY INFERENCES

There is a dire need to develop customized crop and soil-specific fertilizer materials for predominant cropping and farming systems prevalent in different agro-ecologies. Synchronous supply of nutrients as per the demand of the crop should be supplied not only to offset the nutrient deficiencies in a cropping system but also to reduce the environmental pollution caused by applied fertilizers at the same time. Good fertility management should lower the risk of soil erosion by generating a more robust and vigorous crop growth that closes the canopy and covers the soil more quickly. With adequate and balanced fertilization, more biomass is produced which in turn leads to sustained productivity and better soil health. Continuously increasing use of fertilizers, pesticides, and other agrochemicals due to large-scale adoption of high yielding fertilizer-responsive crops and varieties has led to gradual build-up of nutrients and toxic chemicals in soil and groundwater as well as enhanced greenhouse gas emissions into the atmosphere. Second, overdrafting of groundwater from deeper and deeper layers for irrigation purposes has resulted in transporting underground geogenic pollutants to the soil surface. Considering fast contamination of groundwater, soils, and water bodies, it is critical to establish and execute a road map for mitigating environmental pollution so that sustainable development is maintained with least disturbance to the ecosystem. Adequate, balanced, and integrated use of fertilizers, manures, and agrochemicals should be encouraged. To regulate the use of fertilizers and agrochemicals by farmers, policies, and laws may be developed and implemented.

KEYWORDS

- **nitrogenous fertilizer**
- **water quality**
- **soil pollution**
- **groundwater pollution**
- **remediation**

REFERENCES

Aulakh, M. S.; Bijay-Singh. Nitrogen Losses and Fertilizer Nitrogen Use Efficiency in Irrigated Porous Soils. *Nutr. Cycl. Agroecosyst.* **1997,** *47*, 197–212.

Aulakh, M. S.; Garg, A. K.; Kabba, B. S. Phosphorus Accumulation and Leaching, and Residual Effects on Crop Yields from Long-Term Applications in Subtropics. *Soil Use Manage.* **2007,** *23*, 417–427.

Bajwa, M. S.; Bijay-Singh; Parminder-Singh. Nitrate Pollution of Groundwater Under Different Systems of Land Management in the Punjab. In: *Proceedings of 1st Agricultural Science Congress*; National Academy of Agricultural Sciences: New Delhi, 1993; pp 223–230.

Bijay-Singh; Sekhon, G. S. Nitrate Pollution of Groundwater from Farm Use of Nitrogen Fertilizers—A Review. *Agric. Environ* **1979,** *4*, 207–225.

Bijay-Singh; Tiwari, M. K.; Abrol, Y. P. Reactive Nitrogen in Agriculture, Industry and Environment in India. *Indian National Sci. Acad.* **2008**, 42.

Bijay-Singh; Yadvinder-Singh. Rate limiting Steps in Nitrogen Turnover: Ecosystems Based Nitrogen Needs and Actual Use. *Proc. Indian Natl. Sci. Acad.* **1993,** *B59*, 173–182.

Bokhtiar, S. M.; Sakurai, K. Effects of Organic Manure and Chemical Fertilizer on Soil Fertility and Productivity of Plant and Ratoon Crops of Sugarcane. *Arch. Agron. Soil Sci.* **2005,** *51*, 325–334.

Cameron, K. C.; Haynes, R. J. Retention and Movement of Nitrogen in Soils. In: Mineral Nitrogen in the Plant Soil System; Haynes, R. J., Ed.; Academic Press: Orelando, 1976; pp 166–241.

Chen, J.-H. The Combined Use of Chemical and Organic Fertilizers and/or Biofertilizer for Crop Growth and Soil Fertility. In *International Workshop on Sustained Management of the Soil-Rhizosphere System for Efficient Crop Production and Fertilizer Use*, 2006; pp 16–20.

Dinnes, D. L.; Karlen, D. L.; Janes, D. B.; Kaspar, T. C.; Hatfield, J. L.; Colvin, T. S.; Cambardella, C. A. Nitrogen Management Strategies to Reduce Nitrate Leaching in Tile-Drained Midwestern Soils. *Agron. J.* **2002,** *94*, 153–171.

Elrahman, S. H. A.; Mostafa, M. A. M. Applications of Nanotechnology in Agriculture: An Over-View. *Egyptian J. Soil Sci.* **2015,** *55*, 197–214.

El-Yazeid, A. A.; Abou-Aly, H. A.; Mady, M. A.; Moussa, S. A. M. Enhancing Growth, Productivity and Quality of Squash Plants Using Phosphate Dissolving Microorganisms (Bio Phosphor) Combined with Boron Foliar Spray. *Res. J. Agric. Biol. Sci.* **2007,** *3* (4), 274–286.

Freney, J. R. Strategies to Reduce Gaseous Emissions of Nitrogen from Irrigated Agriculture. *Nutr. Cycl. Agroecosyst.***1997,** *48*, 155–160.

Garg, A.; Shukla, P. R.; Kapshe, M. The Sectoral Trends of Multigas Emissions Inventory of India. *Atmos. Environ.* **2006,** *40*, 4608–4620.

Gupta, P. K.; Jha, A. K.; Koul, S.; Sharma, P.; Pradhan, V.; Gupta, V.; Sharma, C.; Nahar Singh. Methane and Nitrous Oxide Emission from Bovine Manure Management Practices in India. *Environ. Pollut.* **2007,** *146*, 219–224.

Hauck, R. D. Slow Release and Bio-Inhibitoramended Nitrogen Fertilizers. In: *Fertilizer Technology and Use*; Engelstad, O. P., Ed., 3rd ed.; SSSA: Madison, 1985; pp 293–322.

IPCC (Intergovernmental Panel on Climate Change) Climate Change. *The Scientific Basis. Contribution of Working Group I to the Third Assessment of the IPCC*; Cambridge University Press: Cambridge, 2001; p 881.

Kumar, M. D.; Shah, T. Groundwater Pollution and Contamination in India: The Emerging Challenge. http:// www.iwmi.cglar.org/wmi-tata/files/pdf/ground-pollute4_FULL_.pdf

Malhi, S. S.; Oliver, E.; Mayerle, G.; Kruger, G.; Gill, K. S. Improving Effectiveness of Seedrow-Placed Urea with Urease Inhibitor and Polymer Coating for Durum Wheat and Canola. *Com. Soil Sci. Plant Anal.* **2003,** *34*, 1709–1727.

Mikkelsen, R. Nanofertilizer and Nanotechnology: A Quick look. *Better Crops* **2008,** *102*, 18–19.

Pandiselvi, T.; Jeyajothiand, R.; Kandeshwari, M. Organic Nutrient Management a Way to Improve Soil Fertility and Sustainable Agriculture—A Review. *Int. J. Adv. Life Sci.* **2017,** *10* (2), 175–181.

Patra, D. D. Absorption of Nitrogen by Japanese Mint from Mineralization and immobilization of N by Soil Microbial Biomass. *Com. Soil Sci. Plant Anal.* **1997,** *28*, 1711–1719.

Patra, D. D.; Kiran, U.; Chand, S. Use of Urea Coated with Natural Products to Inhibit Urea Hydrolysis and Nitrification in Soil. *Biol. Fert. Soil* **2009,** *38*, 617–621.

Patra, D. D.; Sachdev, M. S.; Subbiah, B. V. ^{15}N Studies on the Transference of Legume Fixed N to Associated Cereals in Intercropping Systems. *Biol. Fert. Soil* **1986,** *2*, 167–171.

Prakash, O.; Alva, A. K.; Paramasivam, S. Use of the Urease Inhibitor N-(n-butyl)-Thiophosphoric Triamide Decreased Nitrogen Leaching from Urea in a Fine Sandy Soil. *Water Air Soil Pol.* **1999,** *116*, 587–595.

Saha, B. N.; Saha, S.; Poddar, P.; Tek Chand. Phosphate Fractions as Influenced by Long-Term Phosphorus Fertilization. *Indian J. Agric. Res.* **2013,** *47* (1), 59–65.

Santra, S. C.; Mallick, A.; Samal, A. C. Biofertilizer for Bioremediation in Book '*Recent Trends in Biofertilizers'*; Pati, B. R., Mandal, S. M., Eds., I K International Publishing House Pvt. Ltd., 2015; pp 205–234.

Saxena, S.; D'Souza, S. F. Heavy Metal Pollution Abatement Using Rock Phosphate Mineral. 2005. DOI: 10.1016/j.envint 2005.08.011

Shaviv, A. Advances In Controlled Release Fertilizers. *Adv. Agron.* **2000,** *71*, 1–49.

Singh, B.; Singh, Y.; Sekhon, G. S. Fertilizer-N Use Efficiency and Nitrate Pollution of Groundwater in Developing Countries, 1995. DOI: 10.1016/0169-7722(95)00067-4.

Singh, M. D.; Chirag, G.; Prakash, P. O.; Mohan, M. H.; Prakasha, V. G. Nano-Fertilizers Is a New Way to Increase Nutrients Use Efficiency in Crop Production. *Inter. J. Agric. Sci.* **2017,** *9* (7), 3831–3833.

Usman, M.; Madu, V. U.; Alkali, G. The Combined Use of Organic and Inorganic Fertilizers for Improving Maize Crop Productivity in Nigeria. *Int. J. Sci. Res. Pub.* **2015,** *5*, 1–7.

CHAPTER 9

Methods of Trace Analysis for Persistent Pollutants

CHELLAMUTHU SELVI[1], ESWAR DEEPTHI[2], MARIAPPAN PARAMASIVAM[1], KONAPPAN JAYALAKSHMI[3], SAKTHI AMBOTHI RATHNASAMY[4], and GURUSAMY DHANDAPANI[5]

[1]*Pesticide Toxicology Laboratory, Department of Agricultural Entomology, Tamil Nadu Agricultural University, Coimbatore, Tamil Nadu, India*

[2]*Department of Agriculture, Government of Tamil Nadu, Vellore, Tamil Nadu, India*

[3]*Department of Veterinary medicine, Veterinary College and Research Institute, Orathanadu, Thanjavur, Tamil Nadu, India*

[4]*Department of Plant Biotechnology, Tamil Nadu Agricultural University, Coimbatore, Tamil Nadu, India*

[5]*Department of Botany, Kongunadu Arts and Science College (Autonomous), Bharathiar University, Coimbatore, Tamil Nadu, India*

ABSTRACT

Persistent organic pollutants (POPs) are chemical substances that have drawn considerable international attention over the past several decades; as they possess the capacity to travel wide, their residues continue to persist for a long time in the environment and accumulate in ecosystems. The principal sources of POPs are agricultural practices, industries, and sewage wastes. The importance of monitoring the distribution of POPs in different food and environmental matrices using modern analytical techniques is highly

Environmental Contaminants: Impact, Assessment, and Remediation.
Pritam Ganguly, Jajati Mandal, M. Paramsivam, & Sandip Patra, (Eds.)

acknowledged. This chapter focuses on the analytical methodologies and protocols developed for analyzing POPs in different food and environmental matrices, with emphasis on advanced chromatographic techniques employed in the detection of POPs at trace levels (pg L^{-1} or pg kg^{-1}). The analytical procedures, starting from sampling up to final determination, are discussed, taking into account the latest advances in extraction, preconcentration, and instrumentation for quantification of POPs in various matrices.

9.1 INTRODUCTION

Persistent organic pollutants (POPs) are chemical compounds that have drawn considerable international attention over the past several decades, as they possess the capacity to travel wide, persist for a long term in the environment, and accumulate in ecosystems (Fitzgerald and Wikoff, 2014). The principal sources of POPs are agricultural practices, industries, and sewage wastes. The importance of monitoring the distribution of POPs in different food and environmental matrices using modern analytical techniques is highly acknowledged by the international scientific community and regulatory authorities, to ensure POPs-free environment. Exposure to POPs causes serious health issues in human beings and wild animals, and in turn, the environment (Li et al., 2006; Paramasivam et al., 2012, 2015a). POPs are pervasive, lipophilic, and persistent in nature, and are likely to accumulate in tissues of living organisms (Afful and Anim, 2010; Olisah et al., 2019a). Extensive use and residues of POPs have been reported in the African continent, whereas the use of such compounds has been banned in the developed world (Olisah et al., 2020). POPs, also classified as endocrine disruptors, reduce the efficacy of thyroid hormones and affect the human reproductive system (ATSDR, 2000), and possess carcinogenic properties (El-Shahawi et al., 2010). As a result, they have been known to cause diseases and abnormalities in fish, birds, and mammals (Malarvannan et al., 2020) for the past several decades.

POPs are present in measurable concentrations in almost all the products used in daily life, and hence, monitoring of organic contaminants in the environment has gained an importance in recent times. Development of protocols to monitor the presence of POPs in food and environmental matrices is essential for stringent implementation of food safety standards to ensure consumer protection, and also for quick removal and degradation of their metabolites using remediation techniques. This chapter focuses on the analytical methodologies and protocols developed for analyzing POPs in different food and environmental matrices, with emphasis on advanced chromatographic techniques employed in the detection of POPs at trace levels (pg L^{-1} or pg kg^{-1}). The analytical procedures, starting from sampling up to final determination, are discussed, taking into account the latest advances in extraction, preconcentration, and instrumentation for quantification of POPs in various matrices. For instance, recent progress in sample preparation methods significantly minimizes sample and solvent extraction volumes and even uses solventless analytical procedures in some cases.

9.2 CLASSIFICATION OF POPS

Initially, 12 POPs were identified and classified as "Dirty Dozen," as they were found to function as endocrine disruptors (UNEP, 2001). These include organochlorine pesticides (OCPs), industrial chemicals, unintended byproducts, and others (Olisah et al., 2019; UNEP, 2009; Paramasivam et al., 2012). POPs can be classified into major groups (Table 9.1), such as legacy POPs comprising 23 organochlorine pesticides (OCPs), 76 polychlorinated biphenyls (PCBs), 27 polybrominated diphenyl ethers (PBDEs), three hexabromocyclododecanes (HBCDs), and 13 perfluoroalkyl substances (PFASs), and emerging POPs which include polychlorinated naphthalenes (PCNs) and short-chain chlorinated paraffins (SCCPs) (Choo et al., 2020). Most recently, hexachlorobutadiene (HCBD) and pentachloroanisole (PCA) were included as POPs in Annex C and A of the Stockholm Convention, and they were restricted from production and use in China in 2017, after they were found to cause serious environmental hazards (Rauert et al., 2018). One of the main properties of POPs is halogenation of their aliphatic or aromatic rings, which results in low solubility in water, highly solubility in lipids, reluctance to chemical, and biological and photolytic degradation. In the case of nonhalogenated POPs, their stable chemical structures contribute to their persistence in the environment.

TABLE 9.1 Persistent Organic Pollutants—Stockholm Convention, 2009.

Category	Chemicals
Organochlorine pesticides	Aldrin, chlordane, dieldrin, DDT, endrin, heptachlor, hexachlorobenzene (HCB), mirex, and toxaphene
Industrial chemicals	Pentabromodiphenyl ether, octabromodiphenyl ether, α-hexachlorocyclohexane, β-hexachlorocyclohexane, perfluorooctanesulfonate, hexabromobiphenyl, chlordecone, lindane, and pentachlorobenzene
Unintended byproducts	Polychlorinated dibenzofurans (PCDF) and polychlorinated dibenzo-p-dioxins (PCDD)
Others	Hexachlorobutadiene (HCBD) and pentachloroanisole (PCA)

9.3 SOURCE OF POPS

Of the several POPs present in the environment, including industrial chemicals, pesticides, organic solvents, and detergents, water-soluble pesticides are a matter of great concern (Ollera et al., 2011). The key source of pesticide pollution is improper handling of pesticides at the farm level (Anderson and Yerraguntla, 2002). Pesticides enter into the environment from different sources such as runoff from agricultural fields, improper disposal of wastes, industrial effluents, and chemical spills, and through the process of biomagnification causing significant environmental contamination (Afful and Anim, 2010). Pesticide contamination (especially organochlorines) in the environment has been extensively reported in Asia, Africa, America, and Europe (Elibariki and Maguta, 2017; Wu et al., 2015; Barnhoorn et al., 2015; Lozowicka et al., 2016; Yang et al., 2015; Sajid and Alhooshanib, 2020). Exceedingly high concentration levels have been reported in various matrices including water, eggs, dairy products, meat, meat products, fruits, vegetables, honey, oil, fish, and soil (Guo et al., 2019; Kolani et al., 2016; Bempah et al., 2012; Kishimba et al., 2004). The main entry point of POPs in food is during the food preparation process, where the contaminants from the environment are synthetically introduced into raw food either intentionally or unintentionally (Guo et al., 2019). POPs are toxic in nature and tend to build up in food chains.

Polychlorinated biphenyls (PCBs) possess dielectric, flame-retardant, and coolant properties, and are widely synthesized for use as an insulating material in transformers, motor oil additives, lubricants in paints and inks, and plastics (Ricardo, 2018). Similarly, polybrominated diphenyl ethers (PDBE) and hexabromocyclododecanes (HBCDs) also are been employed as flame-retardants in plastics, electronic equipment, and wood (Alaee, 2003). Compounds like PCDF and PCDD are unintentionally produced from synthesis of other chemicals, solid waste incineration, bleaching of paper pulp, and production of PVC (Ricardo, 2018). The major sources of perfluoroalkyl acids (PFAAs) are automotive and aviation industries, paints and glues, biocides, household products such as wetting and cleaning agents, nonstick coating and packaging materials, and electronics (Sznajder-Katarzynska et al., 2019).

The emerging POPs such as polychlorinated naphthalenes (PCNs) are produced during the manufacturing of chlorine through chlorinated solvents, metallurgical processes (copper ore roasting), and disposal of PCN-containing products (Odabasi et al., 2017), and short-chain chlorinated paraffins (SCCPs) are used as coolants and lubricants in metallurgical industries.

The POPs are transmitted from their production sites into the wind and water and start accumulating over time in ecosystems, causing long-term negative effects.

9.4 METHODS OF ANALYSIS

Analysis of contaminants at trace levels in food and environmental matrices involves (Fig. 9.1): (i) collection of samples, (ii) isolation of analytes from matrix (or) solvent extraction, (iii) removal of interfering compounds, (iv) chromatographic separation, and (v) structural identification using spectroscopic techniques (Yang et al., 2020). POP analysis demands a very low detection limit, and thus, careful sample preparation is necessary to reduce the matrix effect. Methods of sample preparation for the analysis of POPs at low concentrations in food samples involve filtration, adjustment of pH, extraction, and cleanup and enrichment procedures (Dimpe and Nomngongo, 2016).

Steps	Procedures	
1	Sample collection	Food and environmental samples
	↓	
2	Sample homogenization	High volume blade homogenizer
	↓	
3	Extraction	Organic solvents
	↓	
4	Cleanup	Sorbents (PSA and GCB)
	↓	
5	Determination	GC, GC–MS, GC–MS/MS and LC–MS

FIGURE 9.1 Steps in the analysis of POPs from food and environmental matrices.

The selection of a solvent is a crucial step in the analysis method, as the selected solvent must increase the efficiency of the transfer of the target analyte to the solvent while also preserving the properties of microdrop during the process of extraction and being immiscible for efficient extraction of the analyte in the aqueous phase. Similarly, the viscosity of the organic solvent selected must be adequate for the creation of the microdrop and its retention on the microsyringe needle tip while doing the extraction. Over the past few years, the solid-phase extraction (SPE) and liquid-liquid extraction (LLE) methods have been improved leading to the development of quicker and versatile solid-phase and liquid-phase microextraction techniques for the determination of pesticides in food samples. Various extraction and cleanup methods employed for the measurement of POPs in different food and environmental matrices are presented in Tables 9.2 and 9.3.

9.4.1 SAMPLE EXTRACTION METHODS

9.4.1.1 LIQUID–LIQUID EXTRACTION (LLE)

In general, LLE method is an extraction method employed to detect POPs such as PCDD/Fs, PCBs, and OCPs in liquid/aqueous samples like water and milk (Ghidini et al., 2005; Chung and Chen, 2011). The LLE is based on the principle of separation and extraction of compounds in two immiscible liquids, such as water and an organic solvent, depending on their difference in solubilities. This process demands large quantities of organic solvent (Ochiai et al., 2011), and to reduce the consumption of solvent, a revised LLE protocol, the dispersive liquid–liquid microextraction method, has been developed (Wei et al., 2011). Dispersive liquid–liquid microextraction is a simple and low-cost extraction method commonly used for the extraction of compounds, namely PBDEs, PCBs, PAHs, and OCPs from water samples (Zgoła-Grzeskowiak and Grzeskowiak, 2011). Russo et al. (2019) demonstrated the use of liquid–liquid microextraction to extract 15 multiclass pollutants from serum and saliva matrices. The limit of quantification for this method ranged between $2.28 \times 10^{-3}\,\mu g\,mL^{-1}$ for bisphenol A diglycidyl ether and $6.29\,\mu g\,mL^{-1}$ for diethylhexyl phthalate, while the limit of detection varied between $0.068 \times 10^{-3}\,\mu g\,mL^{-1}$ (bisphenol A diglycidyl ether) and $1.031\,\mu g\,mL^{-1}$ (diethylhexyl phthalate).

9.4.1.2 SOLID-PHASE EXTRACTION (SPE)

A second method called solid-phase extraction (SPE) is another classic and commonly used sample preparation method, which works based on "the transfer of target analytes in a liquid, gas, or supercritical fluid matrix to a solid sorbent" (Poole et al., 2016; Paramasivam et al., 2015b). This separation method can be implemented manually or in automatic mode by coupling of SPE with an instrument (chromatography or capillary electrophoresis).

TABLE 9.2 Extraction Techniques and Analytical Methods for Determination of POPs in Environmental Samples.

Type of matrix	Name of the pesticide compounds	Extraction solvent	Extraction and cleanup method	Instrument used for analysis	LOD/LOQ	Recovery and RSD	References
Particulate matter ($PM_{2.5}$)	Hexachlorobutadiene (HCBD), pentachloroanisole (PCA), HCB, PeCB, 1,2,4,5-tetrachloro benzene (TeCB), and 1,2,4-trichlorobenzene (TCB)	–	Ultrasonic solvent extraction and silica gel purification	Gas chromatography-tandem mass spectrometry (GC–MS/MS)	LODs: 1.14×10^{-4} to 2.74×10^{-4} pg m^{-3} LOQs: 3.80×10^{-4} to 9.14×10^{-4} pg m^{-3}	–	Chen et al. (2020)
Soil	α-hexachlorobenzene (BHC), β-BHC, γ-BHC, heptachlor, aldrin, heptachlor epoxide (isomer B), 4,4'DDE, dieldrin, endrin, o,p'-DDD, p,p'-DDD, 2,4'-DDT, 4,4'-DDT, α-endosulfan and β-endosulfan	Hexane: acetone (1:1) mixture	Elution through florisil column (60 cm × 22 mm i.d.) with hexane: acetone (1:1)	GC-ECD	–	–	Gopalan and Chenicherry (2018)
Sediments	Pentachlorobenzene, hexachlorocyclohexanes (α, β, and γ), hexachlorobenzene, DDE/DDD/DDT (o,p' and p,p' isomers) and PCBs (no. 28, 52, 101, 118, 138, 153, and 180)	Dichloromethane		GC–MS	LODs: 0.1 to 0.7 ng mL^{-1}	–	Garcıa-Alonso et al. (2019)
Sediments	12 OCPs (α-HCH, β-HCH, lindane, heptachlor, aldrin, dieldrin, endrin, endosulfan, p,p'-dichloro diphenyldichloroethane (p,p'-DDD), p,p'-DDT, endrin aldehyde and methoxychlor) and eight PCB congeners (2-dichlorobiphenyl (CB-1), 2,3-dichlorobiphenyl (CB-5), 2,4,5-trichlorobiphenyl (CB-29), 2,2',4,4'-tetrachlorobiphenyl (CB-47), 2,2',3',4,6-pentachlorbiphenyl (CB-98), 2,2',4,4',5,6'-hexachlorobiphenyl (CB-154), 2,2',3,3',4,4',6-heptachlorobiphenyl (CB-171), 2,2',3,3', 4,5',6,6'-octachlorobiphenyl (CB-200))	Toluene	Liquid-phase microextraction (LPME)	GC–MS	Method detection limit for each POP analyte ranged from 0.07 to 0.70 ng g^{-1}	Recovery: OCPs: 73 to 111% PCBs: 86 to 110%	Basheer et al. (2005)

TABLE 9.2 *(Continued)*

Type of matrix	Name of the pesticide compounds	Extraction solvent	Extraction and cleanup method	Instrument used for analysis	LOD/LOQ	Recovery and RSD	References
Sediments and water	HCB, α-BHC, β-BHC, γ-BHC, heptachlor, aldrin, trans-chlordane, cis-chlordane, α-endosulfan, p,p'-DDE, dieldrin, o,p'-DDD, endrin, p,p'-DDD, β-endosulfan, p,p'-DDT, methoxychlor (I. S.)	Methanol/acetone mixture ratio 2:3	Supercritical fluid extraction (SFE) and liquid/liquid extraction (LLE)	GC-ECD	–	Recovery: SFE: 98.17 to 134.72% LLE: 84.82 to 1102.83%	Okoya et al. (2013)
Surface water	11 OCPs (α-HCH, β-HCH, γ-HCH, δ-HCH, p,p'-DDT, o,p'-DDT, p,p'-DDD, p,p'-DDE, aldrin, heptachlor, and endrin)	Ethyl acetate, methanol	SPE	GC-Ni-ECD	LOD: 0.04 to 0.3 ng L^{-1}	Recovery: 72.4 to 105%	Liu et al. (2020)
Marine surface waters	OCPs (aldrin, α-HCH, β-HCH, lindane, δ-HCH, dieldrin, α-endosulfan, β-endosulfan, endosulfan sulfate, endrin, endrin aldehyde, heptachlor, heptachlor epoxide (B isomer), methoxychlor, p,p'-DDD, p,p'-DDE, p,p'-DDT, cis-chlordane, transchlordane, and endrin ketone)	Ethylene-vinyl acetate	Coupled of ultrasound-assisted solvent extraction (UASE) and headspace solid-phase microextraction (HS-SPME) as extraction method	Gas chromatography–tandem mass spectrometry (GC–MS/MS)	LODs: 0.01 ng for hexachlorocyclohexane and 0.27 ng for endrin aldehyde	Recovery: >61% RSD: <19%	Cardenas-Soraca et al. (2019)
Water and sediments	α-BHC, HCB, β-BHC, δ-BHC, heptachlor, aldrin, y-chlordane, o,p'-DDE, endosulfan I, dieldrin, o,p'-DDD, endrin, p,p'-DDD, o,p'-DDT, p,p'-DDT, and pentachloronitrobenzene	Dichloromethane	Water- LLE sediments- Soxhlet extraction (SE) and microwave-assisted extraction (MAE)	Gas chromatography-electron-capture detection (GC-ECD)	LOD: Water: 5.5 to 20.6 ng L^{-1} Soil: 0.6 and 2.1 ng g^{-1}	Recovery: 71.03 (dieldrin) to 101.25% In sediments: Soxhlet extraction Recovery: 88.22 to 109.63% (β-BHC) MAE: Recovery: 74.11 to 97.50%	Fatoki and Awofolu (2003)

TABLE 9.2 *(Continued)*

Type of matrix	Name of the pesticide compounds	Extraction solvent	Extraction and cleanup method	Instrument used for analysis	LOD/LOQ	Recovery and RSD	References
Water samples	18 PCBs (PCB-28, PCB-52, PCB-101,PCB-81, PCB-77, PCB-123, PCB-118, PCB-153, PCB-114, PCB-105, PCB-138, PCB-126, PCB-167, PCB-156, PCB-180, PCB-157, PCB-169, and PCB-189 in IUPAC) and 20 organochlorine pesticides (a-BHC, g-BHC, b-BHC, heptachlor, d-BHC, aldrin, heptachlor epoxide, g-chlordane, a-chlordane, endosulfan l, p,p'-DDE, dieldrin, endrin, p,p'-DDD, endosulfan ll, p,p'-DDT, endrin aldehyde, endosulfan sulfate, methoxychlor, and endrin ketone)	–	Stir bar sorptive extraction (SBSE)	SBSE coupled with thermal desorption GC–MS (TD-GC–MS)	LOD: 0.12 to 2.07 ng L^{-1}	Recovery: 64.7 to 111.0% RSD: 0.9 to 17.6%	Feng et al. (2017)
Soil	α-HCH, β-HCH, γ-HCH, δ-HCH, p,p'-DDE, p,p'-DDD, o,p'-DDT, and p,p'-DDT	Acetonitrile	Microwave-assisted magnetic solid-phase extraction (MAE-MSPE)	GC-ECD	LOD: 0.15 to 0.28 ng g^{-1}	Recovery: 71.2 to 102.4% RSD: <10.0%	He et al. (2018)
Tap water, wastewater, soil, tea, and tomato matrices	17 OCPs (α-BHC, γ-BHC, β-BHC, heptachlor, aldrin, heptachlorepoxide, γ-chlordane, α-chlordane, α-endosulfan, p,p'-DDE, dieldrin, endrin, β-endosulfan, p,p'-DDD, endrin aldehyde, p,p'-DDT, and methoxichlor)	n-heptane and toluene	Dual-hollow fiber microextraction (dual-HFμE	LVI-GC–MS(SIM)	LOD: 0.01 to 0.02 μg L^{-1}	Recovery:40.6 to 101.7% RSD: < 16%	Ide and Nogueira (2019)
Environmental and biological matrices (raw milk, soil, underground water, and forage grass)	α-BHC, β-BHC, γ-BHC, δ-BHC, p,p'-DDE, p,p'-DDD, o,p'-DDT, p,p'-DDT	Chloroform-eluted solvent	Hollow fiber–stir bar sorptive extraction (HF-SBSE)	GC–MS	LODs: 0.00013 to 0.0030 μg mL^{-1}	Recovery: 69.0 to 114.1% RSD: <10.2%	Li et al. (2018)

TABLE 9.2 *(Continued)*

Type of matrix	Name of the pesticide compounds	Extraction solvent	Extraction and cleanup method	Instrument used for analysis	LOD/LOQ	Recovery and RSD	References
Sewage sludge soil and water	Hexachlorobenzene, aldrin, dieldrin, endrin, dichlorodiphenyltrichloroethane (DDT), lindane, cis-chlordane, trans-chlordane, heptachlor, and mirex	Acetonitrile	Solid–liquid extraction with low temperature purification in sewage sludge and soil samples Liquid–liquid extraction with low temperature purification for water	GC–MS-SIM	LOQ: Sludge: 2–40 mg kg^{-1}, soil: 1–6 mg kg^{-1}, and water: 0.5 mg L^{-1}	Recovery: 70 to 115% RSD: <13%	Tayane et al. (2018)
Soil	DDT	Hexane	Ultrasonic probe extraction	GC–MS	LOD: 3.5 μg kg^{-1} LOQ: 10 μg kg^{-1}	–	Zhao et al. (2020)
Drinking water	7 OCPs (aldrin, DDT, dieldrin, endrin, endosulfan, heptachlor, and methoxychlor) and 4 pyrethroids (λ-cyhalothrin, cypermethrin, deltamethrin, and permethrin)	Toluene and n-hexane	Single-drop microextraction (SDME)	GC-ECD	LOD: 0.003–0.6 mg L^{-1}	Recovery: 71 to 107%	Carlos et al. (2013)
Surface water, sediments, fish, and aquatic weeds	17 OCP (α, β, γ, δ-BHC, heptachlor, heptachlor epoxide, aldrin, dieldrin, endrin, endrin aldehyde, endosulfan I, endosulfan II, endosulfan sulfate, p,p'-DDE, p,p'-DDT, p,p'-DDD, and methoxychlor (M-8270-14-ASL), and three isomers of α,β,γ-cypermethrin (P-225N) of synthetic pyrethroids)	–	QuEChERS kit	GC–MS	–	–	Arisekar et al. (2018)

TABLE 9.2 *(Continued)*

Type of matrix	Name of the pesticide compounds	Extraction solvent	Extraction and cleanup method	Instrument used for analysis	LOD/LOQ	Recovery and RSD	References
Airborne particles	PCBs (PCB28, 52, 101, 138, 153 and 180), 16-EPA PAHs (naphthalene, acenaphthalene, acenaphthene, fluorene, phenanthrene, anthracene, fluoranthene, pyrene, benzo[a]anthracene, chrysene, benzo[b]flouranthene, benzo[k]fluoranthene, benzo[a]pyrene, dibenzo[a,h]anthracene, benzo[g,h,I] perylene, indeno[1,2,3-cd]pyrene), OCPs compounds, their metabolites and isomers (a-BHC, b-BHC, c-BHC, d-BHC, heptachlor, heptachlor epoxide, dieldrin, aldrin, endrin, endrin aldehyde, endrin ketone, endosulfan I, endosulfan II, endosulfan sulfate, p,p9-DDE, p,p9-DDD, p,p9-DDT, and methoxychlor) were from Accustandard. 1,2,3,4-tetrachloronaphthalene, chrysene-d12, phenanthrene-d10, PCB14, PCB65, PCB166 and a mixture of tetrachlorom-xylene and PCB209 standards	Mixture of n-hexane: petroleum ether (1:1)	Ultrasonic solvent extraction	GC/MSD for PAHs and PCBs ; GC/µ-ECD for OCPs determination	Recovery: 91–96% RSD: 14% for PCBs, for PAHs recoveries (except for naphthalene, 31%) were between 66–90% with RSD 8% OCPs: recoveries were between 67 and 120% with RSD 17%.	–	Aydin et al. (2007)
Arctic ice	48 compounds (polycyclic aromatic hydrocarbons, brominated diphenyl ethers, chlorinated biphenyls, and organochlorinated pesticides)	–	Stir bar sorptive extraction (SBSE)	GC–MS	LODs: organohalogenated compounds: 0.1 to 99 pg L^{-1} PAHs: 102 to 891 pg L^{-1}	–	Lacorte et al. (2009)
Pond, river, and rainwater	PCB-3,84.5,98.5,86.0,101.6	–	SPME fiber based on hollow carbon nanobubbles	GC–MS	LODs: 0.0017 to 0.0042 ng L^{-1}	Recovery: 84.5 to 117.1%	Xu et al. (2020)

TABLE 9.2 *(Continued)*

Type of matrix	Name of the pesticide compounds	Extraction solvent	Extraction and cleanup method	Instrument used for analysis	LOD/LOQ	Recovery and RSD	References
Produced water from oil-based industry	PAHs	–	Headspace solid-phase microextraction technique (HS-SPME)	GC–MS	LOD: 0.2 to 41.4 ng L^{-1}	Recovery: 67.65 to 113.10% RSD: <8.39%	Vasconcelos et al. (2020)
Ambient air pollution	103 pesticides, 17 polycyclic aromatic hydrocarbons (PAHs), and 22 polychlorinated biphenyls (PCBs)	–	Pressurized solvent extraction (PSE) followed by a solid-phase extraction (SPE) and solid-phase microextraction (SPME)	GC–MS/MS LC–MS/MS	LOD: 0.01 to 153.2 ng	Recoveries between 0.01 and 153.2 ng per sampler and between 0.03 and 459.2 ng $sampler^{-1}$ as well as the high recoveries ranged between 76 and 99%	Levy et al. (2020)
Polar snow and ice samples	16 PAHs	–	Solid-phase microextraction (SPME)	GC–MS	LOD 0.1 to1.2 ng L^{-1}	Recovery: 93.7 to 119.7% RSD: 15 to 22%	Arcoleo et al. (2020)

TABLE 9.3 Extraction Techniques and Analytical Methods for Determination of POPs in Food and Other Matrices.

Type of matrix	Name of the pesticide compounds	Extraction solvent	Extraction and cleanup method	Instrument used for analysis	LOD/LOQ	Recovery and RSD	References
Milk, egg, and meat	OCPs (α-HCH, β-HCH, γ-HCH, δ-HCH, dicofol, aldrin, p,p'-DDD, p,p'-DDE, p,p'-DDT, α-endosulfan, β-endosulfan, and endosulfan sulfate)	Acetonitrile	QuEChERS	GC-ECD GC–MS	–	Recovery: Milk: 72 to 108%, Egg: 74 to 101% Meat: 75 to 104.56% RSD: 0.28 to 10.05%	Selvi et al. (2012)
Shellfish and cephalopods	α-BHC, β-BHC, γ-BHC, δ-BHC, aldrin, dieldrin, p,p'-DDE, p,p'-DDD, o,p'-DDT, and p,p'-DDT	Acetonitrile	QuEChERS	Gas chromatograph equipped with a mass spectrometry (GC–MS/MS)	LOD: shellfish: 0.10 to 0.80 ng g^{-1} cephalopods: 0.21 to 0.77 ng g^{-1} LOQ: shellfish: 0.31 to 2.41 ng g^{-1} cephalopods:0.63 to 2.33 ng g^{-1}	Recovery: shellfish: 83.5 to 117.4%, cephalopods: 79.8 to 118.4%, RSD: Shellfish: 0.3 to 27.5%, Cephalopods: 1.2 to 27.9%	Hwang et al. (2020)
Breast milk	α-endosulfan, β-endosulfan, γ-HCH, β-HCH, aldrin dieldrin, HCB, p,p'-DDE, and p,p'-DDT	Hexane: acetone (2:1)	Florisil solid-phase extraction cleanup	GC-ECD	LOD: 0.01 mg kg^{-1} LOQ: 0.01 mg kg^{-1} for p,p'-DDE, p,p'-DDT, α-endosulfan and dieldrin, 0.03 mg kg^{-1} for HCB and γ-HCH and 0.04 mg kg^{-1} for β –HCH	Recovery: 53 to 109%	Burke et al. (2003)

TABLE 9.3 *(Continued)*

Type of matrix	Name of the pesticide compounds	Extraction solvent	Extraction and cleanup method	Instrument used for analysis	LOD/LOQ	Recovery and RSD	References
Placental tissue of the women	HCH and its isomers such as α-HCH, β-HCH, γ-HCH, and δ-HCH, metabolites of DDT (p,p'-DDE, p,p'-DDT, p,p'-DDD and antioxidant enzymes reduced glutathione (GSH)	Formic acid and n-hexane	–	GC-ECD, GC–MS	–	–	Anand et al. (2015)
Fatty biological tissues: fish muscle tissue, coral tissue, and adipose tissue from crocodiles	p,p'-DDE, p,p'-DDT, p,p'-DDD, endosulfan sulfate, α-endosulfan, β-endosulfan, aldrin, dieldrin, endrin, endrin aldehyde, endrin ketone, heptachlor, heptachlor epoxide, methoxychlor, α-HCH, β-HCH, δ-HCH, and γ-HCH	Acetonitrile	QuEChERS	Two-dimensional gas chromatography–time-of-flight mass spectrometry (GC x GC–TOF-MS)	LOD: 0.1 to 2.0 ng g^{-1}	Recovery: 69 to 102% RSD: <10%	Buah-Kwofie and Humphries (2019)
Human serum	OCPs (α-HCH, γ-HCH, o,p'-DDT, p,p'-DDT, o,p'-DDE, p,p'-DDE, o,p'-DDD, and p,p'-DDD)	Dichloromethane	SPE cartridge	Atmospheric pressure gas chromatography–tandem mass spectrometry (APGC–MS/MS)	LODs: 0.22 to 0.90 pg mL^{-1}	Recovery: 84.5 to 118.2% RSD: <20%	Fang et al. (2019)
Honey	All OCPs used: chlorthalonil, chlorpyrifos, icofol, o,p'-DDE, p,p'-DDE, o,p'-DDT, and p,p'-DDD	1,2-dibromoethane	A dispersive liquid–liquid microextraction (DLLME) method	GC/MS	LOD: 0.004 to 0.07 ng g^{-1} LOQ: 0.02 to 0.3 ng g^{-1}	Recovery: 91 to 100%	Mousavi et al. (2016)

TABLE 9.3 *(Continued)*

Type of matrix	Name of the pesticide compounds	Extraction solvent	Extraction and cleanup method	Instrument used for analysis	LOD/LOQ	Recovery and RSD	References
Meat	Aldrin, HCH (α, β, and γ) heptachlorepoxide (HCE) cis and trans isomers, p,p'-DDT, p,p'-DDE, p,p'-DDD, hexachlorobenzene (HCB), endosulfan α and β isomers, endosulfan sulfate, heptachlor, dieldrin, and mirex	Acetonitrile and n-hexane	Accelerated solvent extraction technique	GC-ECD	LOQ: 0.025 $\mu g\ g^{-1}$ fat, except for γ-HCH 0.005 $\mu g\ g^{-1}$	Recovery: 55.5 to 89.9%	Dimitrova et al. (2018)
Packed tea	α-lindane, β-lindane, δ-lindane, heptachlor, heptachlor epoxide, p,p′-DDE, endrin, endosulfan, p,p′-DDD, endosulfan sulfate, p,p′-DDT, endrin ketone, and methoxychlor	n-hexane	Ultrasound-assisted solvent extraction	GC–MS	LODs: 1.4 to 7.2 $ng\ g^{-1}$	Recovery: 86.1 to 100.3% RSD: <16%	Sajid And Alhooshani (2020)
Mussel	eight PCBs, 23 OCPs, and 35 PAHs, including alkyl-PAHs		Ultrasound-assisted cloudy extraction	GC–MS/MS	LODs: for PCBs: 0.012 to 0.058 $\mu g\ kg^{-1}$ OCPs: 0.01 to 0.29 $\mu g\ kg^{-1}$ PAHs: 0.01 to 0.5 $\mu g\ kg^{-1}$	Recovery: PCBs: 101 to 104%, OCPs 99.6 to 106% PAHs: 101 to 110% RSD: PCBs: 0.677 to 2.69% OCPs: 1.14 to 6.60% PAHs: 0.694 to 7.46%	Lee et al. (2020)
Vegetables and fruits	α-BHC, lindane, δ-BHC, aldrin, heptachlor epoxide (B) endosulfan I, dieldrin, and endosulfan II	Acetonitrile	QuEChERS	GC–MS	–	–	Suleiman et al. (2020)

TABLE 9.3 *(Continued)*

Type of matrix	Name of the pesticide compounds	Extraction solvent	Extraction and cleanup method	Instrument used for analysis	LOD/LOQ	Recovery and RSD	References
Fish, shellfish, and muscle of terrestrial animals	15 polybrominated diphenyl ethers (PBDEs) (28, 47, 49, 66, 77, 85, 99, 100, 138, 153, 154, 183, 197, 206, and 209) and three hexabromocyclododecanes (HBCDs) (α, β, and γ)	Ethyl acetate	QuEChERS-like extraction	GC–MS/MS for PBDEs LC–MS/MS for HBCDs	LOQs: 10 pg g^{-1} except BDE 206 and 209 (100 pg g^{-1})	–	Tavoloni et al. (2020)

Abbreviations: Organochlorine pesticides (OCPs); benzene hexachloride (BHC, including α-BHC, β-BHC, γ-BHC, and δ-BHC); 1,1-dichloro-2,2-bis (p-chloro-phenyl)ethylene (p,p′-DDE); 1,1-dichloro-2,2-bis (p-chlorophenyl) ethane (p,p′-DDD); 1,1,1-trichloro-2-(ochlorophenyl)-2-(p-chlorophenyl) ethane (o,p'-DDT); 1,1,1-trichloro-2,2-bis (p-chlorophenyl) ethane (p,p′-DDT); hexabromocyclododecanes (HBCDs); polybrominated diphenyl ethers (PBDEs); polycyclic aromatic hydrocarbons (PAHs); polychlorinated biphenyl (PCBs); LOD, limit of detection; LOQ, limit of quantification; RSDs, relative standard deviations.

In comparison with LLE, SPE utilizes a lesser amount of organic solvent and there is no emulsion formation during the extraction process (Niu et al., 2018). SPE can be used to extract several organic compounds, like pesticides, pharmaceuticals, brominated flame-retardants, industrial additives (Nozal et al., 2005), and polycyclic aromatic hydrocarbons (PAHs) (Andrade-Eiroa et al., 2010), from environmental and food samples. Li et al. (2007) have reported that SPE exhibits high extraction efficiency of PAHs with quantification limits within the range of 0.022–0.67 μg L^{-1}.

9.4.1.2.1 Solid-Phase Microextraction (SPME)

Solid-phase microextraction (SPME) is an eco-friendly solvent-free extraction technique with enhanced sensitivity, reduced matrix effects, and shorter duration for analysis, as sampling, extraction, and enrichment are integrated into a single step (Paramasivam et al., 2015). In SPME, the extraction phase, a fiber coated with a thin polymer, is exposed to the sample and when the coated fiber comes in contact with the sample, adsorption of the analyte begins (Spietelun et al., 2013). Llomparta et al. (2019) demonstrated the use of SPME for the extraction of environmental pollutants, including POPs, pesticides, pharmaceuticals, and personal care products (PPCPs), in environmental matrices. Headspace solid-phase microextraction technique (HS-SPME), an improvement of SPME, is suitable for the extraction of naphthalene, a PAH compound in water of the oil-based industry, at 99.99 μg L^{-1} concentration level (Vasconcelos et al., 2020). Recently, novel SPME fiber consisting of hollow carbon nanobubbles (HCNBs) was established for the analysis of PCBs in water samples with limit of detection (LODs) ranging between 0.0017 and 0.0042 ng L^{-1} and recoveries of analytes ranged from 84.5 to 117.1%. The HCNBs-based SPME presented higher extraction percentage when compared to other commercial fibers (Xu et al., 2020). Arcoleo et al. (2020) proposed helical multiwalled carbon nanotubes (HMWCNTs) as a novel SPME coating as an alternative, and they were evaluated for ultratrace analysis of 16 PAHs in ice and snow samples exhibiting detection limits in the range of 0.1–1.2 ng L^{-1}.

9.4.1.2.2 QuEChERS

The QuEChERS (quick, easy, cheap, effective, rugged, and safe) is a sample extraction and cleaning method, involving use of specific sorbents for dispersive solid-phase extraction (d-SPE) (Selvi et al., 2012; Wei et al., 2015). The most commonly used sorbent is primary secondary amine (PSA) which is effective in removing polar organic acids and graphitized carbon black (GCB), which removes steroids and pigments (Wilkowska and Biziuk, 2011). This method has been widely applied for the detection and extraction of several POPs, including dioxins, perfluoroalkyl substances, polychlorinated biphenyls, and brominated flame-retardants in food and environmental matrices. The various advantages of the QuEChERS methods, when compared to SPE and LLE techniques, are lower solvent consumption, good recoveries, less sample preparation time, very few sample preparation steps, and its cost-effectiveness (Musarurwa et al., 2019). Rial-Berriel et al. (2020) validated a single-step microQuEChERS method requiring low sample volume (250 μl) and using acidified acetonitrile for the quantification of around 360 compounds in blood with recoveries >70%. The developed method was effective for quantification at the sub-ppb level and was used to monitor 3–25 pollutants in 36 barn owls and 112 common kestrels.

Selvi et al. (2012) adopted QuEChERS procedure for the quantification of organochlorine pesticides in meat, milk, and egg samples, where extract cleanup was done using dispersive-SPE with PSA as a sorbent. Covaciu et al. (2020) employed QuEChERS method for extraction of α-BHC, γ-BHC, β-BHC, δ-BHC, heptachlor, aldrin, endrin, endrin aldehyde, heptachlor epoxide, endosulfan I and II, endosulfan sulfate, 4,4'-DDE, 4,4'-DDT, 4,4'-DDD, and methoxychlor in milk. Similarly, Suleiman et al. (2020) employed QuEChERS method for the detection of heptachlor epoxide residues in onion (2.3 mg kg^{-1}) and spinach (2.01 mg kg^{-1}), and endosulfan II (1.43 mg kg^{-1}) in lettuce. Heptachlor epoxide residues detected were also very high in samples such as carrot (0.60 mg kg^{-1}), cabbage (0.72 mg kg^{-1}), and tomatoes (0.52 mg kg^{-1}). The levels of OCPs detected in the fruit and vegetable samples were observed to exceed maximum residue limits (MRLs) prescribed by the Codex (CAC, 2009). However, the widespread adoption of QuEChERS method in analysis of POP faces the challenges such as the difficulty in detecting POPs at trace levels, occurrence of coextractives in final extract, and the requirement of instruments with high resolution for multiresidue analysis (Kim et al., 2019).

9.4.1.3 SOLID–LIQUID EXTRACTION (SLE)

Solid–liquid extractions, both traditional methods such as maceration and percolation and most recent methods, are employed on the principles of diffusion and/or osmosis. This extraction process is performed by bringing the extractant solvents (organic and/or inorganic liquids) in contact with an insoluble solid matrix. The various types of the solid–liquid extraction processes are discussed below.

9.4.1.3.1 Soxhlet Extraction

The conventional Soxhlet extraction is a common solid–liquid extraction method used for the extraction of furans and dioxins, such as PCBs, from various matrices (Tang, 2013), and it is mainly effective for the extraction of organic pollutants in complex environmental matrices, like dust, sediment, and soil, as the Soxhlet ensures contact between the matrix and extraction solvent and improves the recovery of targets (Kunene and Mahlambi, 2020). However, this extraction process requires large amounts of solvent, is time-consuming, and cannot be used to extract thermolabile compounds, as the sample needs to be evaporated after extraction. Fatoki and Awofolu (2003) proved the use of Soxhlet extraction method for the extraction of OCP compounds like α-BHC, HCB, β-BHC, δ-BHC, heptachlor, aldrin, dieldrin, γ-chlordane, endosulfan I, 2,4,-DDE, 2,4'-DDD, 2,4'-DDT, and 4,4'-DD from sediment and water samples. The LOD and recovery of the analytical method were in the range of 5.5–20.6 ng L^{-1} and 88.22–109.63%, respectively.

9.4.1.3.2 Supercritical Fluid Extraction (SFE)

Supercritical fluid extraction (SFE) method includes the usage of gas above the critical point to extract analytes from the matrix. In this method, generally carbon dioxide (CO_2) gas is used as a supercritical fluid. The SFE method solves the problem of solvent disposal, as the extracting gas can be evaporated. This method has been employed by Gilbert-Lopez et al. (2009) and LeDoux (2011) to extract OCPs from egg, butter, meat, and oil products. The drawback of this extraction technique is cleanup steps are mandatory because along with target analytes, several interfering organic compounds are also extracted.

9.4.1.3.3 Pressurized Liquid Extraction (PLE) or Accelerated Solvent Extraction (ASE)

The PLE or ASE technique is an improved version of SFE, which utilizes high temperature and pressure for achieving high extraction from sample matrices. The development of PLE is achieved by adding solvents to SFE as modifiers that mimic classical liquid extractions and avoid the use of extraction gas. This particular method is widely applied for extracting PCDFs, PCBs, PAHs, and PCDDs from fatty foods including fish, egg, and meat samples (Carabias-Martínez et al., 2005). For instance, Covaciu et al. (2020) used the ASE method for extraction of OCP compounds, such as α-BHC, γ-BHC, β-BHC, δ-BHC, heptachlor, heptachlor epoxide, aldrin, dieldrin, endrin, endrin aldehyde, endosulfan I and II, endosulfan sulfate, 4,4'-DDD, 4,4'-DDE, and methoxychlor, from soil and forage samples.

9.4.1.3.4 Microwave-Assisted Extraction (MAE or MASE)

The microwave-assisted extraction (MAE or MASE) technique employs microwaves for heating the solvent and thereby increases the penetration of the solvent into the sample matrix. MAE is a very effective method requiring lesser extraction time and low solvent usage with improved extraction yield. However, to achieve reduced matrix interference and enhanced sensitivity, further enrichment and purification steps are necessary. MAE using electromagnetic field has been adopted for the extraction of organic compounds from various solid matrices (Cui et al., 2017; Keisandokht et al., 2018) and OCPs from food (Chung and Chen, 2011). With the boom in research, the sample pretreatment methods are constantly revised to facilitate automation and devise an environment-friendly method. The new technique called microwave-assisted magnetic solid-phase extraction (MAE-MSPE) is found to be a better alternative to this technique with advantages such as high sensitivity and little matrix interference (Wilkowska and Biziuk, 2010). Studies carried out by He et al. (2018) demonstrated the use of MAE-MSPE for the extraction of OCP compounds, such as p,p'-DDD, p,p'-DDT p,p'-DDE, α-HCH, β-HCH, and γ-HCH, from soil samples with LOD of 0.15–0.28 ng g^{-1}.

9.4.1.3.5 Ultrasonic-Assisted Extraction (UAE)

UAE is a very simple and inexpensive extraction procedure utilizing ultrasound waves to reduce the operating temperature. UAE maximizes the contact between the sample matrix and solvent, thereby saving analysis time and reducing solvent consumption, hence making it a cost-effective and efficient extraction method (Pena et al., 2006). Tadeo et al. (2010) employed

UAE for the determination of OCPs and PAHs in food, and this method can be used for the simultaneous extraction of OCPs, PAHs, and PCBs in airborne particles (Aydin et al., 2007).

A modified version of UAE technique called ultrasound-assisted solvent extraction (UASE) utilizes solvents as the extracting phase. Sajid and Alhooshani (2020) used UASE method for the extraction of OCPs from tea samples. However, UASE combined with headspace SPME (HS-SPME) can be used to achieve higher efficiency and extraction selectivity when compared to using each method alone. Similarly, a novel strategy based on ethylene vinyl acetate (EVA) passive samplers has been devised to detect around 20 ultratrace OCPs concentrations (pg L^{-1}) in marine water under equilibrium conditions (Cardenas-Soraca et al., 2019). The EVA sampling devices have been highly implemented due to its cost-effectiveness and ease of handling. In addition, from this study it was concluded that EVA sampling devices coupled with UASE-HS-SPME and GC–MS/MS offering high selectivity and sensitivity can be used to monitor pollutants at ultratrace levels in marine surface waters. Ultrasound-assisted extraction microscale cell (UAE-MSC) was used to validate and determine the concentration of OCPs and PBDEs in air particulate matter and the recoveries for OCPs and PBDEs ranged between 65.5 to 107.5 and 79.1 to 105.2%, respectively (Beristain-Montiel, 2016).

9.4.1.4 OTHER EXTRACTION METHODS

An extraction method integrating SPE, SPME, and PSE was employed for the ultratrace level determination of pesticides, PAHs, and PCBs in ambient air. This method employed PSE followed by solid-phase extraction (SPE), a preconcentration step based on SPME, and finally, GC/LC coupled to MS or MS/MS for analysis (Levy et al., 2020). In single-drop microextraction (SDME) technique, a microdrop of a water-immiscible solvent is suspended at the tip of a microsyringe needle immersed in an aqueous solvent, which on agitation transfers the analytes from aqueous solution to the organic phase, following which the microdrop is retracted back into the microsyringe and transferred for chromatographic analysis (Carlos et al., 2013). This technique is gaining attention since it integrates extraction and injection steps, drastically reducing the consumption of organic solvent and providing a high concentration factor. Switchable polarity solvent extraction is a new method which employs protonated (solid carbon dioxide or CO_2 gas) low-polarity amines (Fırat et al., 2018) as extraction solvent, which are later deprotonated to extract analytes into separated amine phase (Memon et al., 2017). This method is a green extraction method that utilizes environment-friendly solvents, and the polarity solvent can be easily reverted back into the original nonpolar amine by removing carbon dioxide from the solvent (Mercer and Jessop, 2010). Chormey et al. (2018) successfully applied switchable solvent microextraction for the determination of contaminants such as hormones and pesticides.

9.4.1.5 ULTRATRACE LEVELS METHODS

9.4.1.5.1 Hollow Fiber-Stir Bar Sorptive Extraction (HF-SBSE)

This technique is an amalgamation of SBSE and hollow fiber membrane extraction, and the microporous hollow fiber acts as the carrier and filter. A thin stainless steel wire and HF loaded with microspheres powder act as the magnetic stirrer. OCPs can be extracted with ZrO_2/SiO_2 composite microspheres fabricated by the sol–gel method, which serve as sorbent. Li et al. (2018) adopted HF-SBSE method to determine the movement of OCPs from environment to biological matrices as a crude method by excluding sample cleanup step. Lacorte et al. (2009) optimized stir bar sorptive extraction (SBSE) followed by thermal desorption and GC–MS for the ultratrace analysis of 48 POPs (PAHs, chlorinated biphenyls, and OCPs) at low pg L^{-1} level in arctic ice.

9.4.1.5.2 Hollow Fiber Microextraction (HFμe)

It is a static-based microextraction technique developed by combining modern and miniaturized sample enrichment approaches (Ide and Nogueira, 2018). In this approach, *polypropylene membranes* are loaded with organic solvents along with a Teflon magnetic stirring bar and placed into the sample for the microextraction. Following this, the back-extraction stage is executed in the single liquid desorption (LD) step, where the analytical devices are placed into glass vial inserts with 100 μL of the same organic solvent and transferred for instrumentation. The dual-HFμE technique proved to be safe, cheap, and amenable for the analysis using recent instrumentation technique for routine sample analysis (Ratola et al., 2008). Ide and Nogueira (2019) demonstrated the use of the dual-HFμE technique for extraction of 17 OCPs in soil, tea, tomato, tap water, and wastewater.

9.4.2 SAMPLE CLEANUP

The cleanup process involves preparing the extract for instrument analysis by removing the interfering substances. Recently, several automated cleanup systems have been developed to accelerate the cumbersome cleaning process; conversely, these automated systems are a challenging task for ordinary users who cannot afford due to high investment cost (Kedikoglou et al., 2018). A few widely applied cleanup methods for the detection of POPs in food and environmental matrices are gel permeation chromatography, liquid–liquid portioning, solid-phase extraction, and solid-phase microextraction (Le Doux, 2011).

9.4.3 INSTRUMENTAL ANALYSIS

9.4.3.1 SEPARATION AND DETECTION

There is a need for precise analytical procedures and careful execution, as POPs need to be detected at ultratrace levels such as parts per trillion (ppt) to parts per million (ppm). The measurement of POPs at trace levels helps monitoring the distribution and persistence of the pollutants in various food and environmental matrices. Cardenas-Soraca et al. (2019) have stated the need to develop low-cost and reliable methods with improved analytical sensitivity for the detection of ultratrace concentrations of POPs. However, the complexity of the samples analyzed, the low concentrations of pollutants, and the lack of automated processes with less solvent usage pose a challenge.

Chromatographic techniques when coupled to a detector can separate organic analytes from each other for accurate qualitative and quantitative determinations (Coskun, 2016). Generally, gas chromatography (GC) with an electron capture detector (ECD) or a mass spectrometer (MS) is the most commonly used chromatographic technique for POP analysis (Olisah et al., 2020; Milne and Morrow, 2009), especially compounds that vaporize due to high vapor pressure (Sharma et al., 2018). Various analytical instruments used for the measurement of presence of POPs level in a wide range of matrices are presented in Tables 9.2 and 9.3.

9.4.3.2 GAS CHROMATOGRAPHY (GC)

Gas chromatography, a commonly used separation technique, is mainly based on the boiling point of the analytes and their interaction with the stationary phase. POPs are mostly semivolatile in nature, and possess moderate polarities; hence, gas chromatography with mass spectrometry (GC–MS) is the most suitable technique. Zhang et al. (2019) and Tölgyessy et al. (2016) employed gas chromatography coupled with ECD to detect POPs in air and soil. However, poly- and perfluorinated alkylated substances (PFASs) related chemicals can be more efficiently measured using LC–MS/MS method in food (Guo et al., 2019).

9.4.3.3 GAS CHROMATOGRAPHY–MASS SPECTROMETRY (GC–MS)

Gas chromatography with mass spectrometry (GC–MS) allows detection of selective analytes and characterization of unknown compounds based on different mass to charge ratios (Lynch, 2017). The various types of GC–MS based on ionization source are: (i) electron ionization (EI) and (ii) chemical ionization (CI). Negative chemical ionization (NCI) allows detection of halogenated hydrocarbons at lower instrumental detection limits (IDLs) than electronic ionization (Fulara and Czaplicka, 2012; Gonzalez-Gago et al., 2015). Atmospheric pressure chemical ionization (APCI) is a recently developed low-energy ionization technique with enhanced sensitivity (Geng et al., 2016; Tokumura et al., 2018) for detection of PAHs and their nitroderivatives (NPAHs) (Zhang et al., 2019a), and brominated/chlorinated compounds (Portoles et al., 2015). Single quadrupole mass spectrometry (MS) is commonly used mass spectrometry technique but has the disadvantage of matrix interferences, which makes triple quadrupole mass spectrometry (MS/MS) a preferred alternative (Fulara and Czaplicka, 2012). However, studies on efficient instrumental methods to detect target analytes are limited, and it is mandatory to develop an ultrasensitive method for the trace analysis of analytes in food and environmental samples (Zhang et al., 2018; Portoles et al., 2016; Geng et al., 2017).

9.4.3.4 GC-TANDEM MASS SPECTROSCOPY

Tandem mass spectroscopy (MS/MS) is a recent and efficient analytical technique for analysis of biological samples with advantages like high sensitivity and low inference (Lee et al., 2020). Chen et al. (2020) used GC coupled with electron ionization-triple quadrupole mass spectrometry for the trace analysis of HCBD and PCA in $PM_{2.5}$ in the atmospheric environment in Taiyuan between November 2017 and October 2018. The mean concentration of $PM_{2.5}$ was 165 and 65.9 μg m^{-3} during the heating and nonheating season, respectively. Zabielski et al. (2013) established the GC–MS/MS to determine compounds at trace and ultratrace levels with high selectivity and specificity. However, its relatively high cost remains a deterrent.

Atmospheric pressure gas chromatography (APGC), a novel ionization source, based on using a reagent gas as a protonation source to achieve abundant ions, has been recently developed, but the application of APGC as an ion source is rare (Organtini et al., 2015). The potential of the APGC coupled with tandem mass spectrometry (APGC-MS/MS) as an alternative to GC–MS coupled with an EI source needs to be further explored. Fang et al. (2019) used APGC-MS/MS method for the determination of OCPs in serum samples that provided high sensitivity.

9.4.3.5 LIQUID CHROMATOGRAPHY–MASS SPECTROMETRY (LC–MS)

Liquid chromatography coupled with mass spectrometry (LC–MS) has been employed by Barzen-Hanson et al. (2017) and Janda et al. (2019) for the identification and determination of PFASs in water samples and by Yu et al. (2018) to determine OCPs, PCBs, and PFASs in airborne particulate matters. Russo et al. (2019) adopted liquid–liquid microextraction followed by liquid chromatography (LC) with ultraviolet tandem fluorescence detection to determine 15 multiclass pollutants in serum and saliva. Recently, Bugsel and Zwiener (2020) used LC-high-resolution mass spectrometry (LC-HRMS) for identification of unknown contaminants in soil.

9.5 CONCLUSIONS

The chapter briefly summarizes the POPs, their classification, and sources, with a focus on the need for their analysis and quantification. A range of conventional and other modern analytical sample preparation techniques, more sensitive chromatographic techniques for the ultratrace level determination of POPs, and their secondary and tertiary metabolites in different matrices are presented and discussed for their deployment in POPs quantification.

KEYWORDS

- **persistent organic pollutants**
- **analytical methods**
- **trace analysis**

REFERENCES

Afful, S.; Anim, A. K. Spectrum of Organochlorine Pesticide Residues in Fish Samples from the Densu Basin. *Res. J. Environ. Earth Sci.* **2010,** *2*, 133–138.

Alaee, M.; Arias, P.; Sjödin, A. An Overview of Commercially Used Brominated Flame Retardants, Their Applications, Their Use Patterns in Different Countries/Regions and Possible Modes of Release. *Environ. Inter.* **2003,** *29*, 683–689.

Anand, M.; Agarwal, P.; Singh, L.; Taneja, A. Persistent Organochlorine Pesticides and Oxidant/Antioxidant Status in the Placental Tissue of the Women with Full-Term and Pre-Term Deliveries. *Toxicol. Res.* **2015,** *4*, 326.

Anderson, S. W.; Yerraguntla, K. M. Remedial Alternatives for Agricultural Contamination. *Proc. Waste Res. Technol.* **2002,** 312.

Andrade-Eiroa, A.; Leroy, V.; Dagaut, P.; Bedjanian, Y. Determination of Polycyclic Aromatic Hydrocarbons in Kerosene and Bio-Kerosene Soot. *Chemosphere* **2010,** *78*, 1342–1349.

Arcoleo, A.; Bianchi, F.; Careri, M. Helical Multi-Walled Carbon Nanotube-Coated Fibers for Solid-Phase Microextraction Determination of Polycyclic Aromatic Hydrocarbons at Ultra-Trace Levels in Ice and Snow Samples. *J. Chromatogr. A.* **2020,** *1631*, 461589.

Arisekar, U.; Shakila, R. J.; Jeyasekaran, G.; Shalin, R.; Kumar, P.; Malani, A. H.; Rani, V. Accumulation of Organochlorine Pesticide Residues in Fish, Water, and Sediments in the Thamirabarani River System of Southern Peninsular India. *Environ. Nanotechnol. Monitor. Manag.* **2018**. https://doi.org/10.1016/j.enmm.2018.11.003.

Aydin, M. E.; Ozcan, S.; Tor, A. Ultrasonic Solvent Extraction of Persistent Organic Pollutants from Airborne Particles. *CLEAN (Weinh)*, **2007,** *35* (6), 660–668.

Basheer, C.; Obbard, J. P.; Lee, H. K. Analysis of Persistent Organic Pollutants in Marine Sediments Using a Novel Microwave Assisted Solvent Extraction and Liquid-Phase Microextraction Technique. *J. Chromatogr. A.* **2005,** *1068* (2), 221–228.

Barnhoorn I. E. J.; Van Dyk J. C.; Genthe B.; Harding W. R.; Wagenaar G. M.; Bornman M. S. Organochlorine Pesticide Levels in *Clarias gariepinus* from Polluted Freshwater Impoundments in South Africa and Associated Human Health Risks. *Chemosphere* **2015,** *120*, 391–397.

Barzen-Hanson, K. A.; Roberts, S. C.; Choyke, S.; Oetjen, K.; McAlees, A.; Riddell, N.; Discovery of 40 Classes of Per- and Polyfluoroalkyl Substances in Historical Aqueous Film-Forming Foams (AFFFs) and AFFF-Impacted Groundwater. *Environ. Sci. Technol.* **2017,** *51* (4), 2047–2057.

Bempah, C. K.; Buah-Kwofie, A.; Enimil, E.; Blewu, B.; Agyei-Martey, G. Residues of Organochlorine Pesticides in Vegetables Marketed in Greater Accra Region of Ghana. *Food Contr*. **2012,** *25*, 537–542.

Beristain-Montiel, E.; Villalobos-Pietrini, R.; Arias-Loaiza, G. E.; Gomez-Arroyo, S. L.; Amador-Munoz, O. An Innovative Ultrasound Assisted Extraction Micro-Scale Cell Combined with Gas Chromatography/Mass Spectrometry in Negative Chemical Ionization to Determine Persistent Organic Pollutants in Air Particulate Matter. *J. Chromatogr. A.* **2016,** *1477* (16), 100–107.

Buah-Kwofie A.; Humphries, M. S. Validation of a Modified QuEChERS Method for the Analysis of Organochlorine Pesticides in Fatty Biological Tissues Using Two-Dimensional Gas Chromatography. *J. Chromatogr. B*. **2019,** 110585–110592.

Bugsel, B.; Zwiener, C. LC-MS screening of poly- and perfluoroalkyl Substances in Contaminated Soil by Kendrick Mass Analysis. *Anal. Bioanal. Chem.* **2020,** *412*, 4797–4805.

Burke E. R.; Holden, A. J., Shaw, I. C. A Method to Determine Residue Levels of Persistent Organochlorine Pesticides in Human Milk from Indonesian Women. *Chemosphere*. **2003,** *50*, 529–535.

CAC (Codex Alimentarius Commission), *Codex Alimentarius.* Foods Derived from Modern Biotechnology; FAO, 2009.

Carabias-Martinez, R.; Rodriguez-Gonzalo, E.; Revilla-Ruiz, P.; Hernandez-Mendez, J. Pressurized Liquid Extraction in the Analysis of Food and Biological Samples. *J. Chromatogr. A.* **2005,** *1089*, 1–17.

Cardenas-Soraca, D. M.; Tuccab, F. I.; Mardones-Penaa, C. A.; Barra-Rios, R. O. Development of an Analytical Methodology for the Determination of Organochlorine Pesticides by Ethylene-Vinyl Acetate Passive Samplers in Marine Surface Waters Based on Ultrasound-Assisted Solvent Extraction Followed with Headspace Solid-Phase Micro Extraction and Gas Chromatography-Tandem Mass Spectrometry. *J. Chromatogr. A.* **2019,** *1605*, 360341.

Carlos, E. A.; Alves, R. D.; de Queiroz, M. E. L. R.; Neves, A. A. Simultaneous Determination of the Organochlorine and Pyrethroid Pesticides in Drinking Water by Single Drop Microextraction and Gas Chromatography. *J. Braz. Chem. Soc.* **2013,** *24* (8), 1217–1227.

Chen, Y. J.; Zhangb, Y.; Chena, Y.; Lua, Y.; Lic, R.; Dongc, C.; Qia, Z.; Liua G.; Chena Z. F.; Cai. Z. 2020. GC-MS/MS Analysis for Source Identification of Emerging POPs in PM2.5. *Ecotoxicol. Environ. Saf.* **2020,** *193*, 110368.

Choo, G.; Wang, W.; Cho, H. S.; Kim, K.; Park, K.; Oh, J. E. Legacy and Emerging Persistent Organic Pollutants in the Freshwater System: Relative Distribution, Contamination Trends, and Bioaccumulation. *Environ. Int*. **2020,** *135*, 105377.

Chormey, D. S.; Bodur, S.; Baskın, D.; Fırat, M.; Bakırdere, S. Accurate and Sensitive Determination of Selected Hormones, Endocrine Disruptors, and Pesticides by Gas Chromatography–Mass Spectrometry After the Multivariate Optimization of Switchable Solvent Liquid-Phase Microextraction. *J. Sep. Sci.* **2018,** *41* (14), 2895–2902.

Chung, S. W. C.; Chen, B. L. S. Determination of Organochlorine Pesticide Residues in Fatty Foods: A Critical Review on the Analytical Methods and Their Testing Capabilities. *J. Chromatogr. A.* **2011,** *1218*, 5555–5567.

Coskun, O. Separation Techniques: Chromatography. *Northern Clin. Istanbul*. **2016,** *3* (2), 156–160.

Covaciu, F. D.; Floare-Avram, V.; Magdas, D. A.; David, A. P.; Marincas, O. Distribution and Fate of Persistent Organochlorine Pesticides on the Soil-Forage-Milk Chain in Three Transylvanian Farms. *Anal. Lett.* **2020,** 1–15.

Cui, K.; Wen, J.; Zeng, F.; Zhou, X.; Li, S.; Zeng, Z. Determination of Organophosphate Ester Flame Retardants and Plasticizers in Soil Samples by Microwave-Assisted Extraction Coupled with Silica Gel/Alumina Multilayer Solid-Phase Extraction and Gas Chromatography-Mass Spectrometry. *Anal. Methods* **2017,** *9* (6), 986–993.

Dimitrova, R. T.; Stoykova, I. I.; Yankovska-Stefanova T. T.; Yaneva, S. A.; Stoyanchev, T. T. Development of Analytical Method for Determination of Organochlorine Pesticides Residues in Meat by GC-ECD. *Revue Med. Vet.* **2018,** *169* (4–6), 77–86.

Dimpe, K. M.; Nomngongo, P. N. Current Sample Preparation Methodologies for Analysis of Emerging Pollutants in Different Environmental Matrices. *Trends Analyt. Chem.* **2016,** *82*, 199–207.

Elibariki, R.; Maguta, M. M. Status of Pesticides Pollution in Tanzania—A Review. *Chemosphere* **2017,** *178*, 154–164.

El-Shahawi, M. S.; Hamza, A.; Bashammakh, A. S.; Al-Saggaf, W. T. An Overview on the Accumulation, Distribution, Transformations, Toxicity and Analytical Methods for the Monitoring of Persistent Organic Pollutants. *Talanta* **2010,** *80*, 1587–1597.

Fang, J.; Zhao1, H.; Zhang, Y.; He, Y.; Sun, Q.; Xu, S.; Cai, Z. Performance of Atmospheric Pressure Gas Chromatography-Tandem Mass Spectrometry for the Analysis of Organochlorine Pesticides in Human Serum. *Anal. Bioanal. Chem.* **2019,** *411*, 4185–4191.

Fatoki, O. S.; Awofolu, R. O. Methods for Selective Determination of Persistent Organochlorine Pesticide Residues in Water and Sediments by Capillary Gas Chromatography and Electron-Capture Detection. *J. Chromatogr. A.* **2003,** *983*, 225–236.

Feng, L.; Zhang, S.; Zhu, G.; Li, M.; Liu, J. Determination of Trace Polychlorinated Biphenyls and Organochlorine Pesticides in Water Samples Through Large-Volume Stir Bar Sorptive Extraction Method with Thermal Desorption Gas Chromatography. *J. Sep. Sci.* **2017,** *40* (23), 4583–4590.

Fırat, M.; Bodur, S.; Tışlı, B.; Ozlu, C.; Chormey, D. S.; Turak, F.; Bakırdere, S. Vortex-Assisted Switchable Liquid-Liquid Microextraction for the Preconcentration of Cadmium in Environmental Samples Prior to Its Determination with Flame Atomic Absorption Spectrometry. *Environ. Monit. Assess.* **2018,** *190* (7), 393.

Fitzgerald, L.; Wikoff, D. S. Persistent Organic Pollutants. Reference Module in Biomedical Sciences. In: *Encyclopedia of Toxicology*, 3rd ed.; Elsevier, 2014; pp 820–825.

Fulara, I.; Czaplicka, M. Methods for Determination of Polybrominated Diphenyl Ethers in Environmental Samples–Review. *J. Sep. Sci.* **2012,** *35*, 2075–2087.

Garcıa-Alonso, S.; Gomez-Lopez, L. V.; Perez-Pastor, R. M. Optimizing a Simple Procedure to Determine Organochlorine Compounds in Sediment Samples: Practical Considerations. *Anal. Methods* **2019,** *11*, 5810.

Geng, D.; Jogsten, I. E.; Dunstan, J.; Hagberg, J.; Wang, T.; Ruzzin, J.; Rabasa-Lhoret, R.; van Bavel, B. Gas Chromatography/Atmospheric Pressure Chemical Ionization/Mass Spectrometry for the Analysis of Organochlorine Pesticides and Polychlorinated Biphenyls in Human Serum. *J. Chromatogr. A*. **2016,** *1453*, 88–98.

Geng, D.; Kukucka, P.; Jogsten, I. E. Analysis of Brominated Flame Retardants and Their Derivatives by Atmospheric Pressure Chemical Ionization Using Gas Chromatography Coupled to Tandem Quadrupole Mass Spectrometry. *Talanta* **2017,** *162*, 618–624.

Ghidini, S.; Zanardi, E.; Battaglia, A.; Varisco, G.; Ferretti, E.; Campanini, G.; Chizzolini, R. Comparison of Contaminant and Residue Levels in Organic and Conventional Milk and Meat Products from Northern Italy. *Food Addit. Contam.* **2005,** *22*, 9–14.

Gilbert-Lopez, B.; Garcia-Reyes, J. F.; Molina-Diaz, A. Sample Treatment and Determination of Pesticide Residues in Fatty Vegetable Matrices: A Review. *Talanta* **2009,** *79*, 109–128.

González-Gago, A.; Pröfrock, D.; Prange, A. Optimizing GC-ICP-MS for Ultra-Trace Quantification of PBDEs in Natural Water Samples Using Species-Specific Isotope Dilution. *J. Anal. Atom. Spectrom.* **2015,** *30* (1), 180–190.

Gopalan, N. K.; Chenicherry, S. Fate and Distribution of Organochlorine Insecticides (OCIs) in Palakkad Soil, India. *Sustain. Environ. Res.* **2018,** *28*, 179–185.

Guo, W.; Pan, B.; Sakkiah, S.; Yavas, G.; Ge, W.; Zou, W.; Tong, W.; Hong, H. Persistent Organic Pollutants in Food: Contamination Sources, Health Effects and Detection Methods. *Int. J. Environ. Res. Public Health.* **2019,** *16*, 4361.

He, X.; Zhou, Y.; Yang, W.; Li, S.; Liu, T.; Wang, T.; Hou, X. Microwave Assisted Magnetic Solid Phase Extraction Using a Novel Amino-Functionalized Magnetic Framework Composite of Type Fe_3O_4-NH_2@MIL-101 (Cr) for the Determination of Organochlorine Pesticides in Soil Samples. *Talanta* **2018,** *196* (1), 572–578.

Hwang, S. M.; Lee H. U.; Kim J. B.; Chung, M. S. Validation of Analytical Methods for Organochlorine Pesticide Detection in Shellfish and Cephalopods by GC-MS/MS. *Food Sci. Biotechnol.* **2020.** https://doi.org/10.1007/s10068-020-00748-0.

Ide, A. H.; Nogueira, J. M. F. Determination of Hydrophilic UV Filters in Real Matrices Using New-Generation Bar Adsorptive Microextraction Devices. *Separations* **2019,** *6* (4), 45.

Ide, A. H; Nogueira, J. M. F. New-Generation Bar Adsorptive Microextraction (Baμe) Devices for a Better Eco-User-Friendly Analytical Approach–Application for the Determination of Antidepressant Pharmaceuticals in Biological Fluids. *J. Pharma. Biomed.* **2018,** *153*, 126–134.

Janda, J.; Nodler, K.; Brauch, H. J.; Zwiener, C.; Lange, F. T. Robust Trace Analysis of Polar (C2–C8) Perfluorinated Carboxylic Acids by Liquid Chromatography-Tandem Mass Spectrometry: Method Development and Application to Surface Water, Groundwater and Drinking Water. *Environ. Sci. Pollut. Res.* **2019,** *26* (8), 7326–7336.

Kedikoglou, K.; Costopoulou, D.; Vassiliadou, I.; Bakeas, E.; Leondiadis, L. An Effective and Low Cost Carbon Based Clean-Up Method for PCDD/Fs and PCBs Analysis in Food. *Chemosphere* **2018,** *206*, 531–538.

Keisandokht, S.; Haddad, N.; Gariepy, Y.; Orsat V. Screening the Microwave-Assisted Extraction of Hydrocolloids from *Ocimum basilicum* L. Seeds as a Novel Extraction Technique Compared with Conventional Heating-Stirring Extraction. *Food Hydrocoll.* **2018,** *74*, 11–22.

Kim, L.; Lee, D.; Cho, H. K.; Choi, S. D. Review of the QuEChERS Method for the Analysis of Organic Pollutants: Persistent Organic Pollutants, Polycyclic Aromatic Hydrocarbons, and Pharmaceuticals. *Trends Environ. Anal. Chem.* **2019,** *22*, e00063.

Kishimba, M. A.; Henry, L.; Mwevura, H.; Mmochi, A. J.; Mihale, M.; Hellar, H. The Status of Pesticide Pollution in Tanzania. *Talanta* **2004,** *64*, 48–53.

Kolani, L.; Mawussi, G.; Sanda, K. Assessment of Organochlorine Pesticide Residues in Vegetable Samples from Some Agricultural Areas in Togo. *Am. J. Anal. Chem.* **2016,** *7*, 332–341.

Kunene, P. N.; Mahlambi, P. N. Optimization and Application of Ultrasonic Extraction and Soxhlet Extraction Followed by Solid Phase Extraction for the Determination of Triazine Pesticides in Soil and Sediment. *J. Environ. Chem. Eng.* **2020,** *8* (2), 103665.

Lacorte, S.; Quintanab, J.; Taulera, R.; Venturab, F.; Tovar-Sanchezc, A.; Duarte, C. M. Ultra-Trace Determination of Persistent Organic Pollutants in Arctic Ice Using Stir Bar Sorptive Extraction and Gas Chromatography Coupled to Mass Spectrometry. *J. Chromatogr. A.* **2009,** *1216*, 8581–8589.

Le Doux, M. Analytical Methods Applied to the Determination of Pesticide Residues in Foods of Animal Origin: A Review of the Past Two Decades. *J. Chromatogr. A.* **2011,** *1218*, 1021–1036.

Lee, J.; Lee, S. Y.; Park, K. W.; Lim, H. H.; Shin, H. S. Simultaneous Determination of PCBs, OCPs and PAHs in Mussel by Ultrasound Assisted Cloudy Extraction and Gas Chromatography–Tandem Mass Spectrometry. *Food Addit. Contam.* **2020,** 1–14.

Levy, M.; Al-Alamb, J.; Delhommea, O.; Milleta, M. An Integrated Extraction Method Coupling Pressurized Solvent Extraction, Solid Phase Extraction and Solid-Phase Micro Extraction for the Quantification of Selected Organic Pollutants in Air by Gas and Liquid Chromatography Coupled to Tandem Mass Spectrometry. *Microchem. J.* **2020,** *157*, 104889.

Li, J.; Li, H.; Zhang, W. J.; Wang, Y. B.; Su, Q.; Wu, L. Hollow Fiber–Stir Bar Sorptive Extraction and Gas Chromatography–Mass Spectrometry for Determination of Organochlorine Pesticide Residues in Environmental and Food Matrices. *Food Anal. Methods*. **2018,** *11* (3), 883–891.

Li, Q. Q.; Loganath, A.; Chong, S. Y.; Tan, J.; Obbard, J. P. Persistent Organic Pollutants and Adverse Health Effects in Humans. *J. Toxicol. Environ. Health Part A.* **2006,** *69*, 1987–2005.

Li, K.; Li H.; Liu, L.; Hashi, Y.; Maeda, T.; Lin, J. M. Solid-Phase Extraction with C30 Bonded Silica for Analysis of Polycyclic Aromatic Hydrocarbons in Airborne Particulate Matters by Gas Chromatography–Mass Spectrometry. *J. Chromatogr. A.* **2007,** *1154*, 74–80.

Liu, Z.; Zheng, G.; Liu, Z. Organochlorine Pesticides in Surface Water of Jiuxi Valley, China: Distribution, Source Analysis, and Risk Evaluation. *J. Chem.* **2020,** Article ID 5101936, 8 pp. https://doi.org/10.1155/2020/5101936.

Llomparta, M.; Celeiroa, M.; Garcia-Jares, C.; Dagnac, T. Environmental Applications of Solid-Phase Microextraction. *Trends Analyt. Chem.* **2019,** *112*, 1–12.

Lozowicka B.; Kaczynski, P.; Wolejko E.; Piekutin J.; Sagitov A.; Toleubayev K.; Isenova G.; Abzeitova, E. Evaluation of Organochlorine Pesticide Residues in Soil and Plants from East Europe and Central Asia. *Desalination Water Treat*. **2016,** *57* (3), 1310–1321.

Lynch, K. L. Toxicology: Liquid Chromatography Mass Spectrometry. In *Mass Spectrometry for the Clinical Laboratory*; Nair, H., Clarke, W., Eds.; Academic Press: San Diego, 2017; pp 109–130.

Malarvannan, G.; Poma, G.; Covaci, A. Interspecies Comparison of the Residue Levels and Profiles of Persistent Organic Pollutants in Terrestrial Top Predators. *Environ. Res*. **2020,** *183*, 109187.

Memon, Z. M.; Yilmaz, E.; Soylak, M. Switchable Solvent Based Green Liquid Phase Microextraction Method for Cobalt in Tobacco and Food Samples Prior to Flame Atomic Absorption Spectrometric Determination. *J. Mol. Liq.* **2017,** *229*, 459–464.

Mercer, S. M.; Jessop, P. G. "Switchable Water": Aqueous Solutions of Switchable Ionic Strength. *Chem. Sus. Chem.* **2010,** *3* (4), 467–470.

Milne, G. L.; Morrow, J. D. Measurement of Biological Materials. In *Clinical and Translational Science*; Robertson, D., Williams, G. H., Eds.; Academic Press: San Diego, 2009; pp 69–86.

Mousavi, M. M.; Mousavi, Nemati M.; Nabili, A. A. A.; Mahmoudpour, M.; Arefhosseini, S. Application of Dispersive Liquid–Liquid Microextraction Followed by Gas Chromatography/Mass Spectrometry as Effective Tool for Trace Analysis of Organochlorine Pesticide Residues in Honey Samples. *J. Iran. Chem. Soc.* **2016**. DOI: 10.1007/s13738-016-0939-2.

Musarurwa, H.; Chimuka, L., Pakade, V. E.; Tavengwa, N. T. Recent Developments and Applications of QuEChERS Based Techniques on Food Samples During Pesticide Analysis. *J. Food Compos. Anal.* **2019,** *84*, 103314.

Niu, Z.; Zhang, W.; Yu C.; Zhang J.; Wen Y. Recent Advances in Biological Sample Preparation Methods Coupled with Chromatography, Spectrometry and Electrochemistry Analysis Techniques. *Trends Analyt. Chem.* **2018,** *102*, 123–146.

Nozal, M. J.; Bernal, J. L.; Jiménez, J. J.; Martin, M. T.; Bernal, J. Determination of Azolic Fungicides in Wine by Solid-Phase Extraction and High-Performance Liquid Chromatography–Atmospheric Pressure Chemical Ionization–Mass Spectrometry. *J. Chromatogr. A.* **2005,** *1076*, 90–96.

Ochiai, N., Ieda, T.; Sasamoto, K.; Takazawa, Y.; Hashimoto, S; Fushimi, A; Tanabe, K. Stir Bar Sorptive Extraction and Comprehensive Two-Dimensional Gas Chromatography Coupled to High-Resolution Time-of-Flight Mass Spectrometry for Ultra-Trace Analysis of Organochlorine Pesticides in River Water. *J. Chromatogr A.* **2011,** *1218*, 6851–6860.

Odabasi, M.; Dumanoglu,Y.; Kara, M. Polychlorinated Naphthalene (PCN) Emissions from Scrap Processing Steel Plants with Electric-Arc Furnaces. *Sci. Total Environ.* **2017,** *574*, 1305–1312.

Okoya, A. A.; Ogunfowokan A. O.; Asubiojo O. I.; Torto N. Organochlorine Pesticide Residues in Sediments and Waters from Cocoa Producing Areas of Ondo State, Southwestern Nigeria. *Hindawi Pub. Corp. ISRN Soil Sci.* **2013,** *131647*, 12 p.

Olisah, C.; Adeniji, A. O.; Okoh, O. O.; Okoh, A. I. Occurrence and Risk Evaluation of Organochlorine Contaminants in Surface Water Along the Course of Swartkops and Sundays River Estuaries, Eastern Cape Province, South Africa. *Environ. Geochem. Health.* **2019,** 1–25.

Olisah, C.; Okoh, O. O.; Okoh, A. I. Global Evolution of Organochlorine Pesticides Research in Biological and Environmental Matrices from 1992 to 2018: A Bibliometric Approach. *Emerg. Contam.* **2019a**, *5*, 157–167.

Olisah, C.; Okoh, O. O.; Okoh, A. I. Occurrence of Organochlorine Pesticide Residues in Biological and Environmental Matrices in Africa: A Two-Decade Review. *Heliyon* **2020,** *6*, e03518.

Ollera, S.; Malatoa, J.; Sánchez-Perezb, A. Combination of Advanced Oxidation Processes and Biological Treatments for Wastewater Decontamination—A Review. *Sci. Total Environ.* **2011,** *409* (20), 4141–4166.

Organtini, K. L.; Haimovici, L.; Jobst, K. J.; Reiner, E. J.; Ladak, A.; Stevens, D. Comparison of Atmospheric Pressure Ionization Gas Chromatography-Triple Quadrupole Mass Spectrometry to Traditional High-Resolution Mass Spectrometry for the Identification and Quantification of Halogenated Dioxins and Furans. *Anal. Chem.* **2015,** *87* (15), 7902–7908.

Paramasivam, M.; Jayalakshmi, K.; Chandrasekaran, S.; Kuttalam, S. Organochlorine Pesticide Residues in Food. In: *Handbook of Food Analysis*; Leo, M. L. N., Toldra, F., Eds., 3rd ed.; CRC Taylor & Francis Group, 2015a; pp 133–148.

Paramasivam, M.; Jayalakshmi, K.; Chandrasekaran, S.; Kuttalam, S. Sample Preparation and Extraction Techniques. In: *Handbook of Food Analysis*; Leo, M. L. N., Toldra, F., Eds., 3rd ed.; CRC Taylor & Francis Group, 2015b; pp 415–426.

Paramasivam, M.; Rajeswaran, J.; Chandrasekaran, S.; Kuttalam, S. Pesticide Residues in Mother's Milk. In: *Pesticides: Evaluation of Environmental Pollution*, Hamir, S. R., Leo M. L. N., Eds.; CRC Taylor & Francis Group, 2012; pp 539–567.

Pena, A.; Ruano, F.; Mingorance, M. D. Ultrasound-Assisted Extraction of Pesticides from Olive Branches: A Multifactorial Approach to Method Development. *Anal. Bioanal. Chem.* **2006,** *385*, 918–925.

Poole, C. F.; Mester, Z.; Miro, M.; Pedersen-Bjergaard S.; Pawliszyn, J. Glossary of Terms Used in Extraction (IUPAC Recommendations 2016). *Pure Appl. Chem.* **2016,** *88*, 517–558.

Portoles, T.; Sales, C.; Abalos, M.; Saulo, J.; Abad, E. Evaluation of the Capabilities of Atmospheric Pressure Chemical Ionization Source Coupled to Tandem Mass Spectrometry for the Determination of Dioxin-Like Polychlorobiphenyls in Complex-Matrix Food Samples. *Anal. Chim. Acta.* **2016,** *937*, 96–105.

Portoles, T.; Sales, C.; Gomara, B.; Sancho, J. V.; Beltran, J.; Herrero, L.; Gonzalez, M. J.; Hernandez, F. Novel Analytical Approach for Brominated Flame Retardants Based on the Use of Gas Chromatography-Atmospheric Pressure Chemical Ionization Tandem Mass Spectrometry with Emphasis in Highly Brominated Congeners. *Anal. Chem.* **2015,** *87*, 9892–9899.

Ratola, N.; Alves, A.; Kalogerakis N.; Psillakis, E. Hollow-Fibre Liquid-Phase Microextraction: A Simple and Fast Cleanup Step Used for PAHs Determination in Pine Needles. *Anal. Chim. Acta.* **2008,** *618* (1), 70–78.

Rauert, C.; Harner, T.; Schuster, J. K.; Eng, A.; Fillmann, G.; Castillo, L. E.; Fentanes, O.; Ibarra, M. V.; Miglioranza, K. S.; Rivadeneira, I. M. Air Monitoring of New and Legacy POPs in the Group of Latin America and Caribbean (GRULAC) Region. *Environ. Pollut.* **2018,** *243*, 1252–1262.

Rial-Berriel, C.; Dacal, A. A.; Gonzalez, F.; Pastor-Tiburon, N. M.; Luzardo, O. P. Micro QuEChERS-Based Method for the Simultaneous Biomonitoring in Whole Blood of 360 Toxicologically Relevant Pollutants for Wildlife. *Sci. Total Environ.* **2020,** *736*, 139444.

Ricardo B. Persistent Organic Xenobiotics In *Marine Pollution Sources, Fate and Effects of Pollutants in Coastal Ecosystems*; Elsevier Science: San Diego, 2018; pp 107–135.

Russo, G. F.; Barbato, F.; Mita, D. G.; Grumetto, L. Simultaneous Determination of Fifteen Multiclass Organic Pollutants in Human Saliva and Serum by Liquid Chromatography–Tandem Ultraviolet/Fluorescence Detection: A Validated Method. *Biomed. Chromatogr.* **2019,** *33* (3), e4427.

Sajid, M.; Alhooshani, K. Ultrasound-Assisted Solvent Extraction of Organochlorine Pesticides from Porous Membrane Packed Tea Samples Followed by GC–MS Analysis. *Microchem. J.* **2020,** *152*, 104464.

Selvi, C.; Paramasivam, M.; Rajathi, D. S.; Chandrasekaran, S. Multiresidue Analysis of Organochlorine Pesticides in Milk, Egg and Meat by GC-ECD and Confirmation by GC-MS. *Bull. Environ. Contam. Toxicol.* **2012,** *89* (5), 1051–1056.

Sharma, D.; Rani, R.; Chaturvedi, M.; Yadav, J. P. Anibacterial Efficacy and Gas Chromatography-Mass Spectrometery Analysis of Bioactive Compounds Present in Different Extracts of *Allium sativum*. *Asian J. Pharm. Clin. Res.* **2018,** *11*, 280–286.

Spietelun, A.; Marcinkowski, L.; de la Guardia, M.; Namiesnik, J. Recent Developments and Future Trends in Solid Phase Microextraction Techniques Towards Green Analytical Chemistry. *J. Chromatogr. A.* **2013,** *1321*, 1–13.

Suleiman, F.; Nuhu, A. A.; Omoniyi, K. I.; Yashim, Z. I. Determination of Organochlorine Pesticide Residues in Some Vegetables and Fruit by QuEChERS Techniques and Gas Chromatography/Mass Spectrometry. *FUDMA J. Sci. (FJS)* **2020,** *4* (2), 365–370.

Sznajder-Katarzynska, K; Surma, M; Cieślik, I. A. Review of Perfluoroalkyl Acids (PFAAs) in Terms of Sources, Applications, Human Exposure, Dietary Intake, Toxicity, Legal Regulation, and Methods of Determination. *J. Chem.* **2019,** Article ID 2717528, 20 p.

Tadeo, J. L.; Sanchez-Brunete, C.; Albero, B.; Garcia-Valcarcel, A. I. Application of Ultrasound-Assisted Extraction to the Determination of Contaminants in Food and Soil Samples. *J. Chromatogr. A.* **2010,** *1217*, 2415–2440.

Tang, H. P. O. Recent Development in Analysis of Persistent Organic Pollutants Under the Stockholm Convention. *TrAC Trends Anal. Chem.* **2013,** *45*, 48–66.

Tavoloni, T.; Stramenga, A.; Stecconi, T. Single Sample Preparation for Brominated Flame Retardants in Fish and Shellfish with Dual Detection: GC-MS/MS (PBDEs) and LC-MS/MS (HBCDs). *Anal Bioanal. Chem.* **2020,** *412*, 397–411.

Tayane, C. R.; Mesquita, R. R.; Santos, A. P.; Cacique, L. J.; De Sa, Silverio F. O.; Pinho, G. P. Easy and Fast Extraction Methods to Determine Organochlorine Pesticides in Sewage Sludge, Soil, and Water Samples Based at Low Temperature. *J. Environ. Sci. Health B.* **2018,** *53* (3),199–206.

Tokumura, M.; Miyake, Y.; Wang, Q.; Nakayama, H.; Amagai, T.; Ogo, S.; Kume, K.; Kobayashi, T.; Takasu, S.; Ogawa, K. Methods for the Analysis of Organophosphorus Flame Retardants-Comparison of GC-EI-MS, GC-NCI-MS, LC-ESI-MS/MS, and LC-APCI-MS/MS. *J. Environ. Sci. Heal. A.* **2018,** *53*, 475–481.

Tölgyessy, P.; Miháliková, Z.; Matulová, M. Determination of Selected Chlorinated Priority Substances in Fish Using QuEChERS Method with Dual dSPE Clean-Up and Gas Chromatography. *Chromatographia* **2016,** *79* (21–22), 1561–1568.

UNEP, Stockholm Convention on Persistent Organic Pollutants (POPs). In: Stock. *Conv. Persistent Org. Pollut.* **2009** (accessed on 20 Oct 2020).

United Nations Environment Programme (UNEP). Final act of the Conference of Plenipotentiaries on the Stockholm Convention on Persistent Organic Pollutants. UNEP/POPS/CONF/4, 2001 (accessed on 20 Oct 2020).

Vasconcelos, P. H. M.; Camelo, A. L. M.; de Lima, A. C. A. Chemometric Tools Applied to Optimize a Fast Solid-Phase Microextraction Method for Analysis of Polycyclic Aromatic Hydrocarbons in Produced Water. *Environ. Sci. Pollut. Res.* **2020,** 1–10.

Vasconcelos, P. H. M.; Camelo, A. L. M.; de Lima, A. C. A. Chemometric Tools Applied to Optimize a Fast Solid-Phase Microextraction Method for Analysis of Polycyclic Aromatic Hydrocarbons in Produced Water. *Environ. Sci. Pollut. Res. Int.* **2020,** *28* (7), 8012–8021.

Wei, J.; Cao, J.; Tian, K.; Hu, Y.; Su, H.; Wan, J.; Li, P. Trace Determination of Five Organophosphorus Pesticides by Using QuEChERS Coupled with Dispersive Liquid–Liquid Microextraction and Stacking Before Micellar Electrokinetic Chromatography. *Anal. Methods* **2015,** *7* (14), 5801–5807.

Wei, S.; Leong, M.; Li, Y.; Huang, S. Development of Liquid Phase Microextraction Based on Manual Shaking and Ultrasound-Assisted Emulsification Method for Analysis of Organochlorine Pesticides in Aqueous Samples. *J. Chromatogr. A.* **2011,** *1218*, 9142–9148.

Wilkowska, A.; Biziuk, M. Determination of Pesticide Residues in Food Matrices Using the QuEChERS Methodology. *Food Chem.* **2011,** *125* (3), 803–812.

Wilkowska, A. M.; Biziuk, M. Rapid Method for the Determination of Organochlorine Pesticides and PCBs in Fish Muscle Samples by Microwave-Assisted Extraction and Analysis of Extracts by GC-ECD. *J. AOAC Int.* **2010,** *93*, 1987–1994.

Wu, Q.; Leung, J. Y. S.; Yuan, X.; Huang, X.; Li, H.; Huang, Z.; Li, Y. Biological Risk, Source and Pollution History of Organochlorine Pesticides (OCPs) in the Sediment in Nansha Mangrove, South China. *Mar. Pollut. Bull.* **2015,** *96* (1–2), 57–64.

Xu, L.; Huang, S.; Liu, Y.; Wei, S.; Chen, G.; Gong, Z. Hollow Carbon Nanobubbles-Coated Solid-Phase Microextraction Fibers for the Sensitive Detection of Organic Pollutants. *Anal. Chim. Acta.* **2020,** *1097*, 85–93.

Yang, X.; Leng, X.; Qi, Y.; Zhang, J.; Jiang, R.; Li, W.; Zhong, H. Monitoring of Adsorption and Transfer of Organochlorines in Soybean Seeds and Sprouts with Mass Spectrometric Imaging. *Anal. Chim. Acta.* **2020,** *1130*, 10–19.

Yang, Y. Y.; Toor, G. S.; Williams, C. F. Pharmaceuticals and Organochlorine Pesticides in Sediments of an Urban River in Florida, USA. *J. Soils Sediments.* **2015,** *15* (4), 993–1004.

Yu, N.; Guo, H.; Yang, J.; Jin, L.; Wang, X.; Shi, W. Non-Target and Suspect Screening of Per- and Polyfluoroalkyl Substances in Airborne Particulate Matter in China. *Environ. Sci. Technol.* **2018,** *52* (15), 8205–8214.

Zabielski, P.; Ford, G. C.; Persson, X. M.; Jaleel, A.; Dewey, J. D.; Nair, K. S. Comparison of Different Mass Spectrometry Techniques in the Measurement of L-[Ring- (13)C (6)] Phenylalanine Incorporation into Mixed Muscle Proteins. *J. Mass Spectrom.* **2013,** *48* (2), 269–275.

Zgola-Grzeskowiak, A.; Grze´skowiak, T. Dispersive Liquid-Liquid Microextraction. *Trends Analyt. Chem.* **2011,** *30*, 1382–1399.

Zhang, H.; Shen, Y.; Liu, W.; He, Z., Fu, J.; Jiang, G.; Cai, Z.; A Review of Sources, Environmental Occurrences and Human Exposure Risks of Hexachlorobutadiene and Its Association with Some Other Chlorinated Organics. *Environ. Pollut.* **2019,** *253*, 831–840.

Zhang, H.; Watts, S.; Philix, M. C.; Snyder, S. A.; Ong, C. N. Occurrence and Distribution of Pesticides in Precipitation as Revealed by Targeted Screening Through GC-MS/MS. *Chemosphere* **2018,** *211*, 210–217.

Zhang, Y.; Chen, Y.; Li, R.; Chen, W.; Song, Y.; Hu, D.; Cai, Z. Determination of $PM_{2.5}$–Bound Polyaromatic Hydrocarbons and Their Hydroxylated Derivatives by Atmospheric Pressure Gas Chromatography-Tandem Mass Spectrometry. *Talanta* **2019a**, *195*, 757–763.

Zhao, J.; Vanderburgt, S.; Santos, R. M.; Chiang, Y. W. Process Intensification of Dichlorodiphenyltrichloroethane Detection Methods for Determining Trace Concentrations in Soils. *Sus. Chem.* **2020,** *1*, 63–74.

CHAPTER 10

Phytoremediation: A Novel Strategy to Decontaminate Environmental Pollution

POOJA RANI[1], NILANJAN CHATTOPADHAYAYA[1], RAJEEV PADBHUSHAN[1], and UPENDRA KUMAR[2]

[1]Department of Soil Science and Agricultural Chemistry, Bihar Agricultural University, Sabour, Bhagalpur, India

[2]ICAR-National Rice Research Institute (NRRI), Cuttack, Odisha, India

ABSTRACT

Pollution is one of the major evils that deteriorates the environment in an adverse way. Many physical as well as chemical ways of remediation of contaminants are known, but these are not safe for the environmental health. These processes are very costly and can affect the quality of the environment. Sometimes, a secondary pollution may generate after application of these techniques. This chapter indicates about the most suitable, eco-friendly, and green emerging technology of decontamination. This technique does not harm the health of living organisms. This chapter outlines the novel strategies of phytoremediation techniques. The phytoremediation techniques have a special mechanism of mitigating the environmental pollution. Hyperaccumulators play an important role in extracting heavy metals above the underground parts of the plant. Microorganisms are very efficient at the technique of sequestering toxic metals. Plants have the mechanism to convert the toxic contaminants into volatile compounds and convert them in a less toxic form. Phytoremediating plants have the potential to fix the harmful and hazardous contaminants into the soil and thus restrict the translocation and entry of these contaminants in the food chain of living beings. Microbiological and biotechnological approaches also assist this technique to enhance their

Environmental Contaminants: Impact, Assessment, and Remediation.
Pritam Ganguly, Jajati Mandal, M. Paramsivam, & Sandip Patra, (Eds.)

proficiency. By inducing a genetic approach, we can easily modify the efficiency of hyperaccumulators by making them transgenic plants. Phytoremediation is the novel technique to depollute the harmful pollutants from soil, water, and air. Aquatic plants are very efficient in reducing the contaminants. With these strategies of decontamination, environmental pollution can be mitigated without posing any harm. The detailed structure of decontamination technique is designed in this chapter to reduce contamination.

10.1 INTRODUCTION

Various kinds of harmful, toxicant, and hazardous pollutants have been released day by day due to increasing environmental pollution (Xia et al., 2018). These pollutants are nonbiodegradable and can be easily transformed from lower trophic levels to higher trophic levels. The transfer of pollutants from one to another trophic level permits them to enter our food chain, and in this way, these toxic heavy metals can enter easily into our body and cause various diseases (Sorrentino et al., 2018). These heavy metals have the property of bioaccumulation in the living organisms. These heavy metals agonize the biomagnification process and accumulate in the tissues of the body. These harmful contaminants are detrimental not only to human health but also to the biota's health (Alia et al., 2013). The rapid increase in the concentration of heavy metals is worldwide distress. For remediation of these harmful toxic pollutants, a modern and advanced technological approach was established with novel technology. This novel strategy helps to reduce and eradicate the harmful contaminants from the environment. The novel techniques are advanced and improved. Such techniques are useful in understanding the presence of metallic nanoparticles in the atmosphere. These nanoparticles of heavy metals are rising these days and have a very prolonged contribution to deteriorating environmental health (Ebbs et al., 2016). Such compounds show positive impressions in different sectors like agriculture, cosmetics, pharmaceutics, and other consumer products, but they possess negative effects on the environment and affect its quality. In this way, these particles also become targets for decontamination. Different kinds of physical, chemical, and biological techniques for remediation have been classified so far. Traditional approaches include two things, either on-site management or dig and disposed method. But these techniques only overcomes the contamination problem from a particular site and do not mitigate the pollutant permanently. Such method incorporates a severe risk to the environment. An alternative to this method is the technique of soil washing, but this method is very costly and not possible to implement on a wide range. Not only the cost, but the potential of the soil for plant growth also gets deteriorated. Not only the potential, but the biological activities of soil also destroy (Tangahu et al., 2011). Physical and chemical ways of decontamination possess several limitations like labor requirement, high cost of operating, destruction of native soil flora, and changes in the properties of soil. Chemical remediation technique develops secondary pollution in the environment and thus increases the further treatment cost. Chemical treatments are more sticky and difficult to operate.

Biological decontamination methods are more suitable to the environment because they take care of soil microbiological properties and can be operated at a low cost. This novel technique is more adequate and eco-friendly and does not pose a risk to natural organisms. Among the several biological techniques, phytoremediation is the most novel because of its wonderful strategies for decontamination. "Phyto" is a Greek word meaning plant, and Latin word "remedium" denotes remediation, coordinating to make the word phytoremediation. This technique uses the unique properties of a plant to decontaminate the environmental pollution. Different kinds of beautiful processes which occur regularly in a plant, like transpiration, volatilization, bioaccumulation, and root uptake capabilities, are designed in a way to decontaminate the soil pollution, air pollution, and water pollution from the environment. This is a novel and recent technique, which uses green plants for remediation of harmful toxicants. More than 400 plants are recognized as having phytoremediation potential. Not only the traditional plants, but genetically engineered plants are more popular in decontaminating the environmental pollution because desirable genes can be incorporated in the plants by genetic engineering to make the technique of phytoremediation more advance and powerful. Through this technique, it is easy to handle the biotic and abiotic stress in the environment (Ibanez et al., 2015). Thus, it was recognized that phytoremediation can be used in different ecosystems for the remediation of harmful toxicants from the environment, and it is easy to operate at a very low cost. Let us move toward the magnificent novel strategies to understand the technique in a more reliable way.

10.2 WHY REMEDIATION: A NEED-BASED APPROACH

Industrialization and the way of lifestyle have negatively affected the environment. This is because of the development of harmful contaminants and lack of proper dumping facilities. No proper treatment is available to secure environment from these contaminants. This environmental pollution leads to several serious and harmful impacts on the living beings and makes their survival hazardous and difficult. Among all these, the most severe impact is done by heavy metals. These heavy metals pose a serious environmental risk by posing harmful effects on living organisms by disrupting their life chains and indulging harmful metals in their food chain. These possess fatal and lethal impacts on human beings (Ahmad et al., 2011, 2015; Qadir et al., 2014). Releasing heavy metals in the environment creates an imbalance in normal life of fauna, flora, and human beings. The harmful waste materials should undergo proper treatment before being disposed of. Phytoremediation is found to be more proficient because plants are more efficient in decontaminating harmful contaminants; hence, it is an effective method to decontaminate harmful contaminants (Kamran et al., 2014). This is the most economic and cost-effective technology to mitigate the environmental pollution (Dixit et al., 2015). Phytoremediation consists of different strategies to decontaminate the environmental pollution. These are novel approaches to mitigate contamination.

10.3 NOVEL STRATEGIES OF PHYTOREMEDIATION

Phytoremediation has different techniques which are operated by different physiological processes and used to decontaminate the harmful contaminants. Phytoremediation techniques consist of:

- Phytofiltration
- Phytostabilization
- Phytoextraction
- Phytovolatilization
- Phytotransformation

10.3.1 PHYTOFILTRATION

Phytofiltration can be categorized into three different subtechniques:

- Rhizofiltration: It is the technique of phytoremediation, which uses roots of plant for decontamination purpose.
- Blastofiltraton: In this technique, seedlings are used for remediation of environmental pollution.
- Caulofiltration: This technique comprises the use of excised plant shoots.

This technique of mitigation adsorbs the harmful contaminants from the contaminated wastewater surface. Along with the remediation, this method helps to prevent the leaching of these contaminants into the groundwater resources. This technique is conducted in an in situ environment. The three main principles lying with rhizofiltration are absorption, concentration, and precipitation. By dint of these techniques, plant roots can absorb the contaminants from the water and remediate the polluted water sources.

In biofiltration technique, after germination there is an improvement in the surface to volume ratio and the seedlings are capable to adsorb the metal in a huge quantity. This clearly depicts the ability of the seedlings to decontaminate the wastewater (Krishna et al., 2012). It is reported after performing an experiment with melon, okra, castor, and moringa seeds for observing the potential of blastofiltration. It was observed that in the experiment of 72 h, the metal contaminant was found to be decreased about 96–99%. It was detected that castor and okra seeds were proficient, while the seeds of moringa have 100% efficiency of remediating heavy metals. The heavy metal removed was Cd from the contaminated site. The scientist who conducted this experiment finally concluded that the seeds having minor economic importance may help to remediate the heavy metals by using green technology in the next generation (Udokop, 2016). It was also reported that the extracts of *Moringa oleifera* seeds in the aqueous medium were used for metal uptake from the polluted wastewater. The metal uptake was 70, 93, 95, and 76% for cadmium, lead, copper, and chromium, respectively (RaviKumar and Sheeja, 2013). Papaya seeds also have the blastofiltration property, as they have a potential to uptake zinc from the heavily contaminated wastewater. Results indicate with the fact that the uptake increases as the contact time increases (Ong et al., 2012). Not only the papaya, but mango seeds are also known to remediate the toxicity of Pb, Cu, and Cd from the aqueous solution. The eradication efficiencies were from 85 to 100% for these metals (Pareck et al., 2002). Using seeds in blastofiltration has proved to be a very efficient technique, as there is no limitation in developing seeds since seeds can be grown in any condition, dark or light. The prepared cultures require only water and air, and thus it is a very inexpensive technique (Krishna et al., 2012). Plants seeds are easy to manage and available easily at a low cost.

The highest potential of removing contaminants from the water resources rises with the use of caulofiltration. An experiment was conducted on this technique with aquatic plant *Ipomoea aquatica*, which was introduced in lead-contaminated wastewater. The concentration of lead was about 20 mg/L. Further demonstration showed the eradication efficiency of the plant.

The metal accumulated in the roots and the bottom part of stem of the plant. The adventitious roots along with the lateral branches raised the potential of *Ipomoea aquatic* in sequestrating the lead contaminant in water sources (Chanu and Gupta, 2016). Another study was conducted with *Stevia rebaudiana* to observe the prospective for confiscating the heavy metals Se, Cu, As, and Al (Hajara et al., 2014). When the expunged stems were dipped in the solution concentrated with these metals, a significant amount of metals was accumulated in the leaves (Mesjasz-Przybyłowicz et al., 2004). In context with rhizofiltration, it has been observed that the technique has enormous applications in mitigating the harmful contaminants from the aquatic plants (Dhir et al., 2009; Dushenkov and Kapulnik, 2000; Olguín and Sánchez-Galván, 2012; Rai, 2008). Apart from these aquatic macrophytes, some terrestrial plants have also been recognized for the property of sequestering heavy metals through the technique of rhizofiltration by using root biofilters designed by the microorganisms, and they efficiently precipitate the metals by absorbing and concentrating them (Salt et al., 1995). Root exudates play an important role in precipitating metals by lowering the pH in the rhizosphere. Indian mustard, sunflower, and several grasses have the potential to remove metals from the liquid media or aqueous sources (Dushenkov et al., 1995). Among these all, Indian mustard has been found to be more efficient in concentrating these metals. They can sequester a 13 to 563-fold concentration above the initial one (Dushenkov et al., 1995). Aquatic macrophytes chosen for the rhizofiltration must have a rapid growth rate with well-developed roots system, for example, *Eichhornia crassipes.* These plants have a great potential to develop high biomass and are designed naturally with well-developed fibrous root system (Liao and Chang, 2004). These plants are adaptable to different water conditions and are capable of growing in harsh conditions. They can easily sequester the metals from water resources and accumulate these metals in the roots. In this way, *Eichhornia crassipes* is considered as an ideal plant for rhizofiltration (Liao and Chang, 2004). A study conducted on *Eichhornia crassipes* revealed that Cu, Ni, Pb, and Zn can be extracted and accumulated by this aquatic macrophyte efficiently in the roots, which is three to 15 times greater than shoots (Liao and Chang, 2004). *Hydromistia stolonifera*, *Pistia stratiotes*, *Salvinia herzogii*, and *E. crassipes* are efficient macrophytes accumulating Cd (Maine et al., 2001). Another most efficient aquatic macrophyte was *Azolla pinnata* used to extract the contaminants from the aqueous resources. In the study, it was found that the uptake potential of lead increases as the nutrient media decreases. It was concluded that the power of accumulation is directly proportional to the time of exposure. After the fourth day, it was observed that *Azolla pinnata* had accumulated the lead concentrations by 83% (Thayaparan et al., 2013). To understand the process of rhizofiltration, another experiment was conducted with the plant *Carex pendula*, a wetland plant. This plant is also capable of removing lead from the aquatic water bodies. Aquatic plants having rhizofiltration ability must have tolerance capacity against metals, ability to produce high biomass, and have less translocating potential from roots to shoots (Terry and Banuelos, 2000). Less translocation potential from roots to shoots is a major characteristic for rhizofiltration, as it reduces the risk of entering heavy metals into the food chain (Dushenkov and Kapulnik, 2000).

10.3.2 PHYTOSTABILIZATION

Phytostabilization is one of the strategies among phytoremediation techniques, which reduces the mobility of heavy metals in the soil. As the name defines, this technique provides stability to the contaminant in the soil and thus restricts the movement of heavy metals. This technique involves the leaching and runoff prevention, and converts the high toxicant heavy metal into the less bioavailable form through precipitation in the rhizosphere (Nwoko, 2010). This technique is used to fix the heavy metals in the soil, sediment, sludge, and water, and prevent the transfer of heavy metals in the underground water resources (Ghosh and Singh, 2005; Singh, 2012). This is accomplished by the sorption capacity of roots, their complexation ability, precipitation, and metal variance reduction by roots. The technique also depends upon the chemical characteristics of metals. Chemical structures and characteristics determine the phytostabilization efficiency of heavy metals. An experiment was conducted between *Typha domingensis* and *Phragmites australis* to compare the phytostabilization efficiency, and it was revealed that both plants can stabilize Hg and As, respectively, in the residue (Bonanno, 2013) and do not show efficiency in phytostabilizing different heavy metals other than these. Phytostabilization does not possess remediation efficiency permanently. The heavy metals contaminating the environment cannot be reduced permanently by using this technique. Metals acquire the property of inactivation after treatment by phytostabilization and remain in the soil, plant roots, and sediments (Vangronsveld et al., 2009). But apart from this limitation, one of the best advantages with phytostabilization is its stabilizing ability in the soil or roots. Due to the stabilization, there is no need to treat the aerial part of the plant and thus reduces the risk of transfer of contaminants to the edible parts. This advantage makes the technique more popular and very rapid technique among all the techniques of phytoremediation (Abreu et al., 2012). Phytostabilization includes various applicable advantages for plants. This technique has ability to improve the root system by showing tolerance power against heavy and toxic contaminants, potential to immobilize the pollutants, and less translocation capacity (Kramer, 2005). Forage grasses such as *Sorghum dochna*, *Euchlaena mexicana*, and *Pennisetum americanum* are found efficient in stabilizing Cd and Pb (Zhang et al., 2016). *Juncus effusus* L., a wetland plant is used to phytostabilize the Pb contaminant (Najeeb et al., 2014).

10.3.3 PHYTOEXTRACTION

Phytoextraction is the process of remediating perilous harmful contaminants from the environment by accumulating them in the harvestable parts of plants (Nwoko, 2010). Phytoaccumulation, phytosequestration, and phytoabsorption are the synonyms of phytoextraction. In this technique, roots of the phytoremediating plants sequester and extract the harmful contaminants from the soil and accumulate them in the harvestable parts above the ground (Ali et al., 2013; Sekara et al., 2005; Yoon et al., 2006). Phytoextraction is one of the novel techniques of phytoremediation, which is used most widely and appropriately to decontaminate the metal pollutants. This technique is used to extract the metals and metalloids from the environment. The plants suitable for phytoextraction must be vigorous and fast-growing, having high biomass production capacity. Such plants are known as hyperaccumulators. Efficiency of phytoextraction depends upon several factors like bioavailability of metals, soil physiochemical properties, metal speciation, etc. (Ali et al., 2013; Li et al., 2010). A study was conducted on phytoextraction by using *Ipomea aquatica* to observe the efficiency of phytoextraction, and this plant was found to be proficient in accumulation and translocation of heavy metals in the aboveground harvestable parts of plant, that is, shoots (Wang et al., 2008). One of the most effective plants *Polypogon monspeliensis* is reported for arsenic phytoextracting potential, which extracts the harmful contaminants from soil and water (Ruppert et al., 2013). But the field trials of phytoextraction are limited (Gong et al., 2018). It was also revealed that phytoremediation is very time-consuming and needs a safe and sufficient level of metals in the soil (Deng et al., 2016; Wang et al., 2020). So, therefore, proper approaches are taken to rectify the problem associated with time so that the time taken can be reduced and thus enhance efficiency (Rizwan et al., 2018; Ji et al., 2011).

Many studies have shown that many plants, including alpine penny-cress, rapeseed, Indian mustard (Simmons et al., 2015), sunflowers (Shaheen and Rinklebe, 2015), and willow and poplar trees (Kacálková et al., 2015), are found to be efficient in phytoextracting potential. Hyperaccumulators are very efficient in the phytoextraction. Appropriate strategies should be taken to enhance the phytoextraction efficiency of the hyperaccumulators. These strategies include the addition of ammonium chloride (Yang et al., 2019), incorporation of *B. megaterium* (Jeong et al., 2012; Khan et al., 2009), and application of citric acid (Liu et al., 2008; Yu et al., 2019). Using these techniques not only reduces the time taken by the plants in phytoextraction but also enhances the efficiency of the hyperaccumulators plant. Thus, phytoremediation is known as the eco-friendly, aesthetically pleasing, cost-effective, and metal-extracting technique.

10.3.4 PHYTOVOLATILIZATION

Phytovolatilization is the unique strategy of phytoremediation, which adopts a wonderful mechanism to decontaminate the harmful contaminants in the form of vapor. Harmful and toxicant constituents are converted into the volatile form, and stomata help to release these materials into the atmosphere. Stomata are used for the gaseous exchange in the soil, and in this way, they support to remove the contaminants from the soil (Terry and Ba-nuelos, 1999; Nwoko, 2010; Sarma, 2011). Although this technique is proficient in removing the contaminants, but its use is limited due to its nature of incomplete removal of contaminants from the soil.

In this technique, the plant converts the harmful material into a volatile form and then releases it into the atmosphere, which means the removal of contaminants from the soil and their accumulation in the atmosphere. The accumulated contaminants in the atmosphere again fall back in the soil after rainfall (Gill, 2014). This strategy of phytoremediation creates controversy in decontamination (Padmavathiamma and Li, 2007; Sarma, 2011). Therefore, this technique is suitable to decontaminate the highly volatile elements such as Hg and Se (Sharma et al., 2015; Wang et al., 2012).

10.3.5 PHYTOTRANSFORMATION

As the name indicates, phytotransformation or phytodegradation basically works with the modification of contaminants chemically. This process of decontamination arrests the pollutants from soil, water resources, or sediments, and either transforms the harmful toxic contaminants into inactivate compounds through the process of plant metabolism or degrades them completely or makes them immobilized (Pivetz, 2001; Tangahu et al., 2011) in the roots as well as shoots (Bulak et al., 2014). Enzymes associated with plants roots and microorganisms such as bacteria, fungi, and actinomycetes play an important role in transforming the high toxic contaminants into the less toxic substance (Dobos and Puia, 2009). This technology takes more than one season to work efficiently with some essential physicochemical properties present in the soil.

TABLE 10.1 Strategies of Phytoremediation Techniques

Name	Depiction	Purpose
Phytofiltration	Organic pollutants are broken down with the help of microorganisms	Decontamination
Phytoextraction	Harmful contaminants can be uptaken by plant roots and translocate easily to accumulate in parts above the ground	Decontamination
Phytostabilization	Contaminants are fixed and immobilized in the soil	Stabilization of contaminants
Phytodegradation	Organic compounds degradation by plants	Decontamination
Phytovolatilization	Plants endorse harmful contaminants and transform it into the volatile compound and finally volatilize.	Transfer of pollutants

Source: Adapted from Ali et al., 2013; Vangronsveld et al., 2009

10.4 BIOLOGICAL OUTLOOK OF PHYTOREMEDIATION

10.4.1 MICROBIAL INTERFERENCE

Phytoremediation's biological outlook deals with the several processes arising within the soil. These progressions restructure the beautification of natural interaction between plant rhizosphere and the microbes, translocation mechanism, uptake mechanism, and several rhizospheric processes (Nwoko, 2010). Degradation, compartmentation, and production or activation of plant chelator tangled in the process of storage and transport are some of the important tolerance mechanisms that need to be understood for making phytoremediation more operative. For getting more effective results of phytoremediation, it is most important to understand the biological process within it (Pilon-Smits and Freeman, 2006). The plant rhizospheric environment produces several kinds of organic acids, which are enriched with beneficial nutrients, developing a microenvironment involved in the proliferation and promotion of microbial action (Masciandaro et al., 2013). The collaboration between the plants roots and helpful microorganisms enhances the plant biomass and increases tolerance capacity for metals. It reveals that microbes like bacteria, fungi, and actinomycetes are very important for the phytoremediation. Microflora help out to make phytoremediation more proficient and effective (Gao et al., 2012).

Not only the bacteria and fungi but mycorrhizal association also has an immense role in increasing the proficiency of phytoremediation. It helps to absorb and accumulate the harmful contaminants in the plant tissue (Leung et al., 2013). Mycorrhizal association helps to increase the efficiency of hyperaccumulators. A study revealed that *Brassica juncea* coinoculated with arbuscular mycorrhizal fungi (AMF) enhanced the proline content by 63.47% and alleviated the drought stress situation after treating with sodium chloride (Sarwat et al., 2016). Microorganisms have their mechanism for the treatment of heavy metals, which may increase or decrease the accumulation of metals in the plant depending on the kind of interaction (Sharma et al., 2013).

A study was directed with hyperaccumulators, nonhyperaccumulators, and arbuscular mycorrhizal fungi to check the benefits obtained with AMF inoculation. In this experiment, hyperaccumulator *Cyanodon dactylon* and nonhyperaccumulator *Pteris vittata* were used. Three types of single inoculum and two kinds of mixed inoculum with *Cyanodon dactylon* and *Pteris vittata* were used in this study. These inoculums were introduced to an exposure of 0, 100, and 200 mg kg^{-1} of arsenic. Both the species expressed the enhancement in biomass and the enzyme arsenate reductase in each concentration level of arsenic (Leung et al., 2013). However, some bacteria that are metal resistant can also contribute to the process of phytoremediation by their genes. Their genes can be used for improving the decontamination technique (El-Deeb et al., 2012).

This can be concluded with an experiment, which was performed to test the potential of biotransformation with a heavy metal, that is, chromium. It was found that the proficiency of *Pistia stratiots* and *Eichhornia crassipes* increases in the presence of bacterial strains *Exiguobacterium*, *Bacillus pumilus*, and *Pseudomonas doudoroffii*. The bacterial strains show the beneficial effect due to the presence of root exudates, which enhance the capacity (Ejaz et al., 2013). *P. stratiotes* and *E. crassipes* were used to remove the chromium in the presence of *Exiguobacterium* by treating the wastewater (Ejaz et al., 2013). Roots of the plant play an important role in mitigation of pollutant. It increases the soil structure and efficiency of microbial community, which help in bioavailability and speciation of metal in the soil (Chattopadhyaya et al., 2012; Cosio et al., 2014). Aquatic plants have plasma lemma in their root cells, on which the metals can be absorbed on the wall as a result of passive diffusion or acropetal transportation by the roots of aquatic plants (Choo et al., 2006; Nwoko, 2010). Membrane transport protein is the process by which these contaminants get absorbed from the soil. Thus, biological process marks a unique stage in the process of phytoremediation (Pilon-Smits and Freeman, 2006).

10.4.2 INVOLVEMENT OF CARRIER MATERIAL

The translocation of metal depends upon a unique biological process. This process involves proton pump like ATPase. It consumes energy and is able to generate electrochemical gradients. Proteins are also used as an electrochemical gradient known as cotransporters. Proteins also help to translocate the metals. It reveals that each transport mechanism plays a unique role. As for example, phosphate transporters are used to translocate arsenate, and selenate by sulfate transporters (Nwoko, 2010; Tangahu et al., 2011).

The biological overview is necessary to understand the sequestration of metals inside the cell. Accumulation of metals can occur inside the vacuoles. This extraction is mainly completed by the help of carrier proteins or through mineralization or volatilization (Dhankher et al., 2012). This process of sequestration is not applicable to each metal. There are some metals too, which cannot be sequestered by this way, so those metal ions may be combined in a complex way with the ligands having low molecular weight. These low-molecular-weight ligands may be histidine, nicotinamine, organic acids, or mutagenic acids. These ligands have intracellular cells, which help sequester the metal ions inside the cytoplasm (Haydon and Cobbett, 2007).

10.4.3 PARTICIPATION OF SYNTHETIC CHELATING MATERIAL

To maintain homeostasis is one of the main objectives within the plant cell, which is coordinated by the endorsement and efflux of metals inside. Plant genomes play an important role in encoding metal carriers. These metal carriers are several in numbers and specific to their pattern, systemization, and localization for translocation of metals (Colageno and Guerinot, 2006). A much specified tolerance mechanism of metal was introduced with the help of chelation. In this process, natural or synthetic chelating agents are used to chelate the metal ions having intracellular complexation. These chelating materials or agents are either synthesized naturally in the plant or amplified synthetically in the environment (Guimaraes et al., 2008). Chelating agents support the plant roots for solubilizing these metals. The efficiency of solubilization is so high that it can even solubilize insoluble precipitates (Tangahu et al., 2011). A wide range of phytoextraction can only be achieved by the help of chelate-assisted phytoremediation. Uptake and translocation of metal increase by using chelating agent in the mitigation technology (Alkorta et al., 2004). In soil, metals release from the cationic exchange site and consequently form a metal complex. This way permits the release of the metals in the soil easily. After releasing from the cation exchange sites of soil, the toxic metals can be absorbed by the roots of the plants (Dipu et al., 2012). Citric acid, elemental sulfur, and EDTA (Ethylenediaminetetraacetic acid) are examples of chelating agents. They enhance the bioavailability of metal in the soil (Ali et al., 2013). Selection of a suitable chelating agent is one of the most concerns in the way of choosing strategy for decontamination. Sometimes a chelating agent shows efficient response to a single metal, but it may impact negatively in the presence of two or more metals. It may show confrontational impact. So, picking appropriate chelating agent is very necessary (Gao et al., 2012).

An experiment was conducted to observe the efficiency of chelating agents. Chelators EDTA and ethylene diamine disuccinic acid (EDDS) were compared for extracting lead with plant *Cynara cardunculus* through the technique of phytoextraction. It was revealed that EDTA shows more proficiency in extracting lead through roots as compared to ethylene diamine disuccinic acid (EDDS) (Epelde et al., 2008). Plants produce high biomass when treated with EDTA as compared to EDDS. EDTA is more efficient in phytoextracting lead as compared to EDDS. However, EDDS enhances the biodegradability. On the other hand, it was found to be hazardous when treated with EDTA. EDTA is produced synthetically, and it is nonbiodegradable and prone to leaching into groundwater (Maine et al., 2001).

10.4.4 ASSOCIATION WITH NATURAL CHELATING AGENT

To mitigate the problems associated with synthetic and artificial chelating agents, natural chelators are used with plants to decontaminate harmful toxicants from the environment. Natural chelators like citric acid has biodegradable nature. These natural compounds are not toxic to plants (Smolinska and

Krol, 2012). In a study, it was observed that the concentration of uranium was high in the biomass when plant was treated with citric acid (Schmidt, 2003).

TABLE 10.2 Response of Plants on Heavy Metal Accumulation.

Floras	Polluted areas	Contaminants
Lollium perenne L. (ryegrass)	Soil	Copper
Water hyacinth (*Eichhornia crassipes*)	Water	Arsenic
B. napus L. (canola)	Soil	Cadmium, zinc, and copper
Cajanus cajan (pigeon pea)	Soil	Arsenic and cadmium
Salix spp. (willox)	Water	Cadmium, nickel, and lead
Populas deltoides (Poplar tree)	Soil and water	Organic pollutants
Sorghastrum nutans (Indian grass)	Soil	Organic pollutants
Helianthus annus L. (sunflower)	Soil	Lead and zinc
Ficus microcarpa	Soil	Cadmium and copper
Lemna minor (duckweed)	Soil	Organic pollutants
Zea may (maize)	Soil	Cadmium
Pistia stratiotes	Water	Zinc and cadmium
Spirodela polyrrhiza	Water	Cadmium
Myriophyllum aquaticum	Water	Cadmium and copper
Ludwigina palustris	Water	Cadmium and arsenic
Mentha aquatic	Water	Lead and copper
Centella asiatica	Water	Arsenic and cadmium
C. zizanioides	Soil	Organic pollutants
Cyperus rotundus	Soil	Cadmium and chromium
Ambrosia artemisiifilia	Soil	Lead
Pontederiaceae, *Araceae*, *Lemnaceae*, *Convolvulaceae*, *Scrophulariaceae*, *Hydrocharitaceae*, and *Nelumbonaceae*	Soil	Copper

Source: Adapted from Sumiahadi and Acar, 2018; Muthusaravanan et al., 2018)

10.5 MOLECULAR AND BIOTECHNOLOGICAL APPROACHES TO ENHANCE PHYTOREMEDIATION

For getting more effective results, some genetic modifications are required in the plants so that the induced character may improve the technique of phytoremediation (Doty, 2008; Abhilash et al., 2009; Legault et al., 2017;

Hussain et al., 2018). Genetic modification of microorganisms associated with plants plays an important role in removing toxic and harmful contaminants (Basu et al., 2018; Kuffner et al., 2008; Hussain et al., 2018). Host plants are responsible to improve the prospective of phytoremediation by effectively enhancing the efficiency of its phytoremediating potential through genetic engineering. This process requires bacterial isolation and integration of genes along with enhancing the counts of beneficial microorganisms within the host system. Although genetic engineering is the most prominent and widely used technology these days, some challenges and tasks remain. This can restrict the use of genetic engineering in phytoremediation. Usually, the genetically-modified crop does not grow as fast in the field conditions as in vitro circumstances, and in this way, native rhizospheric beneficial microorganisms are used to outcompete these. The outcompetition results in the fall of their population count and thus restricts the process of degradation of pollutants present in the soil (Gilbertson et al., 2007). It is intricate to sustain the constancy of transferred gene in a host plant. But by using some techniques and processes, we can use it for the further framework of decontamination.

Novel strategies to decontaminate the environmental pollution include genetic modification and approaches. The efficiency of remediation increases with the help of using biotechnological pathway. In nature, mainly two kinds of pollutants exist, organic and inorganic. Inorganic pollutants include heavy metals, and organic pollutants include herbicides, polycyclic aromatic hydrocarbons, explosives, PAHs, and halogenated compounds. To remove these harmful contaminants from the environment, transgenic techniques are incorporated in the phytoremediating plants having high biomass. Novel and innovative strains of genes are acquainting with these plants. These plants are devoid of genes from their presence naturally. These specific genes are obtained by harnessing the bacterial or mammalian genes, which are important for the transportation of harmful contaminants. The genes are encoding especially for metal chelators and enzymes for degrading various resistant and rigid contaminants, metal homeostasis, xenobiotic detoxification, and oxidative stress tolerance.

10.6 AN EMERGING DECONTAMINATION TECHNIQUE: PHYTOMINING

Phytomining is a new plant-based technique assisted with mining and used for recovery of valuable metals. Recovery technique consists of reclamation of soil by extracting metals from ash of contaminated biomass through combustion, and can regenerate revenue. When phytomining is applied to the agricultural field, it is termed as agromining (Sheoran et al., 2009; Mahar et al., 2016; Jiang et al., 2015). Phytomining has eco-friendly benefits over the natural mining process (Ali et al., 2013). In this process, bioenergy generates with low emission of SO_2. This is due to the low amount of sulfur content of bioores. Limitation of phytomining can be achieved by the phytoextraction technique. Phytomining is suitable for heavy metals like Au, Ni, Co, and Tl (Mahar et al., 2016). Phytomining is mainly useful in the restoration of highly degraded soil. It is used for treating disturbed soil. It is widely used for the mine trailing waste and mining sites in the region of tropics. Therefore, it can be said that phytomining is an economic inducement for recovery of low-cost metal and very useful in pollution control.

10.7 CONCLUSION AND FUTURE ASPECTS

Phytoremediation is an emerging technique to mitigate the environmental pollution. This technique of mitigation is based on low cost and produces high amount of dynamic energy through the process of ignition. Mitigation of harmful pollutants should be done by using plants. These plants possess special characteristics like vigorous rapid growth and high biomass production, and are known as hyperaccumulators. Phytoremediation has novel strategies to decontaminate the environmental pollution. These strategies have special mechanisms to remediate the environmental pollution without posing any adverse effect. Phytoremediation is a green technology which is eco-friendly in nature and can be operated at a very low cost. Phytoremediating plants have potential to treat organic and inorganic contaminants from the soil. Phytoremediation is a novel and sustainable approach of mitigating environmental pollution. There are several limitations while operating this technique. It is essential to understand the physiology and chemistry of heavy metals to eradicate them completely. Chelator plays an important role in completion of phytoextraction strategy. Chelating material, either artificial or natural, helps to extract the harmful heavy metals from the soil, and thus plants can accumulate them in the aboveground parts. Between the two kinds of chelating materials, natural chelators are best because they have ability to produce high biomass without deteriorating the quality of soil.

The bright outlook of this technique is that an uneducated or layman person can operate this technique. There is no such complex process that requires high understanding. Molecular and biotechnological approaches also play an important role in the way of phytoremediation. Various bacterial and fungal strains are used to encode the genes required to modify the characteristics of hyperaccumulators. Not only the bacterial and fungal strains but mycorrhizal association also plays an important role in enhancing the efficiency of phytoremediation. Phytomining is one of the important uses of hyperaccumulator plants. In this technique, recovery of metal is done by absorbing it in plants and using as mines.

KEYWORDS

- **phytoremediation**
- **chelates**
- **phytohormones**
- **phytoextraction**
- **phytovolatilization**

REFERENCES

Abreu, C. A. Organic Matter and Barium Absorption by Plant Species Grown in an Area Polluted with Scrap Metal Residue. *Appl. Environ. Soil Sci.* **2012,** 1–7.

Ahloowalia, B. S.; Maluszynski, M.; Nichterlein, K. Global Impact of Mutation-Derived Varieties. *Euphytica* **2004,** *135*, 187–204.

Ahmad, P.; Nabi, G.; Ashraf, M. Cadmium Induced Oxidative Damage in Mustard (*Brassica juncea L.)* Plants can be Alleviated by Salicylic Acid. *South Afr. J. Bot.* **2011,** *77*, 36–44.

Ahmad, P.; Sarwat, M.; Bhat, N. A.;Wani, M. R.; Kazi, A. G.; Tran, L. S. P. Alleviation of Cadmium Toxicity in (*Brassica juncea L.*) by Calcium Application Involves Various Physiological and Biochemical Strategies. *PLoS One* **2015,** *10 (1)*, 0114571.

Ali, H. Phytoremediation of Heavy Metals Concepts and Applications. *Chemosphere* **2013,** *91*, 869–881.

Alkorta, I. Chelate Enhanced Phyotoremediation of Soils Polluted with Heavy Metals. *Rev. Environ. Sci. Biotechnol.* **2004,** *3*, 55–70.

Anawar, H. M.; Garcia-Sanchez, A.; Alam ,M. T. K.; Rahman, M. M. Phytofiltration of Water Polluted with Arsenic and Heavy Metals. *Int. J. Environ. Pollut.* **2008,** *33*, 292–312. https://doi.org/10.1504/IJEP.2008.0194

Bulak, P. Plant Growth Regulators Assisted Phytoextraction. *Biol. Plant.* **2014,** *58*, 1–8.

Chanu, L. B.; Gupta, A. Phytoremediation of Lead using Ipomoea Aquatica Forsk in Hydroponic Solution. *Chemosphere* **2016,** 156, 407–411.

Chattopadhyay, S. Phytoremediation of Mercury and Methyl Mercury Contaminated Sediments by Water Hyacinth (*Eichhornia crassipes*). *Int. J. Phytoremediation.* **2012,** *14*, 142–161.

Chaudhary, H. J. The Potential of the Flora from Different Regions of Pakistan in Phytoremediation: A Review. *Environ. Sci. Pollut. Res. Int.* **2014,** *21*, 801–812.

Colangelo, E. P.; Guerinot, M, L. Put the Metal to the Petal: Metal Uptake and Transport Throughout Plants. *Curr. Opin. Plant Biol.* **2006,** *9 (3)*, 322–330.

Cosio, C. Effects of Macrophytes on the Fate of Mercury in Aquatic Systems. *Environ. Toxicol. Chem.* **2014,** *33*, 1225–1237.

Das, S.; Goswami, S.; Das ,T. A. Copper Hyperaccumulating Plants from Barak Valley, South Assam, India for Phytoremediation. *Int. J. Toxicol. Pharmacol. Res.* **2013,** *5*, 30–32.

Deng, L.; Li, Z.; Wang, J.; Liu, H.; Li, N.; Wu, L.; Hu, P.; Luo, Y.; Christie, P. Long Term Field Phytoextraction of Zinc/Cadmium Contaminated Soil by Sedum Plumbizincicola under Different Agronomic Strategies. *Int. J. Phytoremediation.* **2016,** *18 (2)*, 134–140.

Dhankher, O. P.; Altman A.; Hasegawa, P. M. Biotechnological Approaches for Phytoremediation. *Plant Biotechnol. Agric.* **2012,** 309–328.

Dhir, B.Potential of Aquatic Macrophytes for Removing Contaminants from the Environment. *Crit. Rev. Plant Sci.* **2009,** *39*, 754–781.

Dixit, R.; Wasiullah, M. D.; Pandiyan, K.; Singh, U. B.; Sahu, A.; Shukla, R.; Singh, B. P.; Rai, J. P.; Sharma, P. K.; nLade, H.; Paul, D. Bioremediation of Heavy Metals from Soil and Aquatic Environment: An Overview of Principles and Criteria of Fundamental Processes. *Sustainability* **2015,** *7*, 2189–2212.

Dobos, L.; Puia, C.The Most Important Methods for Depollution of Hydrocarbons Polluted Soils. *Bull. UASMV Agric.* **2009,** *66*, 67–72.

Dushenkov, V.; Kapulnik, Y. In *Phytofiltration of Metals Phytoremediation of Toxic Metals Using Plants to Clean-Up the Environment*, 2000; pp 89–106.

Dushenkov, V. Rhizofiltration: The Use of Plants to Remove Heavy Metals from Aqueous Streams. *Environ.Sci.Technol.* **1995,** *29*, 1239–1245.

Ebbs, S, D. Accumulation of Zinc, Copper, or Ceriumin Carrot (*Daucus carota*) Exposed to Metal Oxide Nanoparticles and Metal Ions. *Environ. Sci. Nano* **2016**.

El-Deeb, B. Molecular Characterization of Endophytic Bacteria from Metal Hyperaccumulator Aquatic Plant (Eichhornia crassipes) and its Role in Heavy Metal Removal. *Geomicrobiol. J.* **2012,** *29*, 906–915.

Epelde, L. Effects of Chelates on Plants and Soil Microbial Community: Comparison of EDTA and EDDS for Lead Phytoextraction. *Sci. Total Environ.* **2008,** *410*, 21–28.

Gao, Y. Metal Resistant Microorganisms and Metal Chelators Synergistically Enhance the Phytoremediation Efficiency of *Solanum nigrum L.* in Cd and Pb Contaminated Soil. *Environ. Technol.* **2012,** *33*, 1383–1389.

Ghosh, M.; Singh, S. P. A Review on Phytoremediation of Heavy Metals and Utilization of it's by Products. *Appl. Ecol. Environ. Res.* **2005,** *3*, 1–18.

Gill, M. Phytoremediation : Green Technology to Clean the Environment. *Int. J.* **2014,** *2*, 879–886.

Guimaraes, M, D, A. Toxic Idadee Tolerânciaao Cádmioem Plantas. *Rev. Trópica–Ciências Agrare Biológicas2*. **2008**, 58–68.

Hajara, E. W. I. Assessment of Heavy Metals Tolerance in Leaves, Stems and Flowers of *Stevia rebaudiana* Plant. *Procedia Environ. Sci.* **2014,** *20*, 386–393.

Harguinteguy, C. A.; Schreiber, R.; Pignata, M. L. Myriophyllum Aquaticum as a Biomonitor of Water Heavy Metal Input Related to Agricultural Activities in the Xanaes River. *Ecol Indic*. **2013,** *27*, 8–16. https ://doi.org/10.1016/j.ecolind.2012.11.018

Haydon, M. J.; Cobbett, C. S. Transporters of Ligands for Essential Metal Ions in Plants. *New Phytol*. **2007,** *174*, 499–506.

Ibanez, S. Transgenic Plants and Hairy Roots: Exploiting the Potential of Plant Species to Remediate Contaminants. *N. Biotechnol.* **2015,** *1871–6784*, 00267–00268.

Imron, M. F.;Kumiawan, A. S.; Wahyudianto. In *Heliyon*, 2019; p 2206.

Jeong, S.; Moon, H.; Nam, K.; Kim, J.; Kim, T. Application of Phosphate Solubilizing Bacteria for Enhancing Bioavailability and Phytoextraction of Cadmium (Cd) from Polluted Soil. *Chemosphere* **2012,** *88*, 204–210.

Ji, P.; Sun, T.; Song, Y.; Ackland, M.; Liu, Y. Strategies for Enhancing the Phytoremediation of Cadmium-Contaminated Agricultural Soils by Solanum Nigrum L. *Environ. Pollut.* **2011,** *159*, 762–768.

Jiang, Y.; Lei, M.; Duan, L.; Longhurst, P. Integrating Phytoremediation with Biomass Valorisation and Critical Element Recovery: A UK Contaminated Land Perspective. *Biomass Bioenergy* **2015,** *83*, 328–339.

Kamran, M. A.; Amna Mufti, R.; Mubariz, N.; Syed, J.H.; Bano, A.; Javed, M. T.; Munis, M. F. H.; Tan, Z.; Chaudhary, H. J. The Potential of the Flora from Different Regions of Pakistan in Phytoremediation: A Review. *Environ. Sci. Pollut. Res. Int.* **2014,** *21*, 801–812.

Kacálková, L. Phytoextraction of Risk Elements by Willow and Poplar Trees. *Int. J. Phytoremediat*. **2015,** *17*, 414–421.

Khan, M. S.; Zaidi, A.; Wani, P. A.; Oves, M. Role of Plant Growth Promoting Rhizobacteria in the Remediation of Metal Contaminated Soils. *Environ. Chem. Lett*. **2009,** *7* (1), 1–19.

Kramer, U. Phytoremediation : Novel Approaches to Cleaning up Polluted Soils. *Curr. Opin. Biotechnol*. **2005,** *16*, 133–141.

Krishna, R. Rhizofiltration: A Green Technology for Remediation of Heavy Metals. *Int. J. Innov. Biosci*. **2012,** *2*, 193–199.

Leung, H. M. Mixed Arbuscular Mycorrhizal (AM) Fungal Application to Improve Growth and Arsenic Accumulation of Pteris Vittata (As hyper- accumulator) Grown in As-contaminated Soil. *Chemosphere* **2013,** *92*, 1367–1374.

Li, J. T. Cadmium Tolerance and Accumulation in Cultivars of a High Biomass Tropical Tree (*Averrhoa carambola*) and its Potential for Phytoextraction. *J. Environ. Qual.* **2010,** *39*, 1262–1268.

Liao, S. W.; Chang, W. L. Heavy Metal Phytoremediation by Water Hyacinth at Constructed Wetlands inTaiwan. *Photogramm. Eng. Remote Sens.* **2004,** *54*, 177–185.

Liu, D.; Islam, E.; Li, T.Q.; Yang, X. E.; Jin, X. F.; Mahmood, Q. Comparison of Synthetic Chelators and Low Molecular Weight Organic Acids in Enhancing Phytoextraction of Heavy Metals by Two Ecotypes of Sedum alfredii Hance. *J. Hazard Mater*. **2008,** *153*, 114–122.

Luo, J.; Cai, L.; Qi, S.; Wu, J.; Gu, X. S. Heavy Metal Remediation with *Ficus microcarpa* Through Transplantation and its Environmental Risks through Field Scale Experiment. *Chemosphere* **2018,** *193*, 244–250.

Mahar, A.; Wang, P.; Ali, A.; Awasthi, M. K.; Lahori, A. H.; Wang, Q.; Li, R.; Zhang, Z. Challenges and Opportunities in the Phytoremediation of Heavy Metals Contaminated Soils: A Review. *Ecotoxicol. Environ. Saf.* **2016,** *26*, 111–121.

Maine, M. A. Cadmium Uptake by Floating Macrophytes. *Water Res*. **2001,** *35*, 2629–2634.

Masciandaro, G. Organic Matter Microorganism Plant in Soil Bioremediation: A Synergic Approach. *Rev. Environ. Sci. Bio/Technol.* **2013,** *12*, 399–419.

Mesjasz Przybyłowicz, J. Uptake of Cadmium, Lead, Nickel and Zinc from Soil and Water Solutions by the Nickel Hyperaccumulator Berkheya coddii. *Acta Biol.Crac.* **2004,** *46*, 75–85.

Mokhtar, H.; Morad, N.; Fizri, F. F. A. Phytoaccumulation of Copper from Aqueous Solutions using *Eichhornia crassipes* and *Centella asiatica*. *Int. J. Environ. Sci. Dev.* **2011,** *2*, 205–210. https ://doi.org/10.7763/IJESD.2011.V2.125

Muthusaravanan, S.; Sivarajasekar, N.; Vivek, J. S.; Paramasivan, T.; Naushad, Mu.; Prakashmaran, J.; Gayathri, V.; Omar, K. A. Phytoremediation of Heavy Metals: Mechanism, Methods and Enhancements. *Environ. Chem. Lett.* **2018,** *16 (4)*, 1339–1359

Najeeb, U. Enhancing the Lead Phytostabilization in Wetland Plant Juncus effusus L. Through Somaclonal Manipulation and EDTA Enrichment. *Arabian J. Chem.* **2014,** *10*, S3310–S3317.

Nwoko, C. O. Trends in Phytoremediation of Toxic Elemental and Organic Pollutants. *Afr. J. Biotechnol.* **2010,** *9*, 6010–6016.

Olguín, E. J.; Sánchez-Galván, G. Heavy Metal Removal in Phytofiltration and Phycoremediation: The Need to Differentiate between Bioadsorption and Bioaccumulation. *New Biotechnol.* **2012,** *30*, 3–8.

Ong, S. T. Papaya (*Carica papaya*) Seed as a Low Costs or Bent for Zinc Removal of Trace Metals with Mango Seed Powder. *Indian J. Chem. Technol.* **2002,** *9*, 540–542.

Padmavathiamma, P. K.; Li, L. Y. Phytoremediation Technology: Hyperaccumulation Metals in Plants. *Water Air Soil Pollut.* **2007,** *184*, 105–126.

Pilon-Smits, E, A.; Freeman, J. L. Environmental Cleanup using Plants :Bio- Technological Advances and Ecological Considerations. *Front. Ecol. Environ.* **2006,** *4*, 203–210.

Pivetz, B. E. Ground Water Issue: Phytoremediation of Contaminated Soil and Ground Water at Hazardous Waste Sites, 2001; pp 1–36.

Qadir, S.; Jamshieed, S.; Rasool, S.; Ashraf, M.; Akram, N. A.; Ahmad, P. Modulation of Plant Growth and Metabolism in Cadmium Enriched Environments. *Rev. Environ. Contamin Toxicol.* **2014,** *229*, 51–88.

Radziemska, M. In *Civil and Environmental Engineering Reports*, 2017; vol *26*, pp 079–089.

Rai, P. K. Heavy Metal Pollution in Aquatic Ecosystems and its Phytoremediation using Wetland Plants: An Ecosustainable Approach. *Int. J. Phytoremediat.* **2008,** *10*, 133–160.

Ravikumar, K.; Sheeja, A. K. Heavy Metal Removal from Water using Moringa Oleifera Seed Coagulant and Double Filtration. *Int. J. Sci. Eng. Res.* **2013,** *4*.

Rizwan, M.; Ali, S.; Rehman, M. Z. U.; Rinklebe, J.; Tsang, D. C. W.; Bashir, A.; Maqbool, A.; Tack, F. M. G.; Ok, Y. S. Cadmium Phytoremediation Potential of *Brassica* Crop Species: A Review. *Sci. Total Environ.* **2018,** *631–632*, 1175–1191.

Ruppert, L.; Lin, Z. Q.; Dixon, R. P.; Johnson, K. A. Assessment of Solid Phase Microfiber Extraction Fibers for them of Volatile Organo Arsinicals Emitted from a Plant–Soil System. *J. Hazard. Mater.* **2013,** *262*, 1230–1236.

Salt, D. E. Phytoremediation: A Novel Strategy for the Removal of Toxic Metals from the Environment using Plants. *Nat. Biotechnol.* **1995,** *13*, 468–474.

Sarma, H. Metal Hyper Accumulation in Plants : A Review Focusing on Phytoremediation Technology. *J. Environ. Sci. Technol.* **2011,** *4*, 118–138.

Sarwat, M.; Hashem, A.; Ahanger, M. A.; AbdAllah, E. F.; Alqarawi, A. A.; Alyemeni, M. N.; Gucel, S. Mitigation of NaCl Stress by Arbuscular Mycorrhizal Fungi through the

Modulation of Osmolytes, Antioxidants and Secondary Metabolites in Mustard (*Brassica juncea* L.) Plants. *Front. Plant Sci.* **2016,** *7*, 869.

Sekara, A. Cadmium and Lead Accumulation and Distribution in the Organs of Nine Crops: Implications for Phytoremediation. *Pol. J. Environ. Stud.* **2005,** *14*, 509–516.

Shahandeh, H.; Hossner, L. R. Plant Screening for Chromium Phytoremediation. *Int. J. Phytoremediation.* **2000,** *2*, 31–51. https ://doi.org/10.1080/15226 51000 85000 29.

Shaheen, S. M., Rinklebe, J. Phytoextraction of Potentially Toxic Elements by Indian Mustard, Rapeseed, and Sunflower from a Contaminated Riparian Soil. *Environ. Geochem. Health.* **2015,** *37* (6), 953–967.

Sharma, R. Rhizosphere Biology of Aquatic Microbes in Order to Access their Bioremediation Potential Along with Different Aquatic Macrophytes. *Recent Res. Sci. Technol.* **2013,** *5*, 29–32.

Sharma, S. Phytoremediation: Role of Terrestrial Plants and Aquatic Macrophytes in the Remediation of Radionuclides and Heavy Metal Contaminated Soil and Water. *Environ. Sci. Pollut. Res.* **2015,** *22*, 946–962.

Sheoran,V.; Sheoran, A. S.; Poonia, P. Phytomining: A Review. *Min. Eng.* **2009,** *22*, 1007–1019.

Simmons, R. W. Towards Practical Cadmium Phytoextraction with *Noccaea caerulescens. Int. J. Phytoremediation.* **2015,** *17*, 191–199.

Singh, S. Phytoremediation: A Sustainable Alternative for Environmental Challenges. *Int. J. Green. Herb. Chem.* **2012,** *1*, 133–139.

Smolinska, B.; Krol, K. Leaching of Mercury During Phytoextraction Assisted by EDTA, KI and Citric Acid. *J. Chem. Technol. Biotechnol.* **2012,** *87*, 1360–1365.

Sorrentino, M. C.; Capozzi, F.; Amitrano, C.; Giordano, S.; Arena, C.; Spagnuolo, V. Performance of Three Cardoon Cultivars in an Industrial Heavy Metal-Contaminated Soil: Effects on Morphology, Cytology and Photosynthesis. *J. Hazard. Mater.* **2018,** *351*, 131–137.

Subhasini, V.; Swamy, A. V. V. S. Phytoremediation of Cadmium and Chromium Contaminated Soils by *Cyperus Rotundus* L. *Am. Int. J. Res. Sci. Technol. Eng. Math.* **2014,** *6*, 97–101.

Sumiahadi, A.; Acar, R. A Review of Phytoremediation Technology: Heavy Metals Uptake by Plants *Earth Environ. Sci.* **2018,** *142*, 012023.

Tangahu, B. V. A Review on Heavy Metals (As, Pb, and Hg) Uptake by Plants through Phytoremediation. *Int. J. Chem. Eng.* **2011,** 1–31.

Terry, N.; Banuelos, G, S. In *Phytoremediation of Contaminated Soil and Water*; CRC Press, 1999; p 408.

Terry, N.; Banuelos, G. In *Phytoremediation of Contaminated Soil and Water*; CRC Press: Boca Raton, FL, 2000.

Thayaparan, M. Rhizofiltration of Pb by Azolla pinnata. *Int. J. Environ. Sci.* **2013,** *3* (6), 3.

Udokop,U. Studies on the Blastofiltration Potentials of Selected Plant Seeds for the Treatment of Lead and Cadmium Contaminated Water Samples. In *Chemistry Africa*, 2016; vol 6, p 37.

Un Nisa, W.; Rashid, A. Potential of Vetiver (*Vetiveria zizanioides* L.) Grass in Removing Selected Pahs from Diesel Contaminated Soil. *Pak. J. Bot.* **2015,** *47*, 291–296.

Vangronsveld, J. Phytoremediation of Contaminated Soils and Groundwater: Lessons from the Field. *Environ. Sci. Pollut. Res.* **2009,** *16*, 765–794.

Wang, J. Remediation of Mercury Contaminated Sites–A Review. *J. Hazard. Mater.* **2012,** *221*, 1–18.

Wang, L.; Hou, D.; Cao, Y.; Ok, Y. S.; Tack, F. M. G.; Rinklebe, J.; O'Connor, D. Remediation of Mercury Contaminated Soil, Water, and Air: A Review of Emerging Materials and Innovative Technologies. *Environ. Int.* **2020,** *134*, 105–281.

Xia, H.; Liang, D.; Chen, F.; Liao, M.; Lin, L.; Tang, Y.; Lv, X.; Li, H.; Wang, Z.; Wang, X.; Wang, J.; Liu, L.; Ren, W. Effects of Mutual Intercropping on Cadmium Accumulation by the Accumulator Plants *Conyza canadensis*, *Cardamine hirsuta*, and *Cerastium glomeratum*. *Int. J. Phytoremediation*. **2018,** *20*, 855–861.

Xu, W.; Lu, G.; Dang, Z.; Liao, C.; Chen, Q.; Yi, X. Uptake and Distribution of Cd in Sweet Maize Grown on Contaminated Soils: A Field-Scale Study. *Bioinorg. Chem. Appl.* **2013.**

Yang, W.; Dai, H.; Skuza, L.; Wei, S. Strengthening Role and the Mechanism of Optimum Nitrogen Addition in Relation to *Solanum nigrum* L. Cd Hyperaccumulation in Soil. *Ecotoxiol. Environ. Saf.* **2019,** *182*, 109–444.

Yoon, J. Remediation of Pb, Cu, and Zn in Native Plants Growing on a Contaminated Florida Site. *Sci.Total Environ.* **2006,** *368*, 456–464.

Yu, G.; Liu, J.; Long, Y.; Chen, Z.; Sunahara, G. I.; Jiang, P.; You, S. Y.; Lin, H.; Xiao, H. Phytoextraction of Cadmium-Contaminated Soils: Comparison of Plant Species and Low Molecular Weight Organic Acids. *Int. J. Phytoremediation*. **2019,** *22*, 1–9.

Zhang, Z. Phytostabilization of Acidic Soils with Heavy Metal Contamination using Three Forage Grasses in Combination with Organic and Inorganic Amendments. *Soil Sediment Contam. Int. J.* **2016,** *25*, 1–10.

Zayed, A.; Gowthaman ,S.; Terry, N. Phytoaccumulation of Trace Elements by Wetland Plants: I. Duckweed. *J. Environ. Qual.* **1998,** *27*, 715–721.

Zurayk, R.; Sukkariyah, B.; Baalbaki, R.; Ghanem, D. A. Ni Phytoaccumulationin *Mentha aquatica* L. and *Mentha sylvestris* L. *Water Air Soil Pollut*. **2002,** 139, 355–364. https :// doi.org/10.1023/A:1015840601 761.

CHAPTER 11

The Headway of Nanotechnology-Mediated Approaches Toward Environmental Clean-up

KINGSUK DAS[1,2+], DEBOMITA DEY[3+], and PRITHUSAYAK MONDAL[4]

[1]Department of Agricultural Biotechnology, Assam Agricultural University, Jorhat, Assam, India

[2]Department of Agricultural Biotechnology, Faculty of Agriculture, Bidhan Chandra Krishi Viswavidyalaya, Mohanpur, Nadia, West Bengal, India

[3]Department of Agricultural Chemistry & Soil Science, Faculty of Agriculture, Bidhan Chandra Krishi Viswavidyalaya, Mohanpur, Nadia, West Bengal, India

[4]Regional Research Station (Terai Zone), Uttar Banga Krishi Viswavidyalaya, Pundibari, Cooch Behar, West Bengal, India

[+]Equal contribution

ABSTRACT

The surging pollution in the environment is one of the burning issues that our planet is facing today. The different environmental segments such as air, water, soil, etc., are depraved with various pollutants and contaminants in the form of particulate matter, heavy metals, synthetic chemicals, industrial effluents, sewage-sludge, oil spills, toxic gases, and many others. Although various conventional remediating approaches are already in action, still researchers worldwide are looking for unique strategies to deal with this

Environmental Contaminants: Impact, Assessment, and Remediation.
Pritam Ganguly, Jajati Mandal, M. Paramsivam, & Sandip Patra, (Eds.)

menace. Simultaneously, a new technological wing has evolved that deals with manipulating matters at nanoscale. Materials fabricated via nanotechnological approaches have proved to be more potent in environmental monitoring and cleanup. Nanomaterials owing to their unique shape as well as size (enlarged surface area) show excellent reactivity toward physicochemical processes such as adsorption, catalysis, and sensing. This book chapter encompasses the progress of nanotechnology-mediated approaches toward environmental clean-up, in particular, applications of various nanomaterials to treat emerging hazardous wastes in air, water, and soil with reduced cost and energy and improved efficiency.

11.1 INTRODUCTION

The escalating pollution of our environs brewing from drastic urbanization, unrestricted industrialization, and callous lifestyle changes of people is indisputably one of the grave concerns of the modern society (Khin et al., 2012). These pollutants include particulate matter, heavy metals, synthetic pesticides, fertilizers, industrial effluents, sewage-sludge, oil spills, toxic gases, and various organic compounds (Vaseashta et al., 2007; Khan and Ghoshal, 2000). Several conventional methods and scientific strategies (e.g., absorption, adsorption, filtration, chemical reactions, photocatalysis, etc.) have been being explored for eradication of these contaminants from different environmental segments such as air, water, and soil (Guerra et al., 2018). Owing to the complexity in physico-chemical properties (e.g., high volatility, low reactivity, etc.) of the mixture of contaminants, the effective capture, and their subsequent degradation are very challenging. The technological advancements are nowadays being directed toward monitoring, detection, and cleanse the contaminants from different environmental compartments. In this framework, the onset of nanotechnology has unveiled immense scope and opportunities for environmental decontamination through the fabrication of desired nanomaterials with unusual features.

Nanotechnology can be considered as a cluster of different fledgling technologies that functions on ultra-small size (not more than 100 nm) to build gadgets, objects, and arrangements with radically unique characteristics and actions by manipulating the size, shape, and other attributes of the materials (Mondal et al., 2017). The spanking growth of nanotechnological approaches has attained a lot of attention in environmental applications due to excellent chemical reactivity/selectivity of nanoparticles having large surface-to-volume ratios. Moreover, the unique surface chemistry of nanomaterials confers them to be attached or engrafted with various functional groups, which can ultimately target pollutant molecules for proper redressing. The altered physico-chemical properties of fabricated nanomaterials can cater outstanding functionalities/selectivities those were found missing in their bulk counterparts. The enriched chemistry of surface modification coupled with the adjustable physical parameters of the nanomaterials present convincing benefits over traditional approaches for addressing environmental contamination (Yunus et al., 2012). The choice of fabricated nanomaterials employed for environmental remediation is also a vital issue. Attention must be paid so that the materials chosen for environmental remediation should have safe toxicological profiles. Biodegradability of nanomaterials is another principal concern as it builds up customer acceptance and satisfaction toward using the technology, in the belief that it can be less hazardous to the environment and become green and safe alternative to conventional methods. Moreover, facile synthesis, cost effectiveness, target-specific capture, recyclability, and regeneration capacity are the major challenges to be taken care of while fabricating advanced nanomaterials for environmental clean-up.

This chapter provides an insight into various facts on the current nanotechnological approaches and their crucial roles to reconcile the essential urge to track and treat the resulting pollutants with minimal cost, lesser energy, and higher efficiency.

11.2 THE CONCEPT, STATUS, AND SUITABILITY OF NANOTECHNOLOGICAL INTERVENTIONS TO THE ENVIRONMENTAL ISSUES

Nanotechnology is the engineering and scientific manipulation of functional systems at the atomic, molecular, and supramolecular scale. According to the definition established by the National Nanotechnology Initiative (NNI), "Nanotechnology is the understanding and control of matter at dimensions of roughly 1 to 100 nm, where unique phenomena enable novel applications" (Allhoff et al., 2010). The United States Food and Drug Administration (USFDA) also describes nanomaterials as "materials that have at least one dimension in the range of approximately 1 to 100 nm and exhibit dimension-dependent phenomena" (USFDA, 2011). Likewise, The International Organization for Standardization (ISO) has referred to nanomaterial as a "material with any external nanoscale dimension or having internal nanoscale surface structure" (ISO, 2010). Similarly, the European Union (EU) Commission explained the term nanomaterial by "a manufactured or natural material that possesses unbound, aggregated or agglomerated particles where external dimensions are between 1 and 100 nm size range" (Potocnik et al., 2011). Based on these definitions, different nano-related terms such as nanofibers, nanoplates, nanowires, nanopowders, nanoclusters, nanocrystals, etc., have been generated (Fig. 11.1).

In 1959, Richard Feynman, the famous American physicist and Nobel Prize laureate, brought in the concept of nanotechnology during his talk in

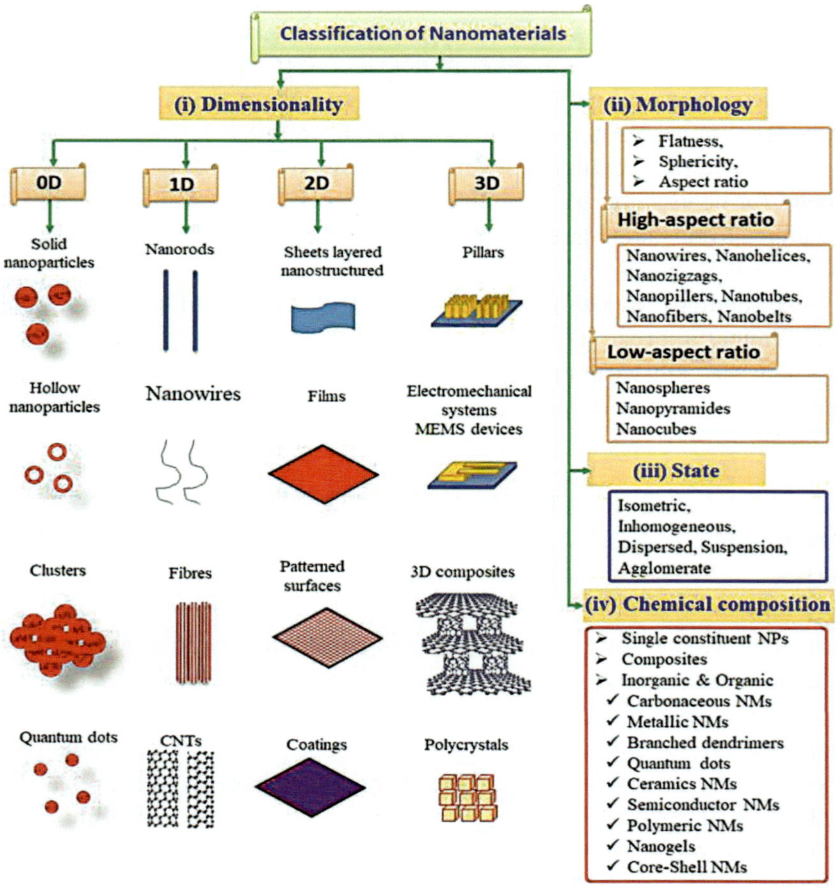

FIGURE 11.1 Different types of nanomaterials.

Source: Reprinted with permission from Saleh (2020). Copyright 2020 Elsevier Inc.

the annual meeting of the American Physical Society (APS) at the California Institute of Technology (Caltech), where he shed light on the possibility of synthesis via direct manipulation of atoms (Feynman, 1960). However, after several years, a Japanese scientist named Norio Taniguchi, defined and applied the term "nanotechnology" for the first time as: "nanotechnology mainly consists of the processing of separation, consolidation, and deformation of materials by one atom or one molecule" (Taniguchi et al., 1974).

Two main strategies or approaches are generally employed for the development of nanomaterials – Top-down and Bottom-up (Iqbal et al., 2012). Nanoscale entities are developed from large and bulky substances without atomic-level control in the top-down method whereas in the bottom-up method, nanomaterials are manufactured from the bottom (i.e., atom-by-atom or molecule-by-molecule) by physical and chemical mechanisms through desired manipulation of self-assembly of atoms and molecules in controlled manner.

Modern agriculture is bestowed with nanotechnology, which can be utilized for supplying novel tools for solving future troubles regarding food as well as energy demand with a sustainable approach. In recent times, it has flourished into an invaluably competent technology for environmental decontamination. With the aim of conserving our environment, a new field of environmental nanotechnology has developed with plenty of prospects. The scope and opportunity of environmental nanotechnology can be further expanded by the controlled adjustments of the nanoscale materials with regard to morphology (particle size and shape), microstructure (pore and surface), and composition (heterojunction and doping). Bulk environmental materials in their nanosizes, can offer fascinating features, for example, optical, thermal, magnetic, electrical, and electrochemical properties toward adsorptive and/or catalytic performances (Hochella et al., 2019). Therefore, efforts are constantly being thrust to obtain engineered nanomaterials combining the principles of nanotechnology with physico-chemical alteration of material surfaces that can overcome the lacunas associated with remedial measures (Kamat and Meisel, 2003; Pandey and Fulekar, 2012). Some nanomaterials, despite being advantageous over bulk ones, are intrinsically unstable under normal conditions, hence need special attention during fabrication at the nanoscale. Agglomeration and polydispersity of nanoparticles are other concerns that need to be prevented as much as possible. Besides, the possible toxicity of metal-based nanoparticles and their residues and recovery costs from the target site may limit their utility in remediation process. Therefore, a sound perception of the material settings, fabrication mechanism, and protocol optimization for superior performance is imperative to develop ideal nanomaterials capable of addressing environmental problems.

Environmental nanotechnology addresses every aspects of environmental management, ranging from interception to mitigation. The total environmental clean-up procedure comprises of three steps – (1) pollution/contamination prevention, (2) sensing and detection, and (3) treatment and remediation. There are several possible routes, whereby a pollutant can be alleviated, such as absorption, adsorption (physisorption and chemisorption), filtration, photocatalysis, and chemical reactions, as illustrated in Figure 11.2 (Tong et al., 2012; Zhao et al., 2011; Ng et al., 2013). A variety of nanomaterials has been utilized for environmental remediation. However, there is no established categorization regarding their specific types and definite uses. From the available literature survey, it has been found that there are three principal types of nanomaterials (i.e., inorganic, carbon-based, and polymer based) exploited for intended purposes whose classifications and applications will be highlighted in the following segments.

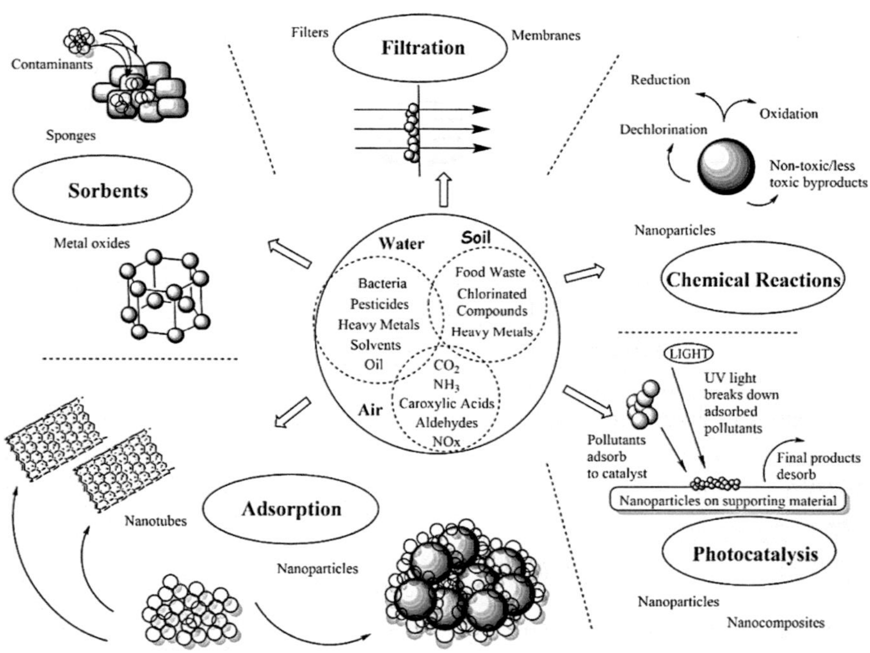

FIGURE 11.2 Environmental remediation approaches.

Source: Reprinted with permission from Guerra et al., 2018.

11.3 NANOTECHNOLOGICAL APPROACHES FOR AIR REMEDIATION

Air pollution, in broad sense, is nothing but alteration of the natural constituents in the atmosphere, which is mainly triggered by the inclusion of different genre of substances varying from physical, chemical, or biological in nature and those materials have broadly geogenic, biogenic, or in some cases anthropogenic sources. It is one of the key pollution problems entire world is facing today (Daly, 2007). The exposure to the poor air has negative impacts on the vegetation and living beings (especially humans) by presumably causing several types of respiratory and cardiovascular diseases. Our atmosphere is heavily loaded with various gaseous pollutants like carbon monoxide (CO), carbon dioxide (CO_2), chlorofluorocarbons (CFCs), hydrocarbons, volatile organic compounds (VOCs), and nitrogen oxides (N_XO_Y). These major air pollutants present in environment are represented in Figure 11.3 along with their relative sizes.

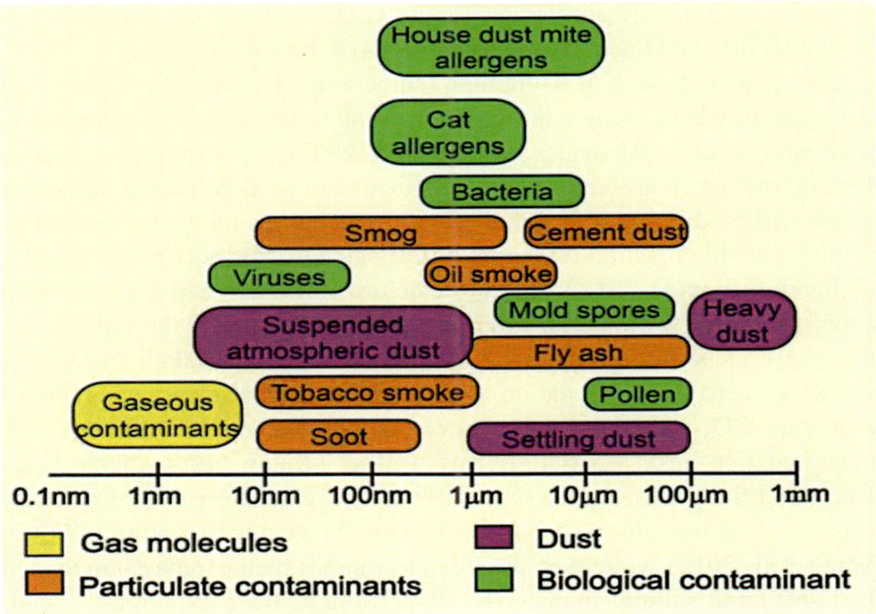

FIGURE 11.3 Type of air contaminants.

Source: Reprinted with permission from Mohamed, 2017.

The global air pollution can be summarized into two different types: outdoor air pollution and indoor air pollution.

11.3.1 OUTDOOR AIR POLLUTION

Global warming is considered as the most culprits in this aspect as it causes abrupt variations inside atmosphere as well as in land and water resources across the world. Greenhouse gases (GHGs) are supposed to be the most unswerving suppliers of the global warming. Many technologies have been developed for monitoring and elimination of these gases from atmosphere and minimization of their risks on human and other environmental segments. Nanotechnology has emerged now as a smart technology to mitigate air pollution by introducing the boons of nanomaterial properties; those can be applied as catalysts membranes, adsorbents, and sensors (Zhao, 2009). Various techniques were planned to control gaseous discharge either by separation or capturing. These include several processes like filtration, adsorption on solids, absorption in liquids, or a coalescence of all the above-mentioned techniques (Cheung et al., 2013). However, the adsorption process of nanoscale materials has been demonstrated to be the most effective and lucrative technique considering the nanomaterials' broad surface area, which ultimately accelerates the process of adsorption, and enhances the availability and regeneration capabilities of the nanomaterials.

The main drawback of using high temperature adsorbents involves easy aggregation, which creates the problem of sintering over the carbonation/calcination cycles (Abanades and Alvarez, 2003). To avert these circumstances, calcium (Ca)-based nano-adsorbents are predominantly utilized to adsorb carbon dioxide (CO_2) at elevated temperatures taking the advantage of the reversible reaction (carbonation) of calcium oxides (Ca_xO_y). Coating of titanium dioxide (TiO_2) on nano calcium carbonate can also avert the sintering issue of the material and capture CO_2 efficaciously through adsorption phase technique (Wang et al., 2013). Additionally, alkali metals (Li, Na, K, Cs, etc.) in conjugation with treated nanoparticles were reported to capture CO_2 at higher temperature, for instance, nano-sized citric acid pretreated rice husk ash (CRHA) as well as lithium ortho-silicate based (CRHA-Li_4SiO_4) adsorbents showed excellent characteristics of sintering resistance and was able to adsorb CO_2 at an elevated temperature of 700°C (Wang et al., 2014). Moreover, the alkali nanotubes such as potassium titanate (K-Ti-NT) and sodium titanate (Na-Ti-NT)) have been explored to detain CO_2 at comparatively low temperatures (not exceeding 200°C) (Upendar et al., 2012). Carbon-based nanomaterials (CNMs) are another example of low temperature adsorbents for CO_2. They are extensively used owing to their broad surface area and strong responsiveness against alteration of pore structure (Upendar et al., 2012). Functionalized carbon nanotubes (CNTs) have also catered significant advantages of detaining CO_2 and accelerating adsorption performance even in the presence of moisture that was found to compete with CO_2 for adsorption to the active sites (Jänchen et al., 2007).

Another successful nanotechnological approach in GHG treatment includes conversion or degradation of methane (CH_4) and nitrogen oxides (N_xO_y) via numerous catalytic processes. For example, nickel nanoparticles acted as vital catalysts in production of hydrogen through thermal degradation of methane (Wang and Lua, 2012). Similarly, TiO_2 coated with stainless steel web net showed excellent performance in photocatalytic CH_4 decomposition (Merajin et al., 2013). Moreover, titanate nanotubes (TNTs) as well as their derivatives and by-products have been extensively testified for the photocatalytic oxidation of N_xO_y. Nanosized materials have also been employed to eradicate SO_2 discharge to the environment mainly by two mechanisms – first, by desulfurization of fossil fuel (Saleh et al., 2015), and second, by direct removal of these compounds from the release source by numerous know-hows, especially by different adsorption techniques (Wu et al., 2011) and catalytic reactions (Koutsopoulos et al., 2006; Rodriguez et al., 2010).

Table 11.1 displays ample examples of nanoscale materials those were utilized for the treatment of various types of pollutants dispersed in the atmosphere.

11.3.2 INDOOR AIR POLLUTION

In the indoor air environment, volatile organic compounds (VOCs) are the serious pollutants that need to be controlled and eradicated as they are supposed to cause several respiratory ailments (Hauptmann et al., 2004; Solomon et al., 2008). Out of the VOCs, formaldehyde (HCHO), an important precursor in the manufacturing of industrial resins, is the most abundant airborne carbonyl chemical. Various techniques have been employed to remove formaldehyde such as photocatalysis and physical and chemical adsorption (Nuasaen et al., 2014). Photochemical reactions require UV light and possibly generate ozone and other carcinogenic gases via secondary mechanisms (Miyawaki et al., 2012). According to Lee et al. (2010a, 2010b), the electrospun polyacrylonitrile (PAN)-based carbon nanofiber (CNF) membrane with custom-made microporosity and profuse nitrogenous functional groups exhibited significantly strong adsorption of HCHO even at

TABLE 11.1 Treatment of Different Air Pollutants by Nanomaterials.

Materials	Types of nanoparticles	Target pollutants	Removal mechanism	References
Carbon nanotubes	SWCNTs and MWCNTs	N_xO_y (mixture of NO and NO_2)	When NO and O_2 were passed through CNTs, NO was oxidized to NO_2 followed by adsorption on the surfaces of nitrate species	Long and Yang, 2001; Zhang et al., 2012
	CNTs-APTS	CO_2	Surface of CNTs having abundant amine groups provided with numerous chemical sites for CO_2 adsorption even at low temperature range (20–100°C)	Su et al., 2009
	SWCNTs/ NaClO	Isopropyl vapor	It was achieved by π–π interactions	Hsu and Lu, 2007
	CNTs deposited on quartz filters	VOCs	Physical adsorption by Van der Waals force and chemical adsorption onto adsorbent surface functional groups	Amade et al., 2014
	Si-doped and Boron-doped SWCNTs	CO and CH_2OH gases	Physisorption or Chemisorption, the electronic properties of SWCNT significantly enhanced the gas adsorption	Azama et al., 2017
Fullerene	fullerene B_{40}	CO_2	Strong chemisorption	Dong et al., 2015
	fullerene-like boron nitride nanocage	N_2O	Adsorption and decomposition of N_2O	Esrafili, 2017
Graphene	Graphene oxide/ nanocomposites	CO_2, NH_3, SO_2, H_2S, and N_2	Functional groups on graphene oxides were responsible for the adsorption of gases by promoting synergistic effect between metal centers and surfaces of the graphene oxide	Petit and Bandosz, 2009; Seredych and Bandosz, 2012; Wu et al., 2013
Silica	Si nanoparticle	Atmospheric Pb	The expanded surface and the presence of negatively charged groups (ions) in the Si nanoparticles promoted the rapid Pb capture	Yang et al., 2013c
Zinc oxide	$Zn_{12}O_{12}$ nanocage	CS_2	With the growing number of the CS_2 molecules, the adsorption energy of these compounds decreased as per individual molecule concerned that may be because of the steric repulsion between the CS_2 molecules	Ghenaatian et al., 2013

TABLE 11.1 *(Continued)*

Materials	Types of nanoparticles	Target pollutants	Removal mechanism	References
Magnesium ferrite	$MgFe_2O_4$ nanospheres	SO_2	At the time of adsorption of SO_2, sulfate, and sulfite classes were formed mainly on the surface of $MgFe_2O_4$, and Fe(III) which was partially reduced to Fe(II)	Zhao et al., 2010a
Silver-mesoporous silica nanocomposites	Ag nanoparticles	CO	The amount of silver and dispersion of the catalyst, being directly proportional with the pH value, increased with increasing pH, and ultimately resulted around 98% of total CO oxidation at pH 5.0 and at 70°C temperature	Zhang et al., 2011c

APTS, 3-aminopropyltriethoxysilane; CNTs, carbon nanotubes; MWCNTs, multi-walled carbon nanotubes; SBA-15, Santa Barbara Amorphous-15; SWCNTs, single-walled carbon nanotubes; VOCs, volatile organic compounds.

Source: Reprinted with permission from Mohamed et al., 2017.

low concentrations. Although, air humidity drastically curtailed the lifetime of the nanofiber membrane. To overcome this, manganese oxide (MnO_x) catalysts were dumped on the PAN-ACNFs. The combination of MnO_x with PAN-ACNF was found to be very much effective for providing a remarkable performance in formaldehyde removal mainly in dry and intensely humid circumstances. This was achieved by applying a double-stage removal process – (1) adsorption of HCHO in the PAN-ACNF micropores and (2) oxidative decay by MnO_x nanoparticles deprived of any UV radiation (Miyawaki et al., 2012).

A supplementary example on indoor air pollutants which has received an enormous attention recently is bioaerosols (i.e., airborne collection of biological species like viruses, bacteria, and fungi) having typical feature of rapid spread with airflow, capable of numerous diseases, especially various infections, and allergies (Stark et al., 2003). Air purifier technology can be considered as the most fruitful technique to eliminate bioaersols through ventilation processes using Ag nanoparticles, Cu nanoparticles, CNTs, and natural products as antimicrobial agents. In spite of the enormous efforts made to restore indoor air quality (IAQ), the human exposure to nanomaterial-based filters mainly in indoor environmental conditions may result in several detrimental health consequences, including dermal irritation, peribronchial inflammation and necrosis, and inflammation in the mucous membranes (Lam et al., 2004; Warheit et al., 2007). As the natural antimicrobial substances are thought to produce less toxic effects to humans in comparison with antimicrobial agents containing Ag, CNTs, and metal oxides, and therefore, these natural products are being widely used to elevate the indoor air quality. For example, plant essential oils display promising results for controlling bacterial population, and hence, these are used in ventilation systems of different indoor-contaminated surroundings (Pibiri et al., 2006; Pyankov et al., 2008). Some of the natural nanoparticles (e.g., *Sophora flavescens*) have been placed on the different filter media surface through an aerosol process, still they were unable to keep their spherical shape and coalesced most often on fiber filters under humid environments (Hwang et al., 2012). Based on the previous result, the electro-spraying method, backed by high-intensity electric fields, is applied to produce natural nanoparticles for improving stability and increasing efficacy for the removal of bioaerosols (Jung et al., 2013). Thus, the application of filters built from electro-spraying technique can act as a novel gifted technology for monitoring air quality.

11.4 NANOTECHNOLOGICAL APPROACHES FOR WATER PURIFICATION

As the global water demand is increasing day-by-day, so it becomes more challenging to provide sufficient amount of clean water among people (Water, 2007). As per prediction by World Water Council, around 3.9 billion people by 2030 will be breathing in a "water scarce" area (Lewin et al., 2007). Due to poor sanitization, there is steadily growing infections due to water borne diseases (Ramírez-Castillo et al., 2015). The United States Environmental Protection Agency (US EPA) has classified the water pollution into the subsequent six groups: (1) plant nutrients, (2) biodegradable wastes, (3) heat, (4) sediments, (5) hazardous and toxic chemicals, and (6) radioactive pollutants. Hence, pollutants present in water are composite in nature and these include industrial discharges including heavy metals and diverse anions, organic pollutants, pathogens, etc., (Goyal et al., 2013). These pollutants after mixing with water bodies cause modification of properties of water bodies as they cannot be broken down by natural means.

The wastewater management, conglomerating physical, chemical, and biological methods, usually encompasses of four phases: (1) preliminary, (2) primary, (3) secondary, and (4) tertiary advanced treatment. The techniques normally followed for water cleansing are coagulation and flocculation, sedimentation, filtration, steam distillation, dissolved air flotation, ion exchange, deionization, reverse osmosis, and disinfection (Shon et al., 2007). Substances that are normally utilized in these machineries are coagulants, sediment filters, ion exchangers, activated carbon, ceramics, organic polymers, activated alumina, and many hybrid ingredients (Hotze and Lowry, 2011). Although, these traditional procedures are expensive and might release secondary toxic pollutants in surrounding environments (Gaya and Abdullah 2008). The current advancements in nanotechnological approaches will truly help developing many opportunities to create diverse sources of water supply systems backed by fresh and cost efficient handling competences that can eradicate major constraints incurred in currently used technologies (Qu et al., 2013a).

11.4.1 NANOFILTRATION

Membrane filtration is one of significant method for water purification by eradicating numerous types of contaminants. Reverse osmosis (RO), that is, high-pressure membrane cleaning technique has recently become very wide-spread for wastewater treatment owing to its sound efficacy in eradication of solid waste materials, monovalent and divalent ions, numerous pathogens, etc. Nanofiltration, unlike RO, requires lower energy due to significantly reduced drive pressure (i.e., 7–14 bar). Nanofiltration generates water that encounters extremely rigid specifications regarding water reuse. Because of its high efficiency in elimination of the organic and inorganic entities, viruses and bacteria, the necessity for consequent water disinfection lies minimum.

Nanotechnology bids a vast range of outcomes for different membrane stuffs, including ceramic membranes as well as unique polymeric membranes with anti-fouling smear, new composite membranes combining polymer with various materials such as carbon nanotubes, metal/metallic oxides, zeolites, and aquaporin (Lens et al., 2013). According to Pendergast and Hoek (2011), all membrane materials utilized in nanofiltration can be divided into three groups: nanostructured ceramic membranes, organic–inorganic membranes, and biologically inspired membranes. These biologically inspired membranes possess enough potential to be improved further for fruitful applications.

11.4.2 NANOMATERIALS FOR CATALYSIS AND PHOTOCATALYSIS

Metal and metal oxide nanoparticles due to their highly reactive nature during redox reactions show robust catalytic activities through which pollutant molecules are converted into less toxic and ecologically acceptable final products (Lens et al., 2013). In advanced oxidation process, nano catalysts play excellent role for treating inorganic and organic pollutants by producing extremely responsive radicals that readily react with contaminant molecules. However, excessive cost for supplying the requisite energy (ultrasonicators, UV lamps, ozonators, etc.) often limits its application. Photocatalysis is the most noteworthy oxidation process initiated by adsorption of a photon having sufficiently higher energy required to overcome the interstitial gap between valence band and conduction band of a semiconductor (Fig. 11.4). During illumination of the catalytic surface by photons, electrons moved to the empty conduction shell from the valance shell with the formation of a "hole" in the valance shell with a positive charge. This pairing of electron-hole generates extremely reactive radicals capable of capturing pollutant molecules and their subsequent destruction (Bora and Dutta, 2014).

However, there are certain technical glitches that must be resolved to endow widespread exercise of this technology including catalysts optimization in the utilization of accessible light energy, utilitarian, and more effectual separation of nano-catalysts after treatment and re-application, and enrichment of specific features at the time of chemical reactions. Here also, lofty operating cost for providing required light energy is the main culprit. To make it economically viable, researchers have started concentrating on

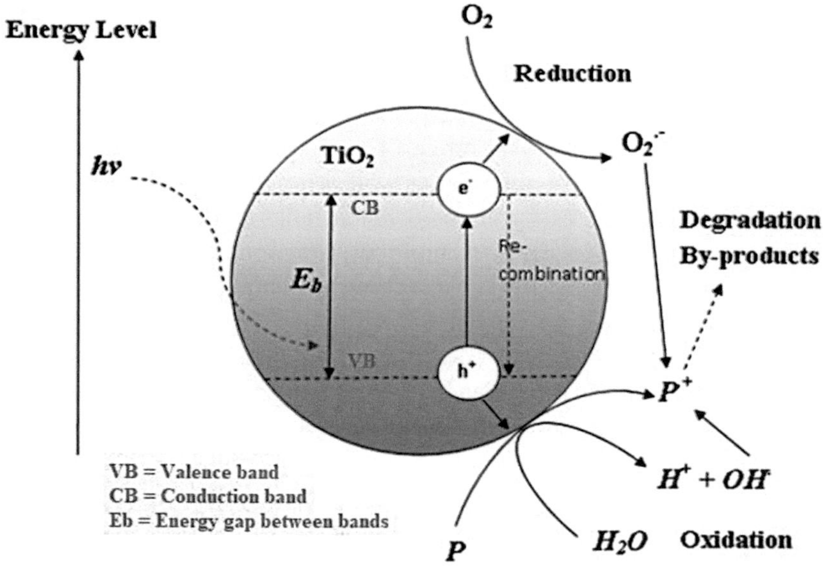

FIGURE 11.4 Photo-degradation of water pollutant (P) by using a semiconductor photocatalyst.

Source: Reprinted with permission from Chong et al., 2010. Copyright 2010 Elsevier Inc.

the utilization of natural energy for photocatalytic methods. Solar energy is a good option for testing in open-air processes. The application of renewable energy in the wastewater treatment launches new prospects for the advancement of effective and commercially feasible technologies (Bora and Dutta, 2014).

11.4.3 NANOMATERIALS FOR WATER DISINFECTION

Certain nanomaterials like carbon nanotubes, silver nanoparticles, titanium dioxide (TiO_2), chitosan, fullerene nanoparticles, etc., are reported to possess excellent antimicrobial activities. These materials cater in water disinfection via different routes such as direct action on different microbial cells and prevention of electron movement through membranes, rupture of cell membranes, cellular oxidation, formation of hydroxyl radicals, and dissolved metal ions responsible for damaging cellular components (Li et al., 2008). Being mild oxidants and relatively inert in water, these nanomaterials hardly produce any harmful by-products. However, there are certain issues that are limiting widespread use of this technique. Direct and strong contact of these nanomaterials with cell membrane or other reactive surfaces of microorganisms are very essential for effective removal of germs. Absence of subsequent (residual) antimicrobial action in the wastewater (such as when chlorine is used) is another disadvantage. As the traditional disinfection procedures require low monetary inputs, nanotechnological approaches will become more pertinent after standardization of nanomaterial development following inexpensive routes.

11.4.4 NANOMATERIALS FOR ADSORPTION OF POLLUTANTS

Large surface area and its multifunctionality (ability to easily bind or react to adjacent atoms or particles) provide nanoparticles with excellent adsorption capacity and long-term stability; however, it sometimes promotes adsorbent degradation using the catalytic properties of nanoparticles. Furthermore, nanoscale materials own unique optical, electrical, mechanical, chemical, and many other properties that enhance their adsorption characteristics for contaminants like heavy metals and various organic pollutants. CNTs proved to the best nanomaterials for adsorption of pollutants. Other than CNTs, metal-based nanoparticles such as silicon (Si), titanium (Ti), and tungsten (W) and metal oxide nanoparticles such as are iron oxides (Fe_xO_y)

are predominantly used for adsorption of different types of heavy metals as well as radionuclides. The adsorption phenomenon is primarily dependent on the electrostatic interaction between the nanoadsorbent surface and the dissolved metals in wastewater. The surface of the nano-adsorbents may be acidic (with anions attracting positive charges) or basic (with cations attracting negative charges) and any alteration in the pH of the solution can drastically affect the strength of this interaction (Qu et al., 2013).

11.4.5 NANOSCALE ZEROVALENT IRON (NZVI)

Remediation of wastewater via nanotech intervention is attaining popularity day-by-day. The zerovalent iron nanoparticles (nZVI) is regarded as the most fascinating nanomaterial because of its trifling manufacture costs, better environmental outcomes, and excellent reactivity with contaminants. This nZVI technology has already been adopted in commercial level, especially for contaminated ground water treatment as it proved to be a brilliant technique that can be used to remove different organic and inorganic pollutants, including numerous pesticides, chlorinated solvents, nitro amines, arsenic, inorganic anions, uranium as well as numerous metals (Crane and Scott, 2012). For both mobile and immobile contaminants, this nZVI injection method can be applied. As there is hardly any research on its negative impact to the environment, great care and emphasis should be given on the behavior of nZVI in the underground system. Detailed knowledge backed by transparent understanding along with anticipated reduction in total expenditure will surely offer nanotechnology a competing space in comparison with the alternative conventional methods.

Below in Table 11.2, miscellaneous nanosized materials those carry out efficient removal of various water pollutants are reported.

11.5 NANOTECHNOLOGICAL APPROACHES FOR SOIL REMEDIATION

An eminent part of the environment is the soil, which is often under threat because of various contaminations ensuing since last a few decades. Rapid growth of industries, chemical disposal of industrial wastes into soil, large utilization of pesticides, fertilizers, and other agrochemicals are responsible for soil contamination. Various heavy metals and metalloids, synthetic pesticides, petroleum, various solvents, etc., are the prominent soil contaminants

TABLE 11.2 Treatment of Water Pollutants by NMs.

Types	Nanomaterials	Target toxicant	References
Metal-based	Ag NPs/Ag ions	*E. coli*	Chou et al., 2005; Xiu et al., 2012; Pal et al., 2007
	TiO_2 NPs	Aromatic hydrocarbons	Da Silva et al., 2016; Gu et al., 2012; Li et al., 2014
	Metal-doped TiO_2	*Staphylococcus aureus*, Rhodamine B, endotoxin, 2-chlorophenol, *E. coli*	Park and Lee, 2014; Sreeja and Vidya, 2016; Younas et al., 2014
	Iron-based	Heavy metals, chlorinated organic solvents	Ebrahim et al., 2016; Guo et al., 2017; Han and Yan, 2016
	Bimetallic NPs	Chlorinated and brominated contaminants	Zhu et al., 2010; Fang et al., 2011; Xie et al., 2014
Silica-based	Carboxylic acid-functionalized mesoporous silica	Cationic dyes, heavy metals	Tsai et al., 2016; Bruzzoniti et al., 2007; Yan et al., 2006
	Amino-functionalized mesoporous silica	Heavy metals	Wang et al., 2015; Nakanishi et al., 2015; Repo et al., 2013
	Mesoporous silica having Thiol-functional group	Heavy metals	Arencibia et al., 2010; Rostamian et al., 2011; Hakami et al., 2012
Graphene	Pristine graphene	Fluoride	Li et al., 2011
	Graphene oxide	Heavy metals, pesticides, pharmaceuticals	Song et al., 2016; Gao et al., 2013; Deng et al., 2017
	ZnO-graphene/ CdS-graphene	Heavy metals	Zhang et al., 2013
Polymer-based	PAMAM dendrimers	Heavy metals	Diallo et al., 2005
	Polymer nanocomposites (PNCs)	Metal ions, dyes, microorganisms	Khare et al., 2016; Mittal et al., 2015; Sun et al., 2015

Source: Reprinted from Guerra et al., 2018.

that affect the soil health and quality leading to deterioration to ecosystem, agriculture and human health, and agriculture, and thus, ultimately threaten the entire environment (Caliman et al., 2011). Therefore, remediation of polluted or contaminated soil has appeared as a serious global challenge. Although sustainable and cost-effective in situ remediation techniques are still lacking; recently, nanotechnology has become an effective, innovative, and promising approach for soil health recovery by means of adsorption, stabilization, redox reaction, conversion, etc., (Fig. 11.5).

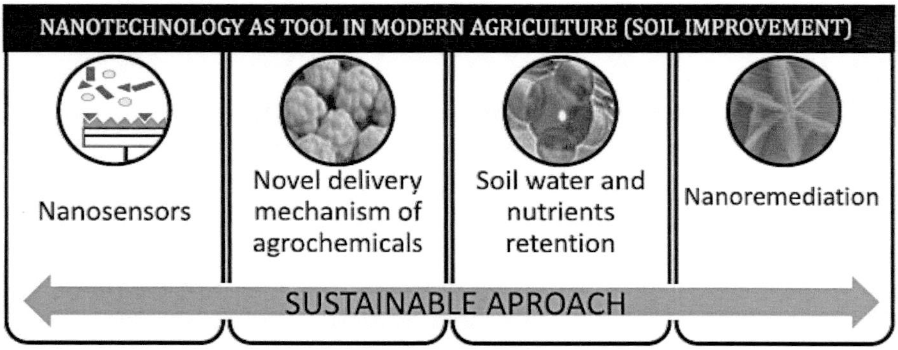

FIGURE 11.5 Technological advances for the improvement of soil quality and reduction of pollution via nanotechnological applications.

Source: Reprinted with permission from Medina-Pérez et al., 2019

11.5.1 SOIL CONTAMINANTS

Soil can be contaminated by various objects like agrochemicals, heavy metals, pesticides, fertilizers, etc. Out of these, heavy metals appear as one of the most notorious and hazardous elements. Generally, pollutant metals can be categorized as: (1) cationic heavy metals; (2) oxyanions, and (3) radionuclides (Table 11.3). Cationic heavy metals, those are persistent, and therefore, increase the chance of bioaccumulation in the environment is usually found as contaminants in soil. These can generate severe toxic effects on human health and the environment. Metal or metalloid oxyanions can act as strong ligands. Like Arsenic (As), an eminent toxic metalloid, which can be presented in soil and groundwater in the form of arsenite [As(III)] and arsenate [As(V)]. Most of the metals or radionuclides are redox-active of which most common pairs include As(V)-As(III), Cr(VI)-Cr(III), Se(VI)-Se(IV), Tc(VII)-Tc(IV), and U(VI)-U(IV). These elements of various oxidation states may show various solubility and toxicity such as As(III), Cr(VI), Se(VI), Tc(VII), and U(VI) exhibit more mobility and toxicity compared to the respective counterparts [Cr = Chromium, Se = Selenium, Tc = Technetium, U = Uranium].

TABLE 11.3 Classification of Pollutant Heavy Metals.

Type of heavy metals	Source	Example	Reference
Cationic heavy metals	Anthropogenic activities like mining, electrical, and electronics manufacturing	Lead (Pb), Cadmium (Cd), and Mercury (Hg)	Xiong et al., 2009; Anirudhan et al., 2008
Oxyanions	Widespread industrial applications	Arsenate, Chromate, and Selenate	MacDonald and Kavanaugh, 1994
Radionuclides	Energy generation via nuclear fission, extraction, processing and utilization of nuclear fuels, and mining activities	Uranium (U) and Technetium (Tc)	O'Loughlin et al., 2003; Wildung et al., 1979

11.5.2 NATURAL AND ANTHROPOGENIC PRESENCE OF NANOMATERIALS IN SOIL AND ENVIRONMENT

Soil is a reservoir of various natural and engineered nanoparticles released from many industrial and environmental applications, which largely affects human health, ecosystem, and various geochemical processes. These nanomaterials are comprised of both natural and anthropogenic nanostructures. Geogenic dust, volcanic smut, various carbonaceous particulate matters and sub-products of combustion processes have been present as the sources of natural nanomaterials (Zosky et al., 2014; Ribeiro et al., 2010). Nanomaterials are present by not only their natural origin, but also manufacturing and transportation processes result in non-intentional release of these elements. Due to the applications of nanotechnology in numerous fields, the entry of nanomaterials into the environment may be resulted through many pathways such as these can be deposited in soil and water by the emissions to the atmosphere, the wastewater treatment may also cause the discharge of nanoparticles in the environment or as a sub-product of water treatments, sewage-sludge, and even accidentally during spills. Generally, air and water act as dispersion medium of the nanoparticles but the soil is the chief recipient of most of the natural or anthropogenic nanomaterials. Despite of knowledge regarding the releasing pathways, some nanomaterials existing in the environment are still unfamiliar (Elsaesser and Howard, 2012).

11.5.2.1 NATURAL NANOPARTICLES IN SOIL

Nanoparticles are produced by various cosmic, earth, and weather-dependent phenomena and these are lifted into the atmosphere through air currents, volcanic eruptions, and strong winds. Some natural nanoparticles are present in soil as clay fractions those are smaller than 2 micrometers, especially the colloidal particles ranging between 1 and 100 nanometers. Organic matters present in the soil often show some similar characteristics of nanoparticles such as surface speciation and electric charge. Besides, humic and fulvic acids also act as electron donors and acceptors during the biodegradation of contaminant compounds (Wu et al., 2013). They take part in transportation of contaminants too.

Another source of natural nanoparticles is deserts, which is the primary and permanent source. Almost about 50% minerals present in aerosols come from deserts (Shi et al., 2005). Weathering of various silicates and microbial degradation of minerals result in the formation of oxides, hydroxides, and oxy-hydroxides of metallic elements such as Fe, Al, and Mn that act as major source of geogenic nanoparticles. Some nanomaterials, present in the environment, are directly linked to the origin, but are well distributed in the environment in various forms. For example, aluminum nanostructures may be presented as gibbsite and boehmite in soil. Besides, microbial processes by bacteria and fungi form manganese nanomaterials. Furthermore, one of the most abundant nanoparticles present in soil is iron hydroxides that help in nutrient absorption, exchange of molecules (Theng and Yuan, 2008).

11.5.2.2 ANTHROPOGENIC NANOPARTICLES IN SOIL

Human activities have developed several modern technologies comprising of new materials and particles even in nanoscale dimension for various applications. The anthropogenic sources of nanomaterials can be classified in two categories: (1) primary and (2) secondary. Nanomaterials from primary sources generally occur owing to mineral exploitation, whereas the secondary sources come from industrial activities (Lungu et al., 2015; Glover et al., 2011). Primary nanomaterials are released from exploitation of fossil fuels, extraction of ferrous and non-ferrous minerals, and exploitation of natural materials for construction. Such nanoparticles of primary sources are less detrimental than those of secondary sources.

Some carbon nanotubes especially graphene which are released due to discharge of products containing these materials like plastics, automotive components, electrodes; have very good thermal and electrical conductivity. Besides, most of the metallic nanomaterials come from cosmetic sector, medicinal and drug industries, bioremediation of polluted soil and water.

About 64% carbonaceous nanoparticles are reported to be present in solid products, whereas the products of the remaining compositions are embedded in liquid.

11.5.3 SPECIFIC EXAMPLES OF NANOPARTICLES FOR SOIL REMEDIATION

Various soil remediation techniques applied globally are in situ, on site, and ex situ techniques. Remediation of contaminated soil and environment through nanotechnology is known as nanoremediation (Machado et al., 2013). Most of the researches are present based on the application of nanotechnology for decontamination of water or aqueous solutions; very few studies are available on nanoremediation of contaminated soil. It has been suggested in many literatures that nanomaterials can adsorb and also degrade pollutants by several mechanisms like adsorption, precipitation, surface complexation, redox reactions, ion exchange, and electrostatic interactions (Shi et al., 2011; Trujillo-Reyes et al., 2014). Polluted soils can be remediated by injecting or spraying colloidal solution of nanoparticles using pressure or gravity. A treatment zone is formed after the injection of nanomaterials in soil and these are remained in suspension.

Broadly, there are two types of nanomaterials applied for the soil remediation and these are discussed in Table 11.4.

TABLE 11.4 Classification of Nanomaterials for Remediation of Contaminated Soil.

Types of nanomaterials	Properties	References
Adsorptive materials	Remove contaminants by the principles of adsorption on both surface and internal structure Very effective for removal of heavy metals such as Cu, Hg, Cd, and Cr. Simple and inexpensive	Caliman et al., 2011; Wei et al., 2013; Shan et al., 2009
Reactive materials	Involve chemical reactions like redox, acid-base, precipitation, ion exchange, dissolution, photocatalysis, etc. Mostly effective in the remediation of aqueous solution or slurry	Caliman et al., 2011; Wei et al., 2013

Most commonly known nanomaterials used for remediation of heavy metals from polluted soils are described below:

11.5.3.1 ZERO-VALENT IRON (ZVI) NANOPARTICLES

Nanoremediation of soil using iron-based nanoparticles has been evolved rapidly as a new approach in various scales in both field and laboratory. It is advantageous to use iron nanoparticles as iron is one of the most abundant metals on the earth and preparation of nanomaterials from iron is cost-effective. Besides, it is devoid of toxicity in elemental form and hence, environment friendly.

Zero-valent iron (ZVI) nanoparticles are generally synthesized under the absence of oxygen by reducing Fe^{2+}/Fe^{3+} by sodium borohydride ($NaBH_4$) in aqueous solutions at circumambient temperature (Wang and Zhang, 1997). The microscopic size of these ZVI nanoparticles helps their easy and fast penetration into contaminated zone. Their expansive surface area allows close and greater contact with contaminants improving decontamination efficiency, and presence of several reactive sites in them enhances their reactivity. Major demerits of such nanoparticles include their rapid oxidation in air due to large surface area and high reactivity. Moreover, they can agglomerate very fast into micron-scale or larger aggregates that would ultimately reduce their mobility and reactivity. To prevent such drawbacks, such nanomaterials can be modified by using various stabilizers, supporting materials, dispersants that would be more beneficial in remediation of contaminated soils than the bare nanomaterials (Li et al., 2006a; Hoch et al., 2008).

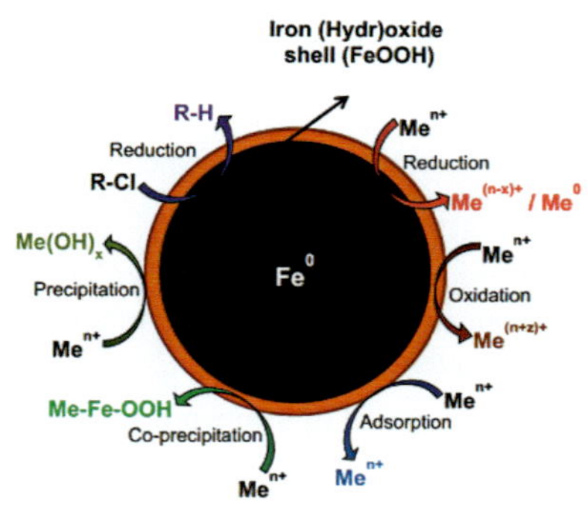

FIGURE 11.6 The core-shell structure of ZVI nanoparticles describing mechanisms related to the removal of contaminants.

Source: Reprinted with permission from O'Carroll et al. (2013). Copyright 2012 Elsevier Inc.

Zero-valent nanoparticles are very effective in treating heavy metals like cadmium (Cd), mercury (Hg), Nickel (Ni), and Lead (Pb) because of their excellent adsorption and reducing capacity. Such nanomaterials with the help of adsorption and reduction are highly engaged in heavy metals elimination by converting them into less toxic elements and thus reducing their availability in the soil (Cundy et al., 2008; Li et al., 2006b). The redox potential (E^0) of the heavy metal is the major determining factor of the reaction mechanism during decontamination actions happened between metal pollutant and ZVI. Those heavy metals (like Zn and Cd) having higher negative E^0 value compared to ZVI particles or similar E^0 value to ZVI nanomaterials are being reported to be adsorbed on the iron shell; whereas, some heavy metals like As, Cr, Cu, etc., having greater positive E^0 value compared to ZVI nanomaterials are being precipitated in soil. Figure 11.6 represents the core-shell structure of ZVI nanoparticles describing mechanisms related to the removal of contaminants (Li and Zhang, 2007; O'Carroll et al., 2013).

11.5.3.2 IRON SULFIDE NANOPARTICLES

Generally, sulfur containing minerals show greater affinity toward mercury (Hg). Among various S-containing elements, iron sulfide (FeS) has gained major importance because of easy availability, higher Hg^{2+} sorption competency, and longstanding stability in anoxic sediments. However, stabilized FeS nanoparticles have been reported to have more stability, mobility, and deliverability in contaminated soil than non-stabilized FeS nanoparticles which are prone to aggregate formation. FeS nanomaterials are observed to be greatly effective in eradication of Hg(II) from contaminated soil, sediment, and water (Bower et al., 2008). Immobilization of Hg can be done by FeS through several mechanisms such as surface complexation, structural incorporation, and precipitation.

$$FeS\,(s) + Hg^{2+} = FeS\text{-}Hg^{2+} \qquad \text{(Surface complexation)}$$

$$FeS\,(s) + xHg^{2+} = [Fe_{1-x}Hg_x]S\,(s) + xFe^{2+}\ (x<1) \qquad \text{(Structural incorporation)}$$

$$FeS\,(s) + Hg^{2+} = HgS + Fe^{2+} \qquad \text{(Precipitation)}$$

Besides the adsorption of Hg, removal of Cr(VI) has also been reported by utilizing CMC-stabilized FeS nanoparticles (Xiong et al., 2009; Wang et al., 2011). However, a magnetic composite of polyacrylamide stabilized FeS/Fe_3O_4 was also observed to remove U(VI) by reducing U(VI) to U(IV) (Shao et al., 2015). Furthermore, several studies have suggested about the efficiency of non-stabilized FeS in adsorptive elimination of various metallic cations like Cu(II), Cd(II), and Pb(II) (Özverdi and Erdem, 2006; Ito et al., 2004).

11.5.3.3 IRON PHOSPHATE NANOPARTICLES

Several earlier studies have reported the capability of some phosphate materials (such as phosphoric acid, natural rock phosphate) to restore heavy metals contaminated soil by immobilizing Cd, Pb, Cu in polluted soil and water (Yang et al., 2001; Basta and McGowen, 2004; Ma et al., 1995; Zhang and Ryan, 1998). However, the use of such phosphate amendments has been restricted in case of field application as such materials produce some precipitates of secondary metal phosphates with low solubility resulting in environmental hazards and lower bioavailability of heavy metals. As iron is less toxic and naturally abundant and iron phosphate nanomaterials are efficient in immobilization of heavy metals, therefore, these nanomaterials are more effective and useful than conventionally used phosphate materials regarding lower leaching of phosphates and long-lasting reactivity as well as following the similar reaction mechanism. In this aspect, Liu and Zhao (2007) prepared a better stabilized vivianite iron phosphate nanoparticles using sodium carboxymethyl cellulose (CMC-Na) following a standardized protocol. Iron phosphate nanomaterials show good reactivity as well as limited soil mobility. Very tiny size (~10 nm) of these nanoparticles favors in better deliverability and easy incorporation into contaminated soil.

11.5.3.4 IRON OXIDES AND IRON-MANGANESE OXIDES NANOPARTICLES

Various metal oxides, especially iron oxides, Iron-Manganese oxides are naturally available in the soil environment. These are very effective in adsorbing and removing heavy metals (especially metal anions such as As and Se oxyanions) from contaminated soils by large surface area, redox reaction, and anion exchange (Goh et al., 2008). Meanwhile, the higher stability, more adsorption capacity, and better binding ability of zirconium–iron oxide (Zr–FeO) make it more effective than FeO for remediation of contaminated soil over a wide pH range. To increase the mobility and particle deliverability, a new and better stabilized Fe-Mn binary oxide nanomaterials have been produced using starch and CMC as stabilizers for removal of As(III) from soil and water (An and Zhao, 2012). It has been reported that, for As(III) and As(V), the Langmuir maximum adsorption capacity were 338 and 272 mg/g, respectively, those were much greater as compared to non-stabilized Fe-Mn materials (Zhang et al., 2007; Chang et al., 2009).

11.5.3.5 ALLOPHANE

An effective natural nanomaterial is allophane, which is hydrous aluminosilicate by composition and known to be a useful sorbent of Cu. It has been observed in laboratory experiment that Cu from contaminated soil has been removed by allophane (Yuan, 2004). Cation-exchange reaction as well as surface complexation between Cu ions and $(OH)Al(OH)_2$ groups of allophane are mainly responsible for the adsorption of Cu to allophane. Soil pH greatly affects the adsorption of Cu in contaminated soil and generally higher pH causes greater adsorption.

11.5.3.6 CARBON BLACK

Carbon black due to its strong affinity toward heavy metals is very effective for the remediation of soils with high heavy metal contamination. Cheng et al. (2014) conducted a greenhouse experiment with ryegrass cultivation to observe the remediation of soil contaminated by Zn and Cu using surface-modified nano-sized carbon black. They reported that the oxidized carbon black with HNO_3 produced some effective functional groups that were involved in the removal of the heavy metals through adsorption, cation-exchange, and surface complexation.

The list of diverse nanomaterials exploited for remediation of heavy metal from soil is provided in Table 11.5.

TABLE 11.5 Various Nanoparticles Utilized to Remove Metal Pollutants from Contaminated Soil.

Nanoparticles	Target metal	Removal efficiency	References
Combined electrokinetics process with nano-scaled ZVI (nZVI)	Cr(VI)	19% total Cr removal from Cr(VI) contaminated clayey soil	Shariatmadari et al., 2009
Stabilized ZVI nanoparticles with carboxymethyl cellulose (CMC)	Cd	Removal of Cd from aqueous soil solution by adsorption technique	Nasiri et al., 2013
Stabilized nZVI particles	Cr	Reduced leachability of Cr by 50%	Xu and Zhao, 2007

TABLE 11.5 *(Continued)*

Nanoparticles	Target metal	Removal efficiency	References
ZVI nanoparticles	As, Cr, Zn, Cu	Reduced concentrations of As and Cr by 98 and 45%, respectively, and also of Cu and Zn in a chromate copper arsenate contaminated soil	Kumpiene et al., 2006
CMC-stabilized ZVI nanoparticles	Cr(VI)	Reduced Cr(VI) and immobilized it	Wang et al., 2010
Starch-stabilized nZVI	Radionuclide perrhenate (ReO_4^-)	>96% reduction in perrhenate concentration	Liu et al., 2013
CMC stabilized FeS nanoparticles	Hg(II)	97% reduction in water leachable Hg concentration and 99% reduction in TCLP (toxicity characteristic leaching procedure) leachability of Hg	Xiong et al., 2009
CMC stabilized Fe-phosphate nanoparticles (vivianite)	Pb(II)	85–95% reduction in TCLP leachability of Pb(II) and 31–47% decrement in physiologically based extraction test (PBET) bioaccessibility of Pb(II) as well as almost 50% decrement in leaching of phosphates	Liu and Zhao, 2007
CMC-stabilized Fe-phosphate nanoparticles (vivianite)	Cu(II)	Reduction in Cu leachability by 63–87%	Liu and Zhao, 2007
FeO nanoparticles	Pb and As	Reduced ammonium acetate (NHO_4Ac)-extractable Pb and As by 83 and 77.3%, respectively	Almaroai et al., 2014
Zirconium-iron oxide (Zr-FeO) nanoparticles	Pb and As	Decreased ammonium acetate (NHO_4Ac)-extractable Pb and As by 65 and 64.7%, respectively	Almaroai et al., 2014
CMC stabilized Fe-Mn nanoparticles	As	91–96% reduction in water leachable As and 94–98% decrement in TCLP leachability of As	Zhang et al., 2007; Chang et al., 2009
CMC and starch stabilized Fe-Mn nanoparticles	Se(IV)	Decrement in TCLP leachability and water extraction test (WET) leachability of Se(IV) by 76 and 71%, respectively	Xie et al., 2015
Starch-stabilized magnetite nanoparticles	As(V)	Reduction in water leachable As(V) by ~93% and TCLP leachability of As(V) by >83%	Liang and Zhao, 2014

11.6 CONCLUSION

The outrageous activities of humans to the ecosystem have hampered the ecological balance and generated high amounts of hazardous pollutants leading to threat to the entire environment. To get rid of such toxicants, special care, and specific efforts have to be made. In this context, nanotechnology has appeared to provide enormous opportunities over conventional technologies. Environmental nanotechnology has brought considerable success in contaminant prevention, detection, monitoring, and treatment aspects using inorganic, carbonaceous, and polymeric nanomaterials. However, more research is required for rational design and well utilization of different nanomaterials maintaining the delicate balance among property, reactivity, and stability. This chapter highlighted the hurdles of nanotech applications in environmental issues. Currently, in many cases nanotechnological approaches are not competitive with the traditional treatment approaches. Mechanistic studies on nano/bio interfaces are also vital to determine the outcome of these nanoscale materials after submission to the various environmental segments for remediative purposes so that these new materials may not become a source of environmental contamination in future. Nevertheless, nanotechnology has open doors of possibilities that can be implemented for environmental cleanup. Finally, it can be concluded with enormous hope that nanotechnology-mediated approaches will definitely come up with more innovative strategies of decontamination from different environmental media, which will contribute to future sustainability as a whole.

KEYWORDS

- **environment**
- **pollution**
- **nanotechnology**
- **contaminants**
- **remediation**

REFERENCES

Abanades, J. C.; Alvarez, D. Conversion Limits in the Reaction of CO_2 with Lime. *Energy. Fuels* **2003,** *17* (2), 308–315.

Allhoff, F.; Lin, P.; Moore, D. In *What is Nanotechnology and Why Does it Matter? From Science to Ethics*; John Wiley and Sons, 2010; pp 3–5. https://download.e-bookshelf.de/download/0000/5994/95/L-G-0000599495-0015302846.pdf

Almaroai, Y. A.; Vithanage, M.; Rajapaksha, A. U.; Lee, S. S.; Dou, Z.; Lee, Y. H.; Sung, J. K.; Ok, Y. K. Natural and Synthesized Iron-Rich Amendments for As and Pb Immobilization in Agricultural Soil. *Chem. Ecol.* **2014**, *30*, 267–279.

An, B.; Zhao, D. Immobilization of As(III) in Soil and Groundwater using a New Class of Polysaccharide Stabilized Fe–Mn Oxide Nanoparticles. *J. Hazard. Mater*. **2012,** *211–212*, 332–341.

Anirudhan, T. S.; Divya, L.; Ramachandran, M. Mercury(II) Removal from Aqueous Solutions and Waste Waters using a Novel Cation Exchanger Derived from Coconut Coir Pith and its Recovery. *J. Hazard. Mater.* **2008,** *157* (2–3), 620–627.

Annan, E.; Agyei-Tuffour, B.; Bensah, Y. D.; Konadu, D. S.; Yaya, A.; Onwona-Agyeman, B.; Nyankson, E. Application of Clay Ceramics and Nanotechnology in Water Treatment: A Review. *Cogent. Eng.* **2018,** *5* (1), 1476017.

Arencibia, A.; Aguado, J.; Arsuaga, J. M. Regeneration of Thiol-Functionalized Mesostructured Silica Adsorbents of Mercury. *Appl. Surf. Sci.* **2010,** *256*, 5453–5457.

Barakat, M. A. New Trends in Removing Heavy Metals from Industrial Wastewater. *Arab. J. Chem.* **2011,** *4* (4), 361–377.

Basta, N. T.; McGowen, S. L. Evaluation of Chemical Immobilization Treatments for Reducing Heavy Metal Transport in a Smelter Contaminated Soil. *Environ. Pollut.* **2004,** *127*, 73–82.

Bawa, R.; Bawa, S. R.; Maebius, S. B.; Flynn, T.; Wei, C. Protecting New Ideas and Inventions in Nanomedicine with Patents. *Nanomedicine NBM* **2005,** *1* (2), 150–158.

Bora, T.; Dutta, J. Applications of Nanotechnology in Wastewater Treatment – A Review. *J. Nanosci. Nanotechnol.* **2014,** *14*, 613–626.

Bower, J.; Savage, K. S.; Weinman, B.; Barnett, M. O.; Hamilton, W. P.; Harper, W. F. Immobilization of Mercury by Pyrite (FeS_2). *Environ. Pollut.* **2008,** *156* (2), 504–514.

Bruzzoniti, M. C.; Prelle, A.; Sarzanini, C.; Onida, B.; Fiorilli, S.; Garrone, E. Retention of Heavy Metal Ions on SBA-15 Mesoporous Silica Functionalised with Carboxylic Groups. *J. Sep. Sci.* **2007,** *30*, 2414–2420.

Caliman, F. A.; Robu, B. M.; Smaranda, C.; Pavel, V. L.; Gavrilescu, M. Soil and Groundwater Cleanup: Benefits and Limits of Emerging Technologies. *Clean Technol. Environ. Policy* **2011,** *13*, 241–268.

Chang, F.; Qu, J.; Liu, H.; Liu, R.; Zhao, X. Fe–Mn Binary Oxide Oncorporated into Diatomite as an Adsorbent for Arsenite Removal: Preparation and Evaluation. *J. Colloid. Interface. Sci.* **2009,** *338* (2), 353–358.

Cheng, J. M.; Liu, Y. Z.; Wang, H. W. Effects of Surface-Modified Nano-Scale Carbon Black on Cu and Zn Fractionations in Contaminated Soil. *Int. J. Phytoremediation.* **2014,** *16*, 86–94.

Cheung, O.; Bacsik, Z.; Liu, Q.; Mace, A.; Hedin, N. Adsorption Kinetics for CO_2 on Highly Selective Zeolites NaKA and Nano-NaKA. *Appl. Energy.* **2013,** *112*, 1326–1336.

Chong, M. N.; Jin, B.; Chow, C. W.; Saint, C. Recent Developments in Photocatalytic Water Treatment Technology: A Review. *Water Res.* **2010**, *44* (10), 2997–3027.

Chou, K. S.; Lu, Y. C.; Lee, H. H. Effect of Alkaline Ion on the Mechanism and Kinetics of Chemical Reduction of Silver. *Mater. Chem. Phys.* **2005,** *94*, 429–433.

Crane, R. A.; Scott, T. B. Nanoscale Zero-Valent Iron: Future Prospects for an Emerging Water Treatment Technology. *J. Hazard. Mater.* **2012,** *211–212*, 112–125.

Cui, X.; Li, C. M.; Bao, H.; Zheng, X.; Zang, J.; Ooi, C. P.; Guo, J. Hyaluronan-Assisted Photoreduction Synthesis of Silver Nanostructures: From Nanoparticle to Nanoplate. *J. Phys. Chem. C.* **2008,** *112* (29), 10730–10734.

Cundy, A. B.; Hopkinson, L.; Whitby, R. L. D. Use of Iron-Based Technologies in Contaminated Land and Groundwater Remediation: A Review. *Sci. Total Environ.* **2008,** *400*, 42–51.

Da Silva, M. B.; Abrantes, N.; Nogueira, V.; Gonçalves, F.; Pereira, R. TiO_2 Nanoparticles for the Remediation of Eutrophic Shallow Freshwater Systems: Efficiency and Impacts on Aquatic Biota under a Microcosm Experiment. *Aquat. Toxicol.* **2016,** *178*, 58–71.

Daly, A.; Zannetti, P. An Introduction to Air Pollution — Definitions, Classifications, and History. In *Ambient Air Pollution.* Zannetti, P., Al-Ajmi, D., Al-Rashied, S., Ed.; The Arab School for Science and Technology (ASST) (http://www.arabschool.org.sy) and The EnviroComp Institute (http://www.envirocomp.org/), 2007.

Deng, C. H.; Gong, J. L.; Zhang, P.; Zeng, G. M.; Song, B.; Liu, H. Y. Preparation of Melamine Sponge Decorated with Silver Nanoparticles-Modified Graphene for Water Disinfection. *J. Colloid Interface Sci.* **2017,** *488*, 26–38.

Diallo, M. S.; Christie, S.; Swaminathan, P.; Johnson, J. H.; Goddard, W. A. Dendrimer Enhanced Ultrafiltration. 1. Recovery of Cu(II) from Aqueous Solutions using PAMAM Dendrimers with Ethylene Diamine Core and Terminal NH_2 Groups. *Environ. Sci. Technol.* **2005,** *39*, 1366–1377.

Ebrahim, S. E.; Sulaymon, A. H.; SaadAlhares, H. Competitive Removal of Cu^{2+}, Cd^{2+}, Zn^{2+} and Ni^{2+} Ions Onto Iron Oxide Nanoparticles from Wastewater. *Desalin. Water Treat.* **2016,** *57*, 20915–20929.

El Saliby, I. J.; Shon, H.; Kandasamy, J.; Vigneswaran, S. Nanotechnology for Wastewater Treatment: In Brief. In *Encyclopedia of Life Support System (EOLSS)*, 2008; vol 7.

Elsaesser, A.; Howard, C. V. Toxicology of Nanoparticles. *Adv. Drug Deliv. Rev.* **2012,** *64*, 129.

Fang, Z.; Qiu, X.; Chen, J.; Qiu, X. Debromination of Polybrominated Diphenyl Ethers by Ni/Fe Bimetallic Nanoparticles: Influencing Factors, Kinetics, and Mechanism. *J. Hazard. Mater.* **2011,** *185*, 958–969.

Feynman, R. P. There's Plenty of Room at the Bottom. *Eng. Sci.* **1960,** *23*, 22–36.

Gao, P.; Ng, K.; Sun, D. D. Sulfonated Graphene Oxide-ZnO-Ag Photocatalyst for Fast Photo Degradation and Disinfection under Visible Light. *J. Hazard. Mater.* **2013,** *262*, 826–835.

Gaya, U. I.; Abdullah, A. H. Heterogeneous Photocatalytic Degradation of Organic Contaminants over Titanium Dioxide: A Review of Fundamentals, Progress and Problems. *J. Photochem. Photobiol. C Photochem. Rev.* **2008,** *9*, 1–12.

Gehrke, I.; Geiser, A.; Somborn-Schulz, A. Innovations in Nanotechnology for Water Treatment. *Nanotechnol. Sci. Appl.* **2015,** *8*, 1.

Glover, R. D.; Miller, J. M.; Hutchison, J. E. Generation of Metal Nanoparticles from Silver and Copper Objects: Nanoparticle Dynamics on Surfaces and Potential Sources of Nanoparticles in the Environment. *ACS Nano.* **2011,** *5(11)*, 8950–8957.

Goh, K. H.; Lim, T. T.; Dong, Z. Application of Layered Double Hydroxides for Removal of Oxyanions: A Review. *Water Res.* **2008,** *42* (6), 1343–1368.

Gu, J.; Dong, D.; Kong, L.; Zheng, Y.; Li, X. Photocatalytic Degradation of Phenanthrene on Soil Surfaces in the Presence of Nanometer Anatase TiO_2 under UV-Light. *J. Environ. Sci.* **2012,** *24*, 2122–2126.

Guerra, F. D.; Attia, M. F.; Whitehead, D. C.; Alexis, F. Nanotechnology for Environmental Remediation: Materials and Applications. *Molecules* **2018,** *23* (7), 1760.

Guo, M.; Weng, X.; Wang, T.; Chen, Z. Biosynthesized Iron-Based Nanoparticles used as a Heterogeneous Catalyst for the Removal of 2,4-dichlorophenol. *Sep. Purif. Technol.* **2017,** *175*, 222–228.

Hakami, O.; Zhang, Y.; Banks, C. J. Thiol-Functionalised Mesoporous Silica-Coated Magnetite Nanoparticles for High Efficiency Removal and Recovery of Hg from Water. *Water Res.* **2012,** *46*, 3913–3922.

Han, Y.; Yan, W. Reductive Dechlorination of Trichloroethene by Zero-Valent Iron Nanoparticles: Reactivity Enhancement Through Sulfidation Treatment. *Environ. Sci. Technol.* **2016,** *50*, 12992–13001.

Hauptmann, M. L. J.; Stewart, P. A.; Hayes, R. B.; Blair, A. Mortality from Solid Cancers Among Workers in Formaldehyde Industries. *Am. J. Epidemiol.* **2004,** *15*, 1117–1130.

Hoch, L. B.; Mack, E. J.; Hydutsky, B. W.; Hershman, J. M.; Skluzacek, J. M.; Mallouk, T. E. Carbothermal Synthesis of Carbon-Supported Nanoscale Zero-Valent Iron Particles for the Remediation of Hexavalent Chromium. *Environ. Sci. Technol.* **2008,** *42* (7), 2600–2605.

Hochella, M. F.; Mogk, D. W.; Ranville, J.; Allen, I. C.; Luther, G. W.; Marr, L. C.; et al. Natural, Incidental, and Engineered Nanomaterials and their Impacts on the Earth System. *Science* **2019,** *363*, eaau8299.

Hotze, M.; Lowry, G. Nanotechnology for Sustainable Water Treatment. In *Sustainable Water*; The Royal Society of Chemistry, 2011; pp 138–164.

Hu, J.; Chen, C.; Zhu, X.; Wang, X. Removal of Chromium from Aqueous Solution by using Oxidized Multiwalled Carbon Nanotubes. *J. Hazard. Mater.* **2009,** *162* (2-3), 1542–1550.

Hutchison, J. E. Greener Nanoscience: A Proactive Approach to Advancing Applications and Reducing Implications of Nanotechnology. *ACS Nano.* **2008,** *2* (3), 395–402.

Hwang, G. B.; Lee, J. E.; Nho, C. W.; Lee, B. U.; Lee, S. J.; Jung, J. H.; Bae, G-N. Short-Term Effect of Humid Airflow on Antimicrobial Air Filters using Sophoraflavescens Nanoparticles. *Sci. Total Environ.* **2012,** *421–422*, 273–279.

Iqbal, P.; Preece, J. A.; Mendes, P. M. Supramolecular Chemistry. In *Nanotechnology: The "Top-Down" and "Bottom-Up" Approaches*; John Wiley & Sons, Ltd.; Chichester, UK, 2012.

ISO, ISO/TS 80004-1:2010, Nanotechnology – Vocabulary – Part 1: Core Terms. International Organization for Standardization: Geneva, Switzerland, 2010. https://www.iso.org/standard/51240.html

Ito, D.; Miura, K.; Ichimura, T.; Ihara, I.; Watanabe, T. Removal of As, Cd, Hg and Pb Ions from Solution by Adsorption with Bacterially-Produced Magnetic Iron Sulfide Particles Using High Gradient Magnetic Separation. *IEEE Trans. Appl. Supercond.* **2004,** *14* (2), 1551–1553.

Jänchen, J.; Möhlmann, D. T. F.; Stach, H. Water and Carbon Dioxide Sorption Properties of Natural Zeolites and Clay Minerals at Martian Surface Temperature and Pressure

Conditions. In *Studies in Surface Science and Catalysis*, Ruren Xu, Z. G. J. C., Wenfu, Y., Eds.; Elsevier: New York, 2007; vol 170, pp 2116–2121.

Jung, J. H.; Lee, J. E.; Bae, G. N. Use of Electrosprayed Sophoraflavescens Natural-Product Nanoparticles for Antimicrobial Air Filtration. *J. Aerosol. Sci.* **2013,** *57*, 185–193.

Kamat, P. V.; Meisel, D. Nanoscience Opportunities in Environmental Remediation. *C. R. Chim.* **2003,** *6*, 999–1007.

Khan, F. I.; Ghoshal, A. K. Removal of Volatile Organic Compounds from Polluted Air. *J. Loss Prev. Process Ind.* **2000,** *13*, 527–545.

Khare, P.; Yadav, A.; Ramkumar, J.; Verma, N. Microchannel-Embedded Metal–Carbon–Polymer Nanocomposite as a Novel Support for Chitosan for Efficient Removal of Hexavalent Chromium from Water under Dynamic Conditions. *Chem. Eng. J.* **2016,** *293*, 44–54.

Khin, M. M.; Nair, A. S.; Babu, V. J.; Murugan, R.; Ramakrishna, S. A Review on Nanomaterials for Environmental Remediation. *Energy Environ. Sci.* **2012,** *5* (8), 8075–8109.

Koutsopoulos, S.; Johannessen, T.; Eriksen, K. M.; Fehrmann, R. Titania-Supported Pt and Pt–Pd Nanoparticle Catalysts for the Oxidation of Sulfur Dioxide. *J. Catal.* **2006,** *238*, 206–213.

Kumpiene, J.; Ore, S.; Renella, G.; Mench, M.; Lagerkvist, A.; Maurice, C. Assessment of Zerovalent Iron for Stabilization of Chromium, Copper, and Arsenic in Soil. *Environ. Pollut.* **2006,** *144*, 62–69.

Lam, C. W. J. J.; McCluskey, R.; Hunter, R. L. Pulmonary Toxicity of Single-Wall Carbon Nanotubes in Mice 7 and 90 Days After Intratracheal Instillation. *Toxicol. Sci.* **2004,** *77*, 126–134.

Lee, B. U. Y. S.; Jung, J. H.; Bae, G. N. Effect of Relative Humidity and Variation of Particle Number Size Distribution on the Inactivation Effectiveness of Airborne Silver Nanoparticles Against Bacteria Bioaerosols Deposited on a Filter. *J. Aerosol Sci.* **2010a,** *41*, 447–456.

Lee, K. J.; Shiratori, N.; Lee, G. H.; Miyawaki, J.; Mochida, I.; Yoon, S. H.; Jang, J. Activated Carbon Nanofiber Produced from Electro Spun Polyacrylonitrile Nanofiber as a Highly Efficient Formaldehyde Adsorbent. *Carbon* **2010b,** *48*, 4248–4255.

Lens, P. N. L.; Virkutye, J.; Jegatheesan, V.; Kim, S. H.; Al-Abed, S. In *Nanotechnology for Water and Wastewater Treatment*, IWA Publishing, 2013.

Lewin, S.; Norman, R.; Nannan, N.; Thomas, E.; Bradshaw, D.; Collaboration, S. A. C. R. A. Estimating the Burden of Disease Attributable to Unsafe Water and Lack of Sanitation and Hygiene in South Africa in 2000. *S. Afr. Med. J.* **2007,** *97* (8), 755–762.

Li, D.; Cui, F.; Zhao, Z.; Liu, D.; Xu, Y.; Li, H.; Yang, X. The Impact of Titanium Dioxide Nanoparticles on Biological Nitrogen Removal from Wastewater and Bacterial Community Shifts in Activated Sludge. *Biodegradation* **2014,** *25*, 167–177.

Li, X. Q.; Elliott, D. W.; Zhang, W. X. Zero-Valent Iron Nanoparticles for Abatement of Environmental Pollutants: Materials and Engineering Aspects. *Crit. Rev. Solid State Mater. Sci.* **2006a,** *31*, 111–122.

Li, L.; Fan, M.; Brown, R. C.; Leeuwen, J. H. V.; Wang, J.; Wang, W.; Song, Y.; Zhang, P. Synthesis, Properties, and Environmental Applications of Nanoscale Iron-Based Materials: A Review. *Crit. Rev. Environ. Sci. Technol.* **2006b,** *36*, 405–431.

Li, Q.; Mahendra, S.; Lyon, D. Y.; Brunet, L.; Liga, M. V.; Li, D.; Alvarez, P. J. J. Antimicrobial Nanomaterials for Water Disinfection and Microbial Control: Potential Applications and Implications. *Water Res.* **2008,** *42* (18), 4591–4602.

Li, X. Q.; Zhang, W. X. Sequestration of Metal Cations with Zerovalent Iron Nanoparticles – A Study with High Resolution X-Ray Photoelectron Spectroscopy (HR-XPS). *J. Phys. Chem. C.* **2007,** *111*, 6939–6946.

Li, Y.; Zhang, P.; Du, Q.; Peng, X.; Liu, T.; Wang, Z.; Xia, Y.; Zhang, W.; Wang, K.; Zhu, H.; et al. Adsorption of Fluoride from Aqueous Solution by Graphene. *J. Colloid Interface Sci.* **2011,** *363*, 348–354.

Liang, Q.; Zhao, D. Immobilization of Arsenate in a Sandy Loam Soil using Starch-Stabilized Magnetite Nanoparticles. *J. Hazard Mater.* **2014,** *271*, 16–23.

Liu, H. F.; Qian, T. W.; Zhao, D. Y. Reductive Immobilization of Perrhenate in Soil and Groundwater Using Starch-Stabilized ZVI Nanoparticles. *Chin. Sci. Bull.* **2013,** *58* (2), 275–281.

Liu, Y.; Wang, X.; Yang, F.; Yang, X. Excellent Antimicrobial Properties of Mesoporous Anatase TiO_2 and Ag/TiO_2 Composite Films. *Microporous Mesoporous Mater.* **2008,** *114* (1-3), 431–439.

Liu, R.; Zhao, D. In Situ Immobilization of Cu(II) in Soils using a New Class of Iron Phosphate Nanoparticles. *Chemosphere* **2007,** *68*, 1867–1876.

Lu, L.; Goia, D. V. Precipitation of Spherical and 'Fiber-Like'silver Particles. *J. Colloid Interface Sci.* **2013,** *407*, 122–127.

Lungu, M.; Neculae, A.; Bunoiu, M.; Biris, C. In *Nanoparticles' Promises and Risks*; Ed.; Springer, International Publishing Switzerland, 2015.

Ma, Q. Y.; Logan, T. J.; Traina, S. J. Lead Immobilization from Aqueous Solutions and Contaminated Soils using Phosphate Rocks. *Environ. Sci. Technol.* **1995,** *29* (4), 1118–1126.

MacDonald, J. A.; Kavanaugh, M. C. Restoring Contaminated Groundwater: An Achievable Goal? *Environ. Sci. Technol.* **1994,** *28* (8), 362A–368A.

Machado, S.; Pinto, S. L.; Grosso, J. P.; Nouws, H. P. A.; Albergaria, J. T.; Delerue-Matos, C. Green Production of Zero-Valent Iron Nanoparticles using Tree Leaf Extracts. *Sci. Total Environ.* **2013,** *445–446*, 1–8.

Medina-Pérez, G.; Fernández-Luqueño, F.; Vazquez-Nuñez, E.; López-Valdez, F.; Prieto-Mendez, J.; Madariaga-Navarrete, A.; Miranda-Arámbula, M. Remediating Polluted Soils using Nanotechnologies: Environmental Benefits and Risks. *Pol. J. Environ. Stud.* **2019,** *28* (3), 1013–1030.

Merajin, M. T.; Sharifnia, S.; Hosseini, S. N.; Yazdanpour, N. Photocatalytic Conversion of Greenhouse Gases (CO_2 and CH_4) to High Value Products using TiO_2 Nanoparticles Supported on Stainless Steel Webnet. *J. Taiwan Inst. Chem. Eng.* **2013,** *44*, 239–246.

Mittal, H.; Maity, A.; Ray, S. S. Synthesis of Co-Polymer-Grafted Gum Karaya and Silica Hybrid Organic–Inorganic Hydrogel Nanocomposite for the Highly Effective Removal of Methylene Blue. *Chem. Eng. J.* **2015,** *279*, 166–179.

Miyawaki, J.; Lee, G-H.; Yeh, J.; Shiratori, N.; Shimohara, T.; Mochida, I.; Yoon, S. H. Development of Carbon-Supported Hybrid Catalyst for Clean Removal of Formaldehyde Indoors. *Catal. Today* **2012,** *185*, 278–283.

Mohamed, E. F. Nanotechnology: Future of Environmental Air Pollution Control. *Environ. Manag. Sustain. Dev.* **2017,** *6* (2), 429–454.

Mondal, P.; Kumar, R.; Gogoi, R. Azomethine Based Nano-Chemicals: Development, In Vitro and In Vivo Fungicidal Evaluation against *Sclerotium rolfsii*, *Rhizoctonia bataticola* and *Rhizoctonia solani*. *Bioorg. Chem.* **2017,** *70*, 153–162.

Nakanishi, K.; Tomita, M.; Kato, K. Synthesis of Amino-Functionalized Mesoporous Silica Sheets and their Application for Metal Ion Capture. *J. Asian Ceram. Soc.* **2015,** *3*, 70–76.

Nasiri, J.; Gholami, A.; Panahpour, E. Removal of Cadmium from Soil Resources using Stabilized Zero- Valent Iron Nanoparticles. *J. Civil Eng. Urban* **2013,** *3*, 338–341.

Ng, L. Y.; Mohammad, A. W.; Leo, C. P.; Hilal, N. Polymeric Membranes Incorporated with Metal/Metal Oxide Nanoparticles: A Comprehensive Review. *Desalination* **2013,** *308*, 15–33.

Nuasaen, S.; Opaprakasit, P.; Tangboriboonrat, P. Hollow Latex Particles Functionalized with Chitosan for the Removal of Formaldehyde from Indoor Air. *Carbohydr. Polym.* **2014,** *101*, 179–187.

O'Carroll, D.; Sleep, B.; Krol, M.; Boparai, H.; Kocur, C. Nanoscale Zero Valent Iron and Bimetallic Particles for Contaminated Site Remediation. *Adv. Water Resour.* **2013,** *51*, 104–122.

O'Loughlin, E. J.; Kelly, S. D.; Cook, R. E.; Csencsits, R.; Kemner, K. M. Reduction of Uranium(VI) by Mixed Iron(II)/Iron(III) Hydroxide (Green Rust): Formation of UO_2 Nanoparticles. *Environ. Sci. Technol.* **2003,** *37*(4), 721–727.

Özverdi, A.; Erdem, M. Cu^{2+}, Cd^{2+} and Pb^{2+} Adsorption from Aqueous Solutions by Pyrite and Synthetic Iron Sulphide. *J. Hazard Mater.* **2006,** *137* (1), 626–632.

Pal, S.; Tak, Y. K.; Song, J. M. Does the Antibacterial Activity of Silver Nanoparticles Depend on the Shape of the Nanoparticle? A Study of the Gram-Negative Bacterium *Escherichia coli*. *Appl. Environ. Microbiol.* **2007,** *73*, 1712–1720.

Pandey, B.; Fulekar, M. H. Nanotechnology: Remediation Technologies to Clean up the Environmental Pollutants. *Res. J. Chem. Sci.* **2012,** *2*, 90–96.

Parisi, C.; Vigani, M.; Rodriguez-Cerezo, E. Agricultural Nanotechnologies: What are the current Possibilities? *Nano. Today* **2015,** *10*, 124–127.

Park, J. Y.; Lee, I. H. Photocatalytic Degradation of 2-Chlorophenol using Ag-Doped TiO_2 Nanofibers and a Near-UV Light-Emitting Diode System. *J. Nanomater.* **2014,** 250803.

Pendergast, M. T. M.; Hoek, E. M. V. A Review of Water Treatment Membrane Nanotechnologies. *Energy Environ. Sci.* **2011,** *4* (6), 1946–1971.

Pibiri, M. C. G. A.; Vahekeni, N.; Roulet, C. A. Indoor Air Purification and Ventilation Systems Sanitation with Essential Oils. *Int. J. Aromather.* **2006,** *16*, 149–153.

Potocnik, J. *Off. J. Eur. Communities Legis.* **2011,** *L275*, 38–40. DOI: 10.3000/19770677.L_2011.275.eng.

Pyankov, O. V. A. I.; Huang, R.; Mullins, B. J. Removal of Biological Aerosols by Oil Coated Filters. *Clean Soil Air Water* **2008,** *36*, 609–614.

Qu, X.; Alvarez, P. J. J.; Li, Q. Applications of Nanotechnology in Water and Wastewater Treatment. *Water Res.* **2013,** *47*, 3931–3946.

Ramírez-Castillo, F. Y.; Loera-Muro, A.; Jacques, M.; Garneau, P.; Avelar-González, F. J.; Harel, J.; Guerrero-Barrera, A. L. Waterborne Pathogens: Detection Methods and Challenges. *Pathogens* **2015,** *4* (2), 307–334.

Repo, E.; Warchoł, J. K.; Bhatnagar, A.; Mudhoo, A.; Sillanpää, M. Aminopolycarboxylic Acid Functionalized Adsorbents for Heavy Metals Removal from Water. *Water Res.* **2013,** *47*, 4812–4832.

Ribeiro, J.; Flores, D.; Ward, C. R.; Silva, L. F. Identification of Nanominerals and Nanoparticles in Burning Coal Waste Piles from Portugal. *Sci. Total Environ.* **2010,** *408* (23), 6032–6041.

Rodriguez, J. A.; Liu, P.; Pérez, M.; Liu, G.; Hrbek, J. Destruction of SO_2 on Au and Cu Nanoparticles Dispersed on MgO(100) and CeO_2(111). *J. Phys. Chem. A* **2010,** *114*, 3802–3810.

Rostamian, R.; Najafi, M.; Rafati, A. A. Synthesis and Characterization of Thiol-Functionalized Silica Nano Hollow Sphere as a Novel Adsorbent for Removal of Poisonous Heavy Metal Ions from Water: Kinetics, Isotherms and Error Analysis. *Chem. Eng. J.* **2011,** *171*, 1004–1011.

Saleh, T. A. Nanomaterials: Classification, Properties, and Environmental Toxicities. *Environ. Technol. Innov.* **2020,** *20*, 101067.

Saleh, T. A.; Shuaib, T. D.; Danmaliki, G. I.; Al-Daous, M. A. Carbon Based Nanomaterials for Desulfurization: Classification, Preparation, and Evaluation. In *Applying Nanotechnology to the Desulfurization Process in Petroleum Engineering*, 2015; pp 154. books.google.com.

Shan, G.; Yan, S.; Tyagi, R. D.; Surampalli, R. Y.; Zhang, T. C. Applications of Nanomaterials in Environmental Science and Engineering: Review. *Pract. Period Hazard Toxic Radioact. Waste Manage.* **2009,** *13*, 110–119.

Shao, D.; Wang, X.; Li, J.; Huang, Y.; Ren, X.; Hou, G.; et al. Reductive Immobilization of Uranium by PAAM FeS/Fe_3O_4 Magnetic Composites. *Environ. Sci. Water Res. Technol.* **2015,** *1* (2), 169–176.

Shariatmadari, N.; Weng, C. H.; Daryaee, H. Enhancement of Hexavalent Chromium [Cr(VI)] Remediation from Clayey Soils by Electrokinetics Coupled with a Nano-Sized Zero-Valent Iron Barrier. *Environ. Eng. Sci.* **2009,** *26*, 1071–1079.

Shi, Z.; Shao, L.; Jones, T. P.; Lu, S. Microscopy and Mineralogy of Airborne Particles Collected During Severe Dust Storm Episodes in Beijing, China. *J. Geophys. Res. Atmos.* **2005,** *110*, Article number D01303.

Shi, L. N.; Zhang, X.; Chen, Z. L. Removal of Chromium (VI) from Wastewater using Bentonite-Supported Nanoscale Zero-Valent Iron. *Water Res.* **2011,** *45* (2), 886–892.

Shon, H. K.; Vigneswaran, S.; Kandasamy, J.; Cho, J. Characteristics of Effluent Organic Matter in Wastewater. In *Water and Waste Water Treatment Technologies*; EOLSS and UNESCO, 2007.

Singh, P. P. Nanotechnology: Greener Approach for Sustainable Environment. In *Nanotechnology in Environmental Science*, 2018; pp 805–824.

Solomon, S. J.; Schade, G.; Kuttippurath, J.; Ladstätter-Weissenmayer, A.; Burrows, J. VOC Concentrations in an Indoor Workplace Environment of a University Building. *Indoor Built. Environ.* **2008,** *17*, 260–268.

Song, B.; Zhang, C.; Zeng, G.; Gong, J.; Chang, Y.; Jiang, Y. Antibacterial Properties and Mechanism of Graphene Oxide-Silver Nanocomposites as Bactericidal Agents for Water Disinfection. *Arch. Biochem. Biophys.* **2016,** *604*, 167–176.

Sreeja, S.; Vidya Shetty, K. Microbial Disinfection of Water with Endotoxin Degradation by Photocatalysis using Ag@TiO_2 Core Shell Nanoparticles. *Environ. Sci. Pollut. Res.* **2016,** *23*, 18154–18164.

Stark, P. C. B. H.; Ryan, L. M.; Milton, D. K.; Gold, D. R. Fungal Levels in the Home and Lower Respiratory Tract Illnesses in the First Year of Life. *Am. J. Respir. Crit. Care Med.* **2003,** *7*, 168–232.

Sun, X. F.; Liu, B.; Jing, Z.; Wang, H. Preparation and Adsorption Property of Xylan/Poly (Acrylic Acid) Magnetic Nanocomposite Hydrogel Adsorbent. *Carbohydr. Polym.* **2015,** *118*, 16–23.

Tan, X.; Fang, M.; Chen, C.; Yu, S.; Wang, X. Counterion Effects of Nickel and Sodium Dodecylbenzene Sulfonate Adsorption to Multiwalled Carbon Nanotubes in Aqueous Solution. *Carbon* **2008,** *46* (13), 1741–1750.

Taniguchi, N.; Arakawa, C.; Kobayashi, T. In *On the Basic Concept of Nano-Technology*, Proceedings of the International Conference on Production Engineering; Tokyo, Japan, 1974.

Theng, B. K.; Yuan, G. Nanoparticles in the Soil Environment. *Elements* **2008,** *4* (6), 395–399.

Tong, H.; Ouyang, S.; Bi, Y.; Umezawa, N.; Oshikiri, M.; Ye, J. Nano-Photocatalytic Materials: Possibilities and Challenges. *Adv. Mater.* **2012,** *24*, 229–251.

Trujillo–Reyes, J.; Peralta–Videa, J. R.; Gardea-Torresdey, J. L. Supported and Unsupported Nanomaterials for Water and Soil Remediation: Are They a Useful Solution for Worldwide Pollution? *J. Hazard. Mater.* **2014,** *280C*, 487–503.

Tsai, C. H.; Chang, W. C.; Saikia, D.; Wu, C. E.; Kao, H. M. Functionalization of Cubic Mesoporous Silica SBA-16 with Carboxylic Acid Via One-Pot Synthesis Route for Effective Removal of Cationic Dyes. *J. Hazard. Mater.* **2016,** *309*, 236–248.

Upendar, K.; Sri Hari Kumar, A.; Lingaiah, N.; Rama Rao, K. S.; Sai Prasad, P. S. Low-Temperature CO_2 Adsorption on Alkali Metal Titanate Nanotubes. *Int. J. Greenhouse Gas Control* **2012,** *10*, 191–198.

USFDA. Considering Whether an FDA-Regulated Product Involves the Application of Nanotechnology; USA: Federal Drug Administration, 2011; [Jan 25; 2016]. https://www.fda.gov/RegulatoryInformation/Guidances/ucm257698.htm

Vaseashta, A.; Vaclavikova, M.; Vaseashta, S.; Gallios, G.; Roy, P.; Pummakarnchana, O. Nanostructures in Environmental Pollution Detection, Monitoring, and Remediation. *Sci. Technol. Adv. Mater.* **2007,** *8*, 47–59.

Wang, X. B.; Liu, J.; Zhao, D. L.; Song, X. J. Preparation of CMC Stabilized FeS Nanoparticles and their Enhanced Performance for Cr(VI) Removal. *Adv. Mater. Res.* **2011,** *287*, 96–99.

Wang, H. Y.; Lua, A. C. Development of Metallic Nickel Nanoparticle Catalyst for the Decomposition of Methane into Hydrogen and Carbon Nanofibers. *J. Phys. Chem. C* **2012,** *116*, 26765–26775.

Wang, Q.; Qian, H.; Yang, Y.; Zhang, Z.; Naman, C.; Xu, X. Reduction of Hexavalent Chromium by Carboxymethyl Cellulose-Atabilized Zerovalent Iron Nanoparticles. *J. Contam. Hydrol.* **2010,** *114* (1–4), 35–42.

Wang, S.; Wang, K.; Dai, C.; Shi, H.; Li, J. Adsorption of Pb^{2+} on Amino-Functionalized Core–Shell Magnetic Mesoporous SBA-15 Silica Composite. *Chem. Eng. J.* **2015,** *262*, 897–903.

Wang, C. B.; Zhang, W. Synthesizing Nanoscale Iron Particles for Rapid and Complete Dechlorination of TCE and PCBs. *Environ. Sci. Technol.* **1997,** *31* (7), 2154–2156.

Wang, K.; Zhao, P.; Guo, X.; Li, Y.; Han, D.; Chao, Y. Enhancement of Reactivity in Li_4SiO_4-Based Sorbents from the Nano-Sized Rice Husk Ash for High-Temperature CO_2 Capture. *Energy Converts Manag.* **2014,** *81*, 447–454.

Wang, Y.; Zhu, Y.; Wu, S. A New Nano CaO-based CO_2 Adsorbent Prepared using an Adsorption Phase Technique. *Chem. Eng. J.* **2013,** *218*, 39–45.

Warheit, D. B. W. T.; Reeda, K. L.; Frerichs, S.; Sayes, C. M. Pulmonary Toxicity Study in Rats with Three Forms of Ultrafine-TiO_2 Particles: Differential Responses Related to Surface Properties. *Toxicology* **2007,** *230*, 90–104.

Water, U. N. Coping with Water Scarcity: Challenge of the Twenty-First Century. Prepared for World Water Day, 2007.

Wei, Q.; Yang, D.; Fan, M.; Harris, H. G. Applications of Nanomaterial-Based Membranes in Pollution Control. *Crit. Rev. Environ. Sci. Technol.* **2013,** *43*, 2389–2438.

Wildung, R. E.; McFadden, K. M.; Garland, T. R. Technetium Sources and Behavior in the Environment. *J. Environ. Qual.* **1979,** *8* (2), 156–161.

Wu, C. M.; Baltrusaitis, J.; Gillan, E. G.; Grassian, V. H. Sulfur Dioxide Adsorption on ZnO Nanoparticles and Nanorods. *J. Phys. Chem. C* **2011,** *115*, 10164–10172.

Wu, C. Y.; Zhuang, L.; Zhou, S. G.; Yuan, Y.; Yuan, T.; Li, F. B. Humic Substance-Mediated Reduction of Iron (III) Oxides and Degradation of 2,4-D by an Alkaliphilic Bacterium, Corynebacterium Humireducens MFC-5. *Microb. Biotechnol.* **2013,** *6* (2), 141–149.

Xie, Y.; Fang, Z.; Cheng, W.; Tsang, P. E.; Zhao, D. Remediation of Polybrominated Diphenyl Ethers in Soil using Ni/Fe Bimetallic Nanoparticles: Influencing Factors, Kinetics and Mechanism. *Sci. Total Environ.* **2014,** *485–486*, 363–370.

Xie, W.; Liang, Q.; Qian, T.; Zhao, D. Immobilization of Selenite in Soil and Groundwater Using Stabilized Fe-Mn Binary Oxide Nanoparticles. *Water Res.* **2015,** *70*, 485–494.

Xiong, Z.; He, F.; Zhao, D.; Barnett, M. O. Immobilization of Mercury in Sediment using Stabilized iron Sulfide Nanoparticles. *Water Res.* **2009,** *43* (20), 5171–5179.

Xiu, Z.; Zhang, Q.; Puppala, H. L.; Colvin, V. L.; Alvarez, P. J. J. Negligible Particle-Specific Antibacterial Activity of Silver Nanoparticles. *Nano Lett.* **2012,** *12*, 4271–4275.

Xu, T., Cai, Y., O'Shea, K. E. Adsorption and Photocatalyzed Oxidation of Methylated Arsenic Species in TiO_2 Suspensions. *Environ. Sci. Technol.* **2007,** *41* (15), 5471–5477.

Xu, Y. H.; Zhao, D. Y. Reductive Immobilization of Chromate in Water and Soil using Stabilized Iron Nanoparticles. *Water Res.* **2007,** *41*, 2101–2108.

Yadav, K. K.; Singh, J. K.; Gupta, N.; Kumar, V. J. J. M. E. S. A Review of Nanobioremediation Technologies for Environmental Cleanup: A Novel Biological Approach. *J. Mater. Environ. Sci.* **2017,** *8* (2), 740–757.

Yan, Z.; Tao, S.; Yin, J.; Li, G. Mesoporous Silicas Functionalized with a High Density of Carboxylate Groups as Efficient Absorbents for the Removal of Basic Dyestuffs. *J. Mater. Chem.* **2006,** *16*, 2347–2353.

Yang, J.; Mosby, D. E.; Casteel, S. W.; Blanchar, R. W. Lead Immobilization using Phosphoric Acid in a Smelter-Contaminated Urban Soil. *Environ. Sci. Technol.* **2001,** *35*, 3553–3559.

Yantasee, W.; Warner, C. L.; Sangvanich, T.; Addleman, R. S.; Carter, T. G.; Wiacek, R. J.; Warner, M. G. Removal of Heavy Metals from Aqueous Systems with Thiol Functionalized Super Paramagnetic Nanoparticles. *Environ. Sci. Technol.* **2007,** *41* (14), 5114–5119.

Yavuz, C. T.; Mayo, J. T.; William, W. Y.; Prakash, A.; Falkner, J. C.; Yean, S.; Natelson, D. Low-Field Magnetic Separation of Monodisperse Fe_3O_4 Nanocrystals. *Science* **2006,** *314* (5801), 964–967.

Younas, H.; Qazi, I. A.; Hashmi, I.; Ali Awan, M.; Mahmood, A.; Qayyum, H. A. Visible Light Photocatalytic Water Disinfection and its Kinetics using Ag-Doped Titania Nanoparticles. *Environ. Sci. Pollut. Res.* **2014,** *21*, 740–752.

Yuan, G. Natural and Modified Nanomaterials as Sorbents of Environmental Contaminants. *J. Environ. Sci. Health A* **2004,** *39*, 2661–2670.

Yunus, I. S.; Harwin Kurniawan, A.; Adityawarman, D.; Indarto, A. Nanotechnologies in Water and Air Pollution Treatment. *Environ. Technol. Rev.* **2012,** *1* (1), 136–148.

Zhang, G.; Qu, J.; Liu, H.; Liu, R.; Wu, R. Preparation and Evaluation of a Novel Fe–Mn Binary Oxide Adsorbent for Effective Arsenite Removal. *Water Res.* **2007,** *41* (9), 1921–1928.

Zhang, P.; Ryan, J. A. Formation of Pyromorphite in Anglesite-Hydroxyapatite Suspensions Under Varying pH Conditions. *Environ. Sci. Technol.* **1998,** *32* (21), 3318–3324.

Zhang, N.; Yang, M. Q.; Tang, Z. R.; Xu, Y. J. CdS-Graphene Nanocomposites as Visible Light Photocatalyst for Redox Reactions in Water: A Green Route for Selective Transformation and Environmental Remediation. *J. Catal.* **2013,** *303*, 60–69.

Zhao, J. Turning to Nanotechnology for Pollution Control: Applications of Nanoparticles. Dartmouth Undergraduate Journal of Science, 2009.

Zhao, X.; Lv, L.; Pan, B.; Zhang, W.; Zhang, S.; Zhang, Q. Polymer-Supported Nanocomposites for Environmental Application: A Review. *Chem. Eng. J.* **2011,** *170*, 381–394.

Zhu, N.; Luan, H.; Yuan, S.; Chen, J.; Wu, X.; Wang, L. Effective Dechlorination of HCB by Nanoscale Cu/Fe Particles. *J. Hazard. Mater.* **2010,** *176*, 1101–1105.

Zielińska, A.; Skwarek, E.; Zaleska, A.; Gazda, M.; Hupka, J. Preparation of Silver Nanoparticles with Controlled Particle Size. *Procedia Chem.* **2009**, *1* (2), 1560–1566.

Zosky, G. R.; Boylen, C. E.; Wong, R. S.; Smirk, M. N.; Gutierrez, L.; Woodward, R. C.; Cook, A. Variability and Consistency in Lung Inflammatory Responses to Particles with a Geogenic Origin. *Respirology* **2014,** *19* (1), 58–66.

CHAPTER 12

Balanced Nitrogen Fertilization in Paddy for Sustainable Production and Environmental Gain

MAINAK GHOSH and GARIMA SINGH

Department of Agronomy, Bihar Agricultural University, Sabour, Bhagalpur, Bihar, India

ABSTRACT

Nitrogen is considered the most important contributor of economic yield and quality of food grains. The use of nitrogen fertilizer in Indian agriculture has grown rapidly in the last few decades but this trend has negatively impacted soil health and productivity across the country. The imbalanced use of nitrogen by farmers has resulted in low nitrogen use efficiency (NUE) and high nitrogen losses threatening the environment. So, for the sustainable development of agricultural systems, a better scientific understanding of nitrogen management in crops as well as improved technologies to increase NUE are to be focused. Many studies have shown that the use of smart N management technologies improves the NUE by synchronizing the fertilizer application rates to the actual crop demand. In these smart technologies, the distribution of N with precise amount and right time is probably the main reason for higher grain yield and partial factor productivity of N. These technologies include various canopy sensors, such as chlorophyll meter, green seeker, leaf color chart, crop circle, and many more which are used to reduce the N loss and to effectively increase the fertilizer NUE for sustainable crop production. The transition from the current agricultural system into a sustainable agricultural system with these resources–use and efficient technologies can be economically profitable as well as environmentally safe.

Environmental Contaminants: Impact, Assessment, and Remediation.
Pritam Ganguly, Jajati Mandal, M. Paramsivam, & Sandip Patra, (Eds.)

12.1 INTRODUCTION

The most important and concerning major nutrient in crop production is nitrogen (N). Agriculture is solely depending on N and also the Indian economy. Report said that India's national average N consumption is not maintained but increasing and from the year 2001–02, it was 113.1 lakh tons to 173.7 lakh tons in 2015–16 (Annual Report 2016–16, Department of Agriculture, Govt. of India). Here, the farmers cannot ignore the N application because the nutrient has a quick result in keeping the crops more attractive. Nitrogen is considered the most important contributor to the economic yield and quality of food grains. The Government of India has undertaken a number of steps for improving soil fertility which is key to the sustainable development of agricultural productivity. Urea is the most widely used N fertilizer in India. About 300 lakh tons of urea per year is required in the Indian market to mitigate its demand. Currently, the actual price of urea in the Indian market is Rs 23,880 per ton of which the Government of India provides a subsidy of Rs. 17,980 (75.3%) and farmers have to pay only Rs 5900. It indicates that the current subsidy burden on the Government of India for urea is approximately Rs. 73,000 crores (Expenditure Budget 2018–19, Ministry of Finance, Government of India). Therefore, saving N without affecting the grain yield is a major challenge to the farmers. Against a recommended N: P: K dose of 4:2:1, as high as 7.2:2.9:1 has been observed to be used in different regions of India (Annual report 2016–17, Department of Agriculture, Govt. of India), which has negatively impacted the soil health and productivity across the country. Imbalance use of plant nutrients also results in pollution of the atmosphere, groundwater, and eutrophication of water bodies. In Asia, farmers are acquainted with excessive N fertilizer use to maintain yield levels but the nutrient use efficiency is declining (Peng et al., 2006; Singh et al., 2010; Ghosh et al., 2013). It demonstrates how N loss and low N use efficiency (NUE) can occur in paddy cultivation systems (Ghosh et al, 2018) and it has also become a threat to the environment worldwide. Considering sustainability, the improvement of NUE is one of the core objectives in Agriculture. Ghosh et al. (2020) suggested that requirement-based N management is required to increase the N-use efficiency and partial factor productivity. Experiments have shown that smart N management technologies in rice could effectively increase grain yield and partial factor productivity of N (Singh et al., 2012) as compared with the common practice of blanket N application. Modern hand-held implements were identified to bridge the gap between the demand and supply of N. The leaf color chart

(LCC) and optical sensors like chlorophyll meter or green seeker are widely used successfully to increase the partial factor productivity and N use efficiencies. The optimum soil plant analysis development (SPAD) threshold for adequate N application in split applications could reduce N loss compared with broadcast applications and thus could increase NUE. Few studies have focused on the N distribution using smart technologies in rice and wheat which ultimately may modify the trimming of N application. The distribution of N with a precise amount and at the right time is probably the main reason for higher grain yield and partial factor productivity of N.

Excessive use of N fertilizer results in decreasing the NUE, N recovery efficiency (REN), and partial factor productivity of applied N (PEN) in rice. Scientists have developed smart technologies to improve fertilizer NUE through site-specific N management (SSNM) and found that they were effective in increasing the NUE of rice (Dobermann et al., 2002; Peng et al., 2006). On average, SSNM could reduce N fertilizer by 32% and could increase the grain yield by 5% compared with the standard recommendation. The crop's growth stages, such as panicle initiation, spikelet differentiation, and heading of rice were found to have a high affinity of N and resulted in a substantial increase in yield and NUE (Ghosh et al., 2013). Considering the sustainability, it is very important to refine the N dose for improving NUE and economic viability with reduced loss of N to the environment. Farmers have practiced to apply 30% of the recommended dose of N with 100% P and K at the time of sowing followed by N splitting equally at active tillering (AT) and panicle initiation (PI) stages, coinciding with irrigation events. Here, during the splitting of N application, it is very important to keep the soil moist for proper N utilization. Therefore, assessment of the growth stage plays a key role in N topdressing. Singh et al. (2012) also advised that NUE can be raised only when the application coincides with stages of high N demand by crops. In India, blanket N dose for rice varying from 120 to 180 kg N ha^{-1} indicates the spatial variation and farmers are practiced to it in more than two splits. The variation also reflects in crop performance during the vegetative growth on chlorophyll content and accordingly greenness of leaves. It was observed and subsequently become admired by farmers to apply an extra dose of N over the recommendation to avoid the risk. Researchers also noticed that 20–35% N can be saved without deteriorating the rice grain yield by the adjustment of N application with demand (Singh et al., 2002; Ghosh et al., 2013). The reason behind the application of more N fertilizer than the demand is that the general recommendation technique is not appropriate or equal for all regions (Singh et al., 2011). The use of Smart technologies considering the leaf chlorophyll content under SSNM provides good N utilization efficiency, as a positive relationship was found between chlorophyll and N content in leaves (Singh et al., 2012; Ghosh et al., 2020). Smart gadgets such as leaf color charts, chlorophyll meter, and greenseeker are really found for guiding real-time N topdressing in rice (Ghosh et al., 2018). SPAD 502 Chlorophyll meter measures the leaf transmittance at wavelengths 650 and 940 nm, and is a non-destructive alternative for leaf nitrogen estimation. This gadget can help precise N management in rice under field conditions with economic benefits to the farmers.

12.2 N USE FOR FOOD SECURITY

Nitrogen fertilization has enabled many developing countries to achieve self-sufficiency in food production through increased crop yield as it is the main yield-determining nutrient. According to Borlaug (2007), global cereal production was increased by 10% in cereal-growing areas since the 1950s. This enormous yield rise per unit of land area was due to the introduction of high-yielding cultivars that were well-responsive to nitrogen fertilizer and irrigation water. It was also noted that although nitrogen fertilizer use has grown approximately sevenfold from the 1950s, around 30–80% of applied nitrogen is lost to surface and groundwaters and into the atmosphere (Goulding et al., 2008). Undesirable consequences of soil and environmental degradation as a result of the excess application of N fertilization beyond crops' demand in the last several decades are quite discernable. Goudriaan et al. (2001) analyzed that in 1990, fertilizer N accounted for 60% of the net primary production. Various tools and technologies are available that take into account the spatial and temporal variability in soil nutrient supply and develop site-specific nutrient management plans. For example, Crop Circle and Green Seeker sensors, the N-Sensor ALS, color digital camera image analysis, chlorophyll fluorescence, reflectance spectra, and chlorophyll measurement (Gu et al., 2017). Under a smallholder production system, the use of a chlorophyll meter and leaf color chart to obtain real-time readings of leaf greenness for directly evaluating the leaf N status serves as the basis of in-season nitrogen management. SPAD meter proved to measure adequate leaf nitrogen status through the chlorophyll content and is used for the application of N fertilizer in the right amount and at the right time (Singh et al., 2012). The leaf color chart (LCC), a relatively cheap tool when compared with other smart technologies, uses different shades of green color to guide the plant's need-based N application. Plant's need-based N management strategies can improve fertilizer use efficiency and thereby reduces the production cost and environmental impact of N losses.

12.3 N FERTILIZATION IN CROP GROWTH AND YIELD

In the history of agriculture, no input has seen as much growth as witnessed in the use of N fertilizer. For each 1 ton increase of cereal grain, the crop will need 20–27 kg N (Prasad et al., 2004). Nitrogen fertilizer has become the major limiting factor for sustainable and profitable crop production. While analyzing the yield data of rice and wheat, Kumar et al. (1998) observed a lower growth rate in Punjab and Haryana pertaining to the sustainability of rice–wheat cropping system as the growth rates are declining. Here, the proper management of N using a canopy sensor can improve crop yield with limited N use (Dobermann and Fairhurst, 2000). Dobermann and Cassman (2002) recorded an increase in grain yield of maize from 3 to 14 t ha^{-1} and rice from 2 to 8 t ha^{-1} with a rise in plant N accumulation from 50 to 300 kg N ha^{-1} and 25 to 200 kg N ha^{-1}, respectively. However, the excess application of N renders to soil infertility, and the crops fail to respond and degrade environmental quality by increasing soil acidification, N leaching, and emissions of ammonia (NH_3) and nitrogen oxide (NO, N_2O, and NO_2) gases, and eutrophication.

An account of N inputs, outputs, and retention in the soil maintains the N balance and identifies dominant processes of N flow in the agro-ecosystem (Ross et al., 2008). Eileen et al. (2018) defined the N balance as the difference between N input and removed in products from an agricultural system. The N balance reductions can be achieved by coordinating the N inputs and N outputs in time and space while maintaining or increasing yields (Cassman et al., 2002; Snyder et al., 2014). Balance application of N fertilizer dependent upon dynamic optimization of N supply to match the N demand of crop in dose and time to achieve a higher agronomic NUE. Hence, N must be need-based for sustainable production (Miao et al., 2011; Zhao et al., 2010; Sui et al., 2013). This optimization of N requires the measurement and prediction of soil-N supply, crop uptake, and variability (Stockdale et al., 1997). Delayed N fertilizer application lowers the N application rates, such modifications in fertilizer management have proved to effectively reduce costs and increase the overall profitability (Sela et al., 2016). Witt and Dobermann (2002) noted that the basal application of N can be skipped without reducing the economic yield has resulted in higher NUE. Sun et al. (2012) reported that an application of N at later growth stages is very useful and benefited the number of spikelets per panicle. Aggarwal et al. (1997) reported 4–10% yield increase in irrigated rice by improved timing of nitrogen application at a fixed total dose; while changing the N dose to the predicted economic optimum rate resulted in additional increases up to 13%. Thus, to achieve maximum yield as well as to minimize N losses to the environment, it is crucial to match the seasonal pattern of N supply to the N demand in rice crops at each stage of development (Spiertz, 2010). N fertilization rate is reported to influence the number of tiller production in grain crops (Wells and Faw, 1978). SPAD-based N application of 30 kg N ha^{-1} at maximum tillering increased wheat yield by 1.0 or 0.5 t ha^{1} when the reading was equivalent to or less than the SPAD value of 32.5 or 42.5, respectively (Singh et al., 2013). In rice–wheat belt application of 150 kg N ha^{-1} is very common in India (or more in rice) and resulted in low recovery efficiency of N (Cassman et al., 1998; Prasad, 1999). It was noticed that the recovery efficiency of N to rice was 31.0, 26.7, and 25.9% at 50, 100, and 150 kg N ha^{-1} (George and Prasad, 1989), and to wheat, it was comparatively higher varying from 40 to 91% (Prasad et al., 1998). Compared with the wheat crop, the losses of N in rice from the soil–plant system are more leading to a lower recovery of N applied to rice than to wheat. The soil residual N after crop harvest is lost to the environment through leaching, denitrification, volatilization, surface runoff, soil erosion, and N_2O emissions (Eickhout et al., 2006; Ross et al., 2008). One option to reduce soil residual N is to increase N use efficiency and the NUE depends on the management practices (Prasad, 2005). It was also observed that the NUE for most of the cereals ranged approximately 30–35% worldwide (Raun and Johnson, 1999). Application of N in small amounts is a proven tool to increase NUE. Splitting of total N in two or three split doses is common in rice or wheat, but the number of splits can be increased in rice crop as N losses are more. Nutrient management in integrated approach and precision farming have shown to be effective to improve NUE (Pierce and Nowak, 1999). Peng et al. (2006) noticed that N application using canopy sensors may reduce the N fertilizer use by 32% and increase grain yield by 5% when compared with farmers' usual fertilizer practices (FFP). Similar results with a yield gain of 0.5 Mg ha^{-1} (12%) compared with FFP along with 10% saving of fertilizer N clearly bring out the positive effect of N management in site-specific nutrient management (Singh et al., 2002; Pathak et al., 2003b). Real-time N management (RTNM) using a SPAD meter gave the same rice yield with 20–30% less fertilizer N use (Singh et al., 2002; Kyaw, 2003; Singh et al., 2010).

Whereas in eastern India, the grain yield of SPAD-based RTNM ranged from 93 to 105% as that of standard N recommendation, but with lower N application rate (Ghosh et al., 2013). The chlorophyll meter readings are highly correlated with stem N (R^2 = 0.80), leaf N (R^2 = 0.94), and grain yield of rice (R^2 = 0.88) (Ghosh et al., 2020). Research indicated that high NUE like Agronomic NUE (AE_N) by 5.26 kg kg^{-1} (63%), N recovery efficiency (RE_N) by 0.10 kg kg^{-1} (59%), and partial factor productivity of applied N (PFP_N) by 7.72 kg kg^{-1} (26%) a considerable amount of fertilizer N saving than the general recommendation with maintaining the grain yield of rice (Khurana et al., 2008). Alam et al. (2006) also supported RTNM through SPAD and LCC and opined that RTNM could obtain higher AE_N (12–21 kg kg^{-1}) as compared with farmers' practice (6–15 kg kg^{-1}). Ghosh et al. (2013) noticed that the AE_N of SPAD-based N management was increased by 37–75% as compared with the general N recommendation because of similar or higher rice grain yields at a lower N rate in RTNM.

12.4 N FERTILIZATION IN PRODUCTIVITY AND SUSTAINABILITY

With the intensification of agriculture and the increasing awareness toward natural resources, sustainability, and the development of environment-friendly N application approaches are being considered. The main objective of sustainability includes Economic profitability, environmental health, and ethical soundness. Nitrogen is the crucial nutrient, and can account for approximately 15–25% of production costs (Lambert and Lowenberg-DeBoer, 2000) in cereal. Good agricultural practices aimed toward the ecological and economical use of resources for sustainable food production can be the guiding principle in achieving sustainability goals. These sustainability targets have to be developed for specific agro-ecological regions and farming systems as one size may not fit all (Gourley et al., 2007; de Klein et al., 2017). Precision agriculture emphasizes on improved management practices (Zhang et al., 2002) to improve profitability in need-based ways of farming leading to sustainability (Mulla, 2013). Precision agriculture employs tools like various soil and crop sensors that have the ability to sense the status of crop N and deliver spatially variable N applications based on crop N demand. The simplest, tool which avails on spectral properties of leaves to guide need-based fertilizer N application in crops under precision N management is the LCC (Shukla et al., 2004; Singh et al., 2010). The LCC is a key ingredient of the SSNM in rice (Buresh et al., 2005) and the strategies for using LCC rice and maize have been established (Singh et al., 2011, 2012). To manage the fertilizer N, Blackmer and Schepers (1995) observed that the crop-based approach might be an improvement over the soil-based approach. Photosynthesis is closely associated with the canopy and leaf N content (Lemaire et al., 2007) and the leaf N status can be determined nondestructively from the spectral reflectance of the canopy (Wood et al., 1992). The NDVI (Normalized difference vegetation index) is also an important measurement to estimate crop N status from remotely sensed data (Tremblay et al., 2009) and used for mid-season N topdressing for rice (Raun et al., 2002). The NDVI from Greenseeker has been found useful in determining management zones (Sharp et al., 2004) and nitrogen status in cereals for improving NUE (Li et al., 2010; Shaver et al., 2010; Erdle et al., 2011). The Crop Circle ACS 470 another sensor is applied in the detection of green biomass and leaf area estimation in cereals (Trotter et al., 2008; Fitzgerald, 2010) and very stable index for estimating winter wheat N status (Erdle et al., 2011). Shiratsuchi et al. (2011) noticed that the vegetation indices calculated with Crop Circle were the best indices for differentiating N rate effects on maize N status. Chlorophyll meters (SPAD) are commonly used for in-season adjustment of N fertilizer at different crop growth stages (Cao et al., 2012). The SPAD index is positively correlated with leaf N concentrations as well as soil N status (Follett et al., 1992). SPAD meter-guided in-season fertilizer N top dressings are developed (Shukla et al., 2004; Cartelat et al., 2005; Spaner et al., 2005; Huang et al., 2008) and is used for accurate and nondestructive prediction of crop N needs for rice (Swain and Sandeep, 2010; Cabangon et al., 2011; Ghosh et al., 2013) and wheat (Zebarth et al., 2007; Singh et al., 2010), leading to greater fertilizer NUE at all yield levels. The detection of a crop N status using chlorophyll meter for determining N application could maintain crop yield with less N fertilizer (Ghosh et al., 2013). Tubana et al. (2008) reported that increase in fertilizer NUE up to 41% in cereal crops using an in-season N fertilization optimization algorithm. The NUE was significantly increased using active sensor-based in-season N management strategies for winter wheat (Li et al., 2009; Singh et al., 2011) and rice (Yao et al., 2012). Greenseeker showed 15% more NUE than that of farmers' recommendation in wheat (Raun et al., 2002). Scharf et al. (2011) noted that sensor-based N application increased partial profit and yield by \$42 ha^{-1} and 110 kg ha^{-1}, respectively, while the N application rate was reduced by 16 kg N ha^{-1} compared with the local farmers' practice.

Walsh et al. (2013) noticed the soil moisture data and suggested that it is useful in improving the NUE by addressing the temporal and spatial N variability. Thus, the soil moisture parameters can help in improving fertilizer N management and increasing NUE. Cui et al. (2008) noticed that soil test-based N management strategy significantly increased economic wheat yield by $144 per hectare, reduced residual nitrate-N content in the top 90 cm soil layer, and N losses by 81 and 118 kg N ha^{-1}, respectively.

12.5 N MANAGEMENT IN ENVIRONMENTAL GAIN

The upsurge in food production came with a price of more external N inputs and more environmental harm (Eickhout et al., 2006). The use of N is mainly determined by economic incentives such as profitability and subsidies, and less by environmental costs. Hunter and colleagues (2017) observed that the sustainable intensification of food production with environmental gains is largely lacking. India has become the third country after the USA and the European Union where agriculture is the main source of nitrogen pollution. Excessive application of N fertilizers resulted in soil acidification, nitrate-N (NO_3-N) contamination of groundwater, and emissions of nitrous oxide (N_2O). and when the NO_3-N problem in the groundwater goes beyond 10 mg L^{-1}, it can be hazardous to humans and animals through drinking water. Emissions of N_2O gas that contribute to global warming, are 300 times more powerful than carbon dioxide in terms of global warming potential (GWP). Intensive farming and a high rate of N fertilizer application for 20 years in China dropped the soil pH by 0.30–0.80 units from the original level (Guo et al., 2010). Long-term application of NH_4-based N fertilizer increased soil acidity (Mahler et al., 1984), resulting in the development of infertile soil that becomes unresponsive to further application of N fertilizers. The N fertilizers also indirectly emit N_2O through NH_3 volatilization and NO_3-N leaching (Venterea et al., 2005). Li et al. (2015) observed that the net GWP and greenhouse gas intensity (GHGI) increased linearly with the N application rates that exceeded crop demand. The N fertilizers should always be applied at optimum rates to reduce net GWP and GHGI while sustaining crop yields. The most prominent way of applying N fertilizer at optimum rates is precision agriculture, which uses hi-tech approaches to apply just the right amount of nitrogen at the right time as per the plant's need. The sensor-based N fertilization eliminates an enormous amount of loss without affecting yields and bringing the global nitrogen application down. There is a need for a changeover of agricultural systems into highly resource-use efficient systems that are profitable, but at the same time ecologically safe and socially acceptable. Therefore, a great leap toward the balanced use of fertilizer N needs to be taken. For a farmer seeking to reduce N losses, the challenge is identifying the changes that can be made in their operation to reduce N imbalance while maintaining productivity and profitability.

12.6 CONCLUSIONS

Nitrogen is the most crucial nutrient and growth factor in rice crop production, economy, and environmental gain. The inadequate nitrogen fertilizer application threatens global food security and causes severe ecological consequences. Therefore, improving the NUE of applied N fertilizer is an effective means of increasing crop productivity to address global food security in a sustainable manner and to decrease environmental N pollution. Smart technologies in agriculture have provided farmers with access to technical knowledge-based N management at the farm scale. The farmers' growing awareness and adoption of NUE-based technologies like SSNM and real-time nitrogen management using canopy sensors, such as chlorophyll meter, green seeker, crop circle, and even simple leaf color chart can be a crucial step toward achieving sustainable agricultural development. The effective increase in NUE in rice through resource use and efficient smart technologies can be economically profitable and environmentally safe. However, some socioeconomic factors related to initial cost and perceived risks and lack of trust in new technologies in farmers are restraining them from adopting improved nutrient management practices. In this regard, seeking solutions in the adoption of new technologies and outreaching the farmers influencing their decision-making for proper nitrogen management needs more attention.

KEYWORDS

- **green seeker**
- **chlorophyll meter**
- **NDVI**
- **N use efficiency**
- **rice**

REFERENCES

Aggarwal, P. K.; Kropff, M. J.; Cassman, K. G.; ten Berge, H. F. M. Simulating Genotypic Strategies for Increasing Rice Yield Potential in Irrigated, Tropical Environments. *Field Crop Res.* **1997,** *51*, 5–17.

Alam, M. M.; Ladha, J. K.; Foyjunnessa; Rahman, Z.; Khan, S. R.; Rashid, H.; Khan, A. H.; Buresh, R. J. Nutrient Management for Increased Productivity of Rice–wheat Cropping System in Bangladesh. *Field Crops Res.* **2006,** *96*, 374–386.

Blackmer, T. M.; Schepers, J. S.; Meyer, G. E. In *Remote Sensing to Detect N Deficiency in Corn*, Proceeding of Site-Specific Management for Agricultural System, 1995; 27–30.

Borlaug, N. E. Sixty-Two Years of Fighting Hunger: Personal Recollections. *Euphytica* **2007,** *157*, 287–297.

Buresh, R. J.; Witt, C.; Ramanathan, S.; Mishra, B.; Chandrasekaran, B.; Rajendran, R. Site-specific Nutrient Management Managing N, P and K for Rice. *Fertil. News* **2005,** *50*, 25–28.

Cabangon, R. J.; Castillo, E. G.; Tuong, T. P. Chlorophyll Meter-based Nitrogen Management of Rice Grown Under Alternate Wetting and Drying Irrigation. *Field Crops Res.* **2011,** *121*, 136–146.

Cao, Q.; Cui, Z.; Chen, X.; Khosla, R.; Dao, T. H.; Miao, Y. Quantifying Spatial Variability of Indigenous Nitrogen Supply for Precision Nitrogen Management in Small Scale Farming. *Precis. Agric.* **2012,** *13*, 45–61.

Cartelat, A.; Cerovic, Z. G.; Goulas, Y.; Meyer, S.; Lelarge, C.; Prioul, J. L.; Barbottin, A.; Jeuffroy, M. H.; Gate, P.; Agati, G.; Moya, I. Optically Assessed Contents of Leaf polyphenolics and Chlorophyll as Indicators of Nitrogen Deficiency in Wheat (*Triticum aestivum* L.). *Field Crops Res.* **2005,** *91*, 35–49.

Cassman, K. G.; Dobermann, A.; Walters, D. T. Agroecosystems, Nitrogen-use Efficiency, and Nitrogen Management. *Ambio* **2002,** *31*, 132–140.

Cassman, K. G.; Peng, S.; Olk, D. C.; Ladha, J. K.; Reichardt, W.; Dobermann, A.; Singh, U. Opportunities for Increased Nitrogen Use Efficiency from Improved Resource Management in Irrigated Rice System. *Field Crops Res.* **1998,** *56*, 7–39.

Cui, Z.; Zhang, F.; Chen, X.; Miao, Y.; Li, J.; Shi, L.; Xu, J.; Ye, Y.; Liu, C.; Yang, Z.; Zhang, Q.; Huang, S.; Bao, D. On-farm Evaluation of An In-season Nitrogen Management Strategy Based on Soil N_{min} Test. *Field Crops Res.* **2008,** *105*, 48–55.

De Klein, C. A. M.; Monaghan R. M.; Alfaro, M.; Gourley, C. J. P.; Oenema, O.; Powell J. M. Nitrogen Performance Indicators for Dairy Production Systems. *Soil Res.* **2017,** *55*, 479–488.

Dobermann, A.; Cassman, K. G. Plant Nutrient Management for Enhanced Productivity in Intensive Grain Production Systems of the United States and Asia. *Plant Soil.* **2002,** *247*, 153–175.

Dobermann, A.; Fairhurst, T. *Rice: Nutrient Disorders and Nutrient Management*; International Rice Research Institute: Manila, Philippines, 2000.

Eickhout, B.; Bouwman, A. P.; van Zeijts, H. The Role of Nitrogen in World Food Production and Environmental Sustainability. *Agric. Ecosyst. Environ.* **2006,** *116*, 4–14

McLellan, E. L.; Cassman, K. G.; Eagle, A. J.; Woodbury, P. B.; Sela, S.; Tonitto, C.; Marjerison, R. D.; van Es, H. M. The Nitrogen Balancing Act: Tracking the Environmental Performance of Food Production. *Bioscience* **2018,** *68*(3), 194–203.

Erdle, K.; Mistele, B.; Schmidhalter, U. Comparison of Active and Passive Spectral Sensors in Discriminating Biomass Parameters and Nitrogen Status in Wheat Cultivars. *Field Crops Res.* **2011,** *124*, 74–84.

Fitzgerald, G. J. Characterizing Vegetation Indices Derived from Active and Passive Sensors. *Int. J. Remote Sens.* **2010,** *31*, 4335–4348.

Follett, R. H.; Follett, R. F.; Halvorson, A. D. Use of a Chlorophyll Meter to Evaluate the Nitrogen Status of Dryland Winter Wheat. *Commun. Soil Sci. Plant Anal.* **1992,** *23*, 687–697.

George, M.; Prasad, R. Studies on the Effect of Legumes on Fertilizer N Nitrification by Rice Using ^{15}N Technique in Rice-based Multiple Cropping System. *Res. Dev. Agri.* **1989,** *6*, 115–118.

Ghosh, M.; Swain, D. K.; Jha, M. K.; Tewari, V. K. Precision Nitrogen Management using Chlorophyll Meter for Improving Growth, Productivity and N Use Efficiency of Rice in Subtropical Climate. *J. Agric. Sci.* (Canada), **2013,** *5*(2), 1–14.

Ghosh, M.; Swain, D. K.; Jha, M. K.; Tewari, V. K. Chlorophyll Meter-based Nitrogen Management of Wheat in Eastern India. *Exp. Agric.* **2018,** *54*(3), 349–362.

Ghosh, M.; Swain, D. K.; Jha, M. K.; Tewari, V. K.; Bohra, A. Optimizing Chlorophyll Meter (SPAD) Reading to Allow Efficient Nitrogen Use in Rice and Wheat Under Rice-wheat Cropping System in Eastern India. *Plant Prod. Sci.* **2020,** *23*(3), 270–285.

Goudriaan, J.; Groot, J. J. R.; Uithol, P. W. J. Productivity of Agro-ecosystems. In: *Terrestrial Global Productivity*; Roy, J., Saugier, B., Mooney, H. A., Eds.; Academic Press: San Diego, USA, 2001; pp 310–313.

Goulding K.; Jarvis, S.; Whitmore A. Optimizing Nutrient Management for Farm Systems. *Philos. T. Roy. Soc. B* **2008,** *363*, 667–680.

Gourley, C. J. P.; Powell, J. M.; Dougherty, W. J.; Weaver, D. M. Nutrient Budgeting as An Approach to Improving Nutrient Management on Australian Dairy Farms. *Aust. J. Exp. Agric.* **2007,** *47*, 1064–1074.

Gu, J.; Chen, Y.; Zhang, H.; Li, Z.; Zhou, Q.; Yu, C.; Kong, X.; Liu L.; Wang, Z.; Yang, J. Canopy Light and Nitrogen Distributions are Related to Grain Yield and Nitrogen Use Efficiency in Rice. *Field Crops Res.* **2017,** *206*, 74–85.

Guo, J. H.; Liu, X. J.; Zhang, Y.; Shen, J. L.; Han, W. X.; Zhang, W. F.; Christie, P.; Goulding, K. W. T.; Vitousek, P. M.; Zhang, F. S. Significant Acidification in Major Chinese Croplands. *Science* **2010,** *327*, 1008–1010.

Huang, J.; He, F.; Cui, K.; Roland, J.; Buresh, B. X.; Gong, W.; Peng, S. Determination of Optimal Nitrogen Rate for Rice Varieties Using a Chlorophyll Meter. *Field Crops Res.* **2008,** *105*, 70–80.

Hunter, M. C.; Smith, R. G.; Schipanski, M. E.; Atwood, L. W.; Mortensen, D. A. Agriculture in 2050: Recalibrating Targets for Sustainable Intensification. *BioScience* **2017,** *67*, 386–391.

Khurana, H. S.; Phillips, S. B.; Singh, B.; Alley, M. M.; Dobermann, A.; Sidhu, A. S.; Singh, Y.; Peng, S. Agronomic and Economic Evaluation of Site-specific Nutrient management for Irrigated Wheat in Northwest India. *Nutr. Cycling Agroecosyst.* **2008,** *82*, 15–31.

Kumar, P.; Joshi, P. K.; Johansen, C.; Asokan, M. Sustainability of Rice–wheat Based Cropping Systems in India. *Econ. Political Wkly.* **1998,** *33*(A), 182–188.

Kyaw, K. K. Plot-specific N fertilizer Management for Improved N-use Efficiency in Rice Based Systems of Bangladesh. In *Ecology and Development*; Vlek, P. L. G., Denich, M., Martius, C., Giesen, N. V. D., Eds.; Series No. 12. Cuvillier, Gottingen, 2003.

Lambert, D. M.; Lowenberg-DeBoer, J. *Precision Farming Profitability Review*; Site Specific Management Center, Purdue University: West Lafayette, 2000.

Lemaire, G.; Jeuffroy, M. H.; Gastal, F. Diagnosis Tool for Plant and Crop N Status in Vegetative Stage. Theory and Practices for Crop N Management. *Eur. J. Agron.* **2007,** *28*, 614–624.

Li, B.; Fan, C. H.; Zhang, H.; Chen, Z. Z.; Sun, L. Y.; Xiong, Z. Q. Combined Effects of Nitrogen Fertilization and Biochar on the Net Global Warming Potential, Greenhouse Gas Intensity, and Net Ecosystem Budget in Intensive Vegetable Agriculture in Southeastern China. *Agric. Ecosyst. Environ.* **2015,** *100*, 10–19.

Li, F.; Miao, Y.; Zhang, F.; Cui, Z.; Li, R.; Chen, X.; Zhang, H.; Schroder, J.; Raun, W. R.; Jia, L. In-season Optical Sensing Improves Nitrogen-use Efficiency for Winter Wheat. *Soil Sci. Soc. Am. J.* **2009,** *73*, 1566–1574.

Li, Y.; Chen, D.; Walker, C. N.; Angus, J. F. Estimating the Nitrogen Status of Crops Using a Digital Camera. *Field Crops Res.* **2010,** *118*, 221–227.

Mahler, R. L.; Harder, R. W. The Influence of Tillage Methods, Cropping Sequence, and N Rates on the Acidification of a Northern Idaho Soil. *Soil Sci.* **1984,** *137*, 52–60.

Miao, Y.; Stewart, B. A.; Zhang, F. Long-term Experiments for Sustainable Nutrient Management in China. A Review. *Agron. Sustain. Dev.* **2011,** *31*, 397–414.

Mulla, D. J. Twenty Five Years of Remote Sensing in Precision Agriculture: Key Advances and Remaining Knowledge Gaps. *Biosyst. Eng.* **2013,** *114*, 358–371.

Pathak, H.; Singh, Y.; Singh, B. Site-specific Nutrient Management for Improving the Productivity of Rice-wheat Cropping System. In *Nutrient Management for Sustainable Rice-wheat Cropping System* Singh, Y., et al., Eds.; National Agricultural Technical Project, Indian Council of Agricultural Research, New Delhi, India, and Punjab Agricultural University, Ludhiana, Punjab, India, 2003; pp 79–98.

Peng, S.; Buresh, R. J.; Huang, J.; Yang, J.; Zou, Y.; Zhong, X.; Wang, G.; Zhang, F. Strategies for Overcoming Low Agronomic Nitrogen Use Efficiency in Irrigated Rice Systems in China. *Field Crops Res.* **2006,** *96*, 37–47.

Pierce, F. P.; Nowak, P. Aspects of Precision Farming. *Adv. Agron.* **1999,** *67*, 1–85.

Prasad, R.; Kumar, D.; Sharma, S.; Gautam, R.; Dwivedi, M. Current Status and Strategies for Balanced Fertilization. *Fertil. News* **2004,** *49*(12), 73–80.

Prasad, R. *A Text Book of Rice Agronomy*; Jain Brothers: New Delhi, India, 1999.

Prasad, R. Rice-wheat Cropping Systems. *Adv. Agron.* **2005,** *86*, 1–85.

Prasad, R.; Rai, R. K.; Sharma, S. N.; Singh, S.; Shivay, Y. S.; Idnani, L. K. Nutrient management. In *Fifty Years of Agronomic Research in India*; Yadav, R. L., Singh, P., Prasad, R., Ahlawat, I. P. S., Eds.; Indian Society of Agronomy: New Delhi, India, 1998; pp 51–85.

Raun, W. R.; Johnson, G. V. Improving Nitrogen Use Efficiency for Cereal Production. *Agron. J.* **1999,** *91*, 357–363.

Raun, W. R.; Solie, J. B.; Johnson, G. V.; Stone, M. L.; Mullen, R. W.; Freeman, K. W.; Thomason, W. E.; Lukina, E. V. Improving Nitrogen Use Efficiency in Cereal Grain Production with Optical Sensing and Variable Rate Application. *Agron. J.* **2002,** *94*, 815–820.

Ross, S. M.; Izaurralde, R. C.; Janzen, H. H.; Robertson, J. A.; McGill, W. B. The Nitrogen Balance of Three Long-term Agroecosystems on a Boreal Soil in Western Canada. *Agric. Ecosyst. Environ.* **2008,** *27*, 241–250.

Scharf, P. C.; Shannon, D. K.; Palm, H. L.; Sudduth, K. A.; Drummond, S. T.; Kitchen, N. R.; Mueller, L. J.; Hubbard, V. C.; Oliveira, L. F. Sensor-based Nitrogen Applications

Out-performed Producer-chosen Rates for Corn in On-farm Demonstrations. *Agron. J.* **2011,** *103*, 1683–1691.

Sela, S.; van Es, H. M.; Moebius-Clune, B. N.; Marjerison, R.; Melkonian, J.; Moebius-Clune, D.; Schindelbeck, R.; Gomes, S. Adapt-N Outperforms Grower-selected Nitrogen Rates in Northeast and Midwestern United States Strip Trials. *Agron. J.* **2016,** *108*, 1726–1734.

Sharp, T. C.; Evans, G.; Salvador, A. In *Weekly NDVI Relationships to Height, Nodes and Productivity Index for Low, Medium, and High Cotton Productivity Zones*, Proceedings of the Beltwide Cotton Conferences, National Cotton Council, Memphis, TN, USA, 2004; Dugger, C. P., Ritcher, D. A. Eds.; 2048.

Shaver, T. M.; Khosla, R.; Westfall, D. G. Evaluation of Two Ground-based Active Crop Canopy Sensors in Maize: Growth Stage, Row Spacing and Sensor Movement Speed. *Soil Sci. Soc. Am. J.* **2010,** *74*, 2101–2108.

Shiratsuchi, L.; Ferguson, R.; Shanahan, J.; Adamchuk, V.; Rundquist, D.; Marx, D.; Slater, G. Water and Nitrogen Effects on Active Canopy Sensor Vegetation Indices. *Agron. J.* **2011,** *103*, 1815–1826.

Shukla, A. K.; Ladha, J. K; Singh, V. K.; Dwivedi, B. S.; Balasubramanian, V.; Gupta, R. K.; Sharma, S. K.; Singh, Y.; Pathak, H.; Pandey, P. S.; Padre, A. T.; Yadav, R. L. Calibrating the Leaf Color Chart for Nitrogen Management in Different Genotypes of Rice and Wheat in a Systems Perspective. *Agron. J.* **2004,** *96,* 1606–1621.

Singh, B.; Singh, V.; Singh, Y.; Thind, H. S.; Kumar, A.; Singh, S.; Choudhary, O. P.; Gupta, R. K.; Vashistha, M. Supplementing Fertilizer Nitrogen Application to Irrigated Wheat at Maximum Tillering Stage using Chlorophyll Meter and Optical Sensor. *Agric. Res.* **2013,** *2*(1), 81–89.

Singh, B.; Singh, Y.; Ladha, J. K.; Bronson, K. F.; Balasubramanian, V.; Singh, J.; Khind, C. S. Chlorophyll Meter and Leaf Color Chart Based Nitrogen Management for Rice and Wheat in Northwestern India. *Agron. J.* **2002,** *94*, 821–829.

Singh, V.; Singh, B.; Singh, Y.; Thind, H. S.; Gupta, R. K. Need based Nitrogen Management Using the Chlorophyll Meter and Leaf Colour Chart in Rice and Wheat in South Asia: A Review. *Nutr. Cycle Agroecosyst.* **2010,** *88*, 361–380.

Singh, V.; Singh, B.; Singh, Y.; Thind, H. S.; Singh, G.; Kaur, S.; Kumar, A.; Vashistha, M. Establishment of Threshold Leaf Colour Greenness for Need-based Fertilizer Nitrogen Management in Irrigated Wheat (*Triticum aestivum* L.) Using Leaf Colour Chart. *Field Crops Res.* **2012,** *130*, 109–119.

Singh, V.; Singh, Y.; Singh, B.; Thind, H. S.; Kumar, A.; Vashistha, M. Calibrating the Leaf Colour Chart for Need Based Fertilizer Nitrogen Management in Different Maize (*Zea mays* L.) Genotypes. *Field Crops Res.* **2011,** *120*, 276–282.

Snyder C. S.; Davidson E. A.; Smith P.; Venterea, R. T. Agriculture: Sustainable Crop and Animal Production to Help Mitigate Nitrous Oxide Emissions. *Curr. Opin. Environ. Sustain.* **2014,** *9*, 46–54.

Spaner, D.; Todd, A. G.; Navabi, A.; McKenzie, D. B.; Goonewardene, L. A. Can Leaf Chlorophyll Measures at Differing Growth Stages be Used as an Indicator of Winter Wheat and Spring Barley Nitrogen Requirements in Eastern Canada? *J. Agron. Soil Sci.* **2005,** *191*, 393–399.

Spiertz, J. H. J. Nitrogen, Sustainable Agriculture and Food Security - A Review. *Agron. Sustain. Dev.* **2010,** *30*, 43–55.

Stockdale, E. A.; Gaunt, J. L.; Vos, J. Soil-plant Nitrogen Dynamics: What Concepts are Required? *Eur. J. Agron.* **1997,** *7*, 145–159.

Sui, B.; Feng, F.; Tian, G.; Hu, X.; Shen, Q.; Guo, S. Optimizing Nitrogen Supply Increases Rice Yield and Nitrogen Use Efficiency by Regulating Yield Formation Factors. *Field Crops Res.* **2013,** *150*, 99–107.

Sun, Y.; Ma, J.; Sun, Y.; Xu, H.; Yang, Z.; Liu, S.; Jia, X.; Zheng, H. The Effects of Different Water and Nitrogen Managements on Yield and Nitrogen Use Efficiency in Hybrid Rice of China. *Field Crops Res.* **2012,** 127 85–98.

Swain, D. K.; Sandip, S. J. Development of SPAD Values of Medium and Long Duration Rice Variety for Site Specific Nitrogen Management. *J. Agron.* **2010,** *9*(2), 38–44.

Tremblay, N.; Wang, Z.; Ma, B.; Belec, C. Vigneault, P. A comparison of Crop Data Measured by Two Commercial Sensors for Variable-rate Nitrogen Application. *Precis. Agric.* **2009,** *10*, 145–161.

Trotter, T.; Frazier, P.; Trotter, M.; Lamb, D. In *Objective Biomass Assessment Using an Active Plant Sensor (Crop Circle), Preliminary Experiences on a Variety of Agricultural Landscapes*. 9th International Conference on Precision Agriculture, Colorado State University, Fort Collins, Colorado, USA, 2008; Kholsa, R., Ed.

Tubana, B. S.; Arnall, D. B.; Holtz, S. L.; Solie, J. B.; Girma, K.; Raun, W. R. Effect of Treating Field Spatial Variability in Winter Wheat at Different Resolutions. *J. Plant Nutr.* **2008,** *31*, 1975–1998.

Venterea, R. T.; Burger, M.; Spokas, K. A. Nitrogen Oxide and Methane Emissions Under Varying Tillage and Fertilizer Management. *J. Environ. Qual.* **2005,** *34*, 1467–1477

Walsh, O. S.; Klatt, A. R.; Solie, J. B.; Godsey, C. B.; Raun, W. R. Use of Soil Moisture Data for Refined Green Seeker Sensor Based Nitrogen Recommendations in Winter Wheat (*Triticum aestivum* L.). *Precis. Agric.* **2013,** *14*, 343–356.

Wells, B.; Faw, W. F. Short-statured Rice Response to Seeding and N Rates. *Agron. J.* **1978,** *70*, 477–480.

Witt, C.; Dobermann, A. Site-specific Nutrient Management Approach for Irrigated, Lowland Rice in Asia. *Better Crops.* **2002,** *16*, 20–24.

Wood, C. W.; Reeves, D. W.; Duffield, R. R.; Edmisten, K. L. Field Chlorophyll Measurements for Evaluation of Corn Nitrogen Status. *J. Plant Nutr.* **1992,** *15*, 487–500.

Yao, Y.; Miao, Y.; Huang, S.; Gao, L.; Ma, X.; Zhao, G.; Jiang, R.; Chen, X.; Zhang, F.; Yu, K.; Gnyp, M.; Bareth, G.; Liu, C.; Zhao, L.; Yang, W.; Zhu, H. Active Canopy Sensor based Precision N Management Strategy for Rice. *Agron. Sustain. Dev.* **2012,** *32*, 925–933.

Zebarth, B. J.; Botha, E. J.; Rees, H. Rate and Time of Fertilizer Nitrogen Application on Yield, Protein and Apparent Efficiency of Fertilizer Nitrogen Use of Spring Wheat. *Can. J. Plant Sci.* **2007,** *87*, 709–718.

Zhang, N.; Wang, M.; Wang, N. Precision Agriculture: A Worldwide Overview. *Comput. Electron. Agric.* **2002,** *36*, 113–132.

Zhao, L.; Lianghunn, W. U.; Yongshan, Li.; Animesh, S.; Zhu, D.; Uphoff, N. Comparisons of Yield, Water Use Efficiency and Soil Microbial Biomass as Affected by the System of Rice Intensification. *Soil Sci. Plant Anal.* **2010,** *41*, 1–12.

CHAPTER 13

Advancement in Pesticide Formulation Technology for Reducing Load of Contaminants

ROMILA AKOIJAM[1], SUSHMITA TH[2], AKOIJAM RATANKUMAR SINGH[3], and SANDIP PATRA[3]

[1]ICAR Research Complex for North Eastern Hill Region, Manipur Centre, Lamphelpat, India

[2]Department of Entomology, Assam Agricultural University, Jorhat, Assam, India

[3]ICAR Research Complex for North Eastern Hill Region, Umiam, Meghalaya, India

ABSTRACT

Pesticide formulation refers to bulk pesticide chemicals, which contain very highly concentrated active ingredients. They are designed to readily use, mixed and sold. Lack of knowledge of know-how leads to improper handling and use. This led to the development of a resurgence of secondary pests, résistance development, health hazards, and the problem of pesticide residues. All pesticides should be approved by the Environment Protection Agency (EPA) and must be tested for human health and the environment thoroughly for their safety. The broad-spectrum organochlorine insecticides have the highest persistence and deteriorate the environment; their residues are still detected in nature. The new evolution of pesticides has led to the discovery of new chemical families that are lowly persistent in nature, environment-friendly, have a low use rate of insecticides and fungicides, and

Environmental Contaminants: Impact, Assessment, and Remediation.
Pritam Ganguly, Jajati Mandal, M. Paramsivam, & Sandip Patra, (Eds.)

less harm to beneficial insects, and the chemical patterns make the environmentally safe formulations. Because of the cheaper price, the conventional insecticide formulations are still being used in developing countries like India. Special precautions are required for the formulations, which give rise to environmental and health concerns and are prone to drift into the air or move off into the water. The development of new chemistry insecticide formulations and nanotechnology is being applied in scientific disciplines by using nanoscale products which are useful for pest control and reduce the risk to plants and animals, and boost plant growth and its exposure to beneficial insects.

13.1 INTRODUCTION

Pesticide is a broad range term which includes the compounds, such as insecticides, fungicides, herbicides, rodenticides, molluscicides, nematicides, and also the plant growth regulators. Pesticides are the regular contaminants in soil, air, water, plants, fish, birds, and animals from beneficial soil microorganisms to insects and other forms of life. Pesticides are generally formulated and packed into ready-to-use forms for marketing. Pesticide formulation is the bulk pesticide chemical which contains highly concentrated active ingredients and transforms into convenient products which are ready to use and for sale to the consumer. The pesticide formulation is first formulated with the required additives and then packaged and labeled in a suitable container. The formulations should be easy to handle by the applicator, biologically effective, and should have more longevity of effectiveness of the formulation in storage. Throughout history, humankind commonly uses crop protection products, such as nonselective in nature, nondegradable, highly synthetic, persistent, and toxic to many forms of life. Due to the increasing population, getting enough food to survive and good quality of life is difficult. Today, more than 50% of our population depends on agriculture, and it is the basic reason to increase food production by controlling weeds, insects, and pathogens using different crop protection products. Climate change makes a good opportunity for several insect pests to attack the crops from the seedling stage to senescence affecting the total worldwide food production (Jeyasankar and Jesudasan, 2005). Therefore, conventional pesticides are being employed indiscriminately to control these pests and to feed the increasing population. Improper use of pesticides has led to the development of resurgence of secondary pests, résistance development, health hazards, and pesticide residues problem (Rahman et al., 2009; Tripura et al., 2017). Moreover, good agricultural practice is necessary to recommend minimum quantities of insecticides to achieve adequate control. Most pesticides cause adverse effects if they are intentionally or accidentally ingested or in contact with the skin for a long time. Fixing the indiscriminate use of pesticides for safeguarding consumers from the hazards of pesticides is important. All pesticides should be approved by the Environment Protection Agency (EPA) and must be tested thoroughly for their safety to human health and the environment. The use of pesticides along with other pest management techniques maintains the quality of the crops during harvest and helps in destroying pests and diseases (Sarwar, 2013).

During the early days, the crop protection products were usually basic poisons which were nonselective in nature, high persistent, and toxic to plants, animals, insects, and human beings giving a rapid breakdown in the environment. Those chemicals were arsenic, lead, organomercurials, organochlorines such as DDT and the introduction of organophosphates. Rachel Carson in her famous book that changed the world "Silent Spring" also mentions the harmfulness of pesticides. Carson named the book "Silent Spring" because according to her vision, if pesticides were used continuously then the world mankind would face a silent world. The book roared about pesticides and gave birth to modern environmentalism through the powers of pen. The new neonics are slowly responsible for mostly toxicity load and it sprung the second Silent Spring. Therefore, the new evolution of pesticides has led to the discovery of new chemical families that are low persistence in nature, environment-friendly, low use rate of insecticides and fungicides, less harm to beneficial insects, and the chemical patterns make the environmentally safe formulations.

Today, a new era of genetically engineered pesticides are designed to eliminate environmental contamination and are user-friendly to combat disease and pests effectively; they can be used on a broader scale by improving plant growth and targeting the specific pest. Different formulations are being introduced which reduce the load of contaminants into the environment and other forms of life. The increasing awareness of the toxic effects of conventional formulations has led to the transformation of a significant trend toward switching over the formulations, which contain petroleum, organic, water-based formulations and eco-friendly constituents (Knowles, 2008). The new advanced pesticide formulation can alter the toxic, nondegradable and also increase the bioefficacy of the products by adopting the latest technologies, such as wettable powder to suspension concentrate, soluble concentrate to microemulsion (ME), emulsifier concentrate to nanoformulation and granule to controlled release formulation which reduces the doses of the pesticides and boost the food quality with minimum pesticide residues (Beestman, 2003).

13.2 HISTORY OF CONVENTIONAL PESTICIDE FORMULATION TECHNOLOGY

The broad-spectrum insecticides, such as DDT, organophosphates, chlordane, and cyclodein compounds were used on a large scale to control pests in the 1940s. During World War II, DDT was used all over the world to put an end to malaria. After World War II, DDT became an effective insect killer and people called it "Atomic Bomb" of pesticides. The residues of pesticides on produce cause deleterious, acute, or long-term effects on the health of consumers. In developing countries, many conventional pesticide formulations are used which are more persistent primarily because they are cheaper. Most of the pesticides were unable to reach the site of action in the target organism because of physical, chemical, and biological factors. Acknowledging the cost and limitation in the design of the conventional pesticide formulation, pesticide scientists have tried to develop new formulation technologies adopting new chemistry with less harmful and improving delivery of pesticides. The use of conventional pesticide formulation is more in developing countries and its exposure is quite common in farmers resulting in illness and increasing medical costs and low productivity.

The organophosphates and organochlorine group of insecticides were available in granular formulation (GR), wettable powder form (WP), emulsifier concentrates (EC), and soluble concentrates (SL). Earlier, the conventional pesticide formulation is divided into solids, liquids, and gases. The solid formulation includes dust or powders, granules, pellets, tablets or baits, wettable powder, dry flowables, seed treatment, and vapor strips. The liquid formulation includes suspension concentrates (SC) (flowables), solutions, emulsifier concentrates, gels, aerosols, ultralow volume concentrates, and ME. The gas formulation includes fumigants sold as liquids or solids. The granular formulations (GR) contain granules which are 250–1,000 microns in size, and because of the larger size, they are being used for direct broadcasting to the field, and this affects the soil, nontarget insects, and causes phytotoxicity to the crop. They are also difficult to mix in a tank causing dust hazards during application (Gilden et al., 2010). The granules do not stick to the foliage and need moisture to release the active ingredient and are not effective in drought conditions. Another formulation in the form of wettable powder (WP) is applied after dilution as a suspension in water. This formulation has larger particles than the droplets produced by emulsifier concentrates, which gives lower biological activity than the other liquid formulation causing phytotoxicity to crops (Gupta, 2004). They are very high-concentration pesticides. The emulsifier concentrates (EC) cause emulsion stability problems after dilution. They cause phytotoxicity to vegetable crops and the active ingredients cause dermal toxicity and can cause fire hazards as the solvent may affect plastics and rubbers during spray applications (Copping, 2004). They also contain organic solvents that are inflammable which endangered human health and pollute the environment. The SL require surfactant wetter for quality wetting on vegetable leaves. They hydrolyze the active ingredients and are corrosive to metals (Zabkiewicz, 2000). Dust formulations are easily drifted off target during application. The formulations make the equipment to get clogged and difficult to apply in damp or humid environments. The gaseous formulation needs the target site to be enclosed or covered to prevent the gas from escaping. The microencapsulated pesticides act as a fumigant and it is injurious to honey bees. They also have long restricted entry.

Hence, special precautions are needed for the formulations, which give rise to environmental concern and are prone to drift into the air or move off into the water. Wildlife, human beings, fishes, birds, and aquatic invertebrates can be affected by different formulations and they should be formulated in a product that can be accepted globally (Mulqueen, 2003). The old conventional formulation technology can stain fabrics, discolor linoleum, dissolve plastic, and burn foliage and the product prices vary substantially according to the active ingredient in specific formulations. The new advanced formulations reduce the acute toxicity of the pesticides, which are safer to farm workers. In order to keep away these problems, newly developed modern formulations are introduced.

13.3 NEW ADVANCES IN PESTICIDE FORMULATION TECHNOLOGY

The ingredients of a pesticide formulation contain both active and inert components. The active ingredient is the main component to control the pest and the inert ingredients comprise carriers or diluents and adjuvants. Today, there has been a dramatic shift from conventional formulations to advanced pesticide formulation due to their environmental advantages (Mathur, 1999). The new pesticide formulation has low mammalian toxicity, less soluble in water, leaching potential is less, the molecular weight ranged from 200 to 450, and they are new chemistry. The new advanced pesticide formulation is discussed below.

1. Water-based dispersion technology
2. Improved dry product technology
3. Controlled released technology
4. Combined or mixed formulation technology
5. Nanotechnology-based pesticide technology

13.3.1 WATER-BASED DISPERSION TECHNOLOGY

The water-based dispersion technology in pesticide formulation is discussed in the following sections:

13.3.1.1 SUSPENSION CONCENTRATES (SC)

They are usually suspended in an oil phase but it is also suspended in water (Mulqueen, 2003). They have high concentrations of insoluble active ingredients and are safe for the operator and the environment. They are usually low-cost and have the ability to enable water-soluble adjuvants to be built in for enhancing biological activity. Nowadays, it has been switched from wettable powder to SC because of non-dusty, easy to measure, and can be poured easily into the spray tank. Example: Fipronil 5 SC, Carbendazim 50 SC.

13.3.1.2 OIL-IN-WATER O/W EMULSION

Nowadays, oil-in-water has great importance because of the need to eliminate volatile organic solvents (VOC) for good handling (Ware, 2004). They can surpass emulsifier concentrates because of cost-effectiveness, they are safe during manufacture, transportation, and use. Their active ingredient also has low water solubility to avoid crystallization problems (Tadros, 1995). Example: Butachlor 50 EW, Cfluthrin 5 EW.

13.3.1.3 SUSPO-EMULSION (SE)

They are the mixtures of SC and oil-in-water which already have surfactants and because of these, they are becoming popular. Suspo-emulsion (SE) is the mixed combination formulation which helps the farmers to apply the correct amount of each component pesticide and also overcome problems of tank incompatibility. As it already presents the surfactant, it prevents flocculation and thickness and also prevents the separation of the dispersed phase (Tadros, 1995). The surfactants which are already present are similar to those of SC. Example: Fenpropimorph 24.5+ Epoxiconazole 8.2 EC (which is not registered in India).

13.3.1.4 MICROEMULSIONS (ME)

They are thermodynamically stable and transparent, and it is the dispersion of two immiscible liquids which are stable over a different temperature range (Hiromoto, 2007). The particle size of ME is usually fine droplet size which is less than 0.05 microns, that is, 50 nm. The surfactant present in ME can be as high as 10–30% but in oil-in-water emulsion, it is only 5% surfactant. Because of their special characteristics, they enhanced biological activity by having relatively low active ingredients, high surfactant content, and good solubilization of the active ingredient. Example: Neemazal 30 MEC.

13.3.1.5 OIL DISPERSION FORMULATION

It is the new recent advanced technology, which allows very efficient and environmentally friendly agrochemical formulations. The active ingredient present can be dispersed in the oil phase which is very effective for water-sensitive or non-soluble active ingredients (Llacer et al., 2012). When they come in contact with water, the formulation can either change into an emulsion or a SE. Usually, the oil phase comprises mineral oils, vegetable oils, or it might be esters of vegetable oils.

13.3.1.6 AQUEOUS FLOWABLES (AF)

They are the aqueous concentrated containing 40–70% W/W suspensions of micronized insoluble active pesticide in water. Before spraying, they are diluted with water in a spray tank in order to get minimum effective pesticide concentration. Because of low viscosity and good fluidity, they can be transferred easily to the spray tank. To achieve better dispersion of the pesticides in water, they require an effective wetting agent (Castro et al., 1998). They also have an insoluble active ingredient, which requires good suspension.

13.3.1.7 SEED TREATMENT FORMULATIONS

They are film-form used for coating plants and other plant seeds. The coating material is prepared by dispersant, wetting agents, film former, pH regulator, antifreeze, defoamer, and other auxiliaries and water (Dayer et al., 2007). They can be directly coated on the seed surface after dilution in order to achieve a protective film with good strength and good permeability. Because of this property, they are known as seed-coating agents. They are the new molecule pesticide formulations saving the most pesticides as they do not require nearly 45 days at the bud stage and seedling stage and also the dosage is very less which requires about 1/50 of field pesticide application (Dipak, 2015). There are various seed treatment formulations which include water flowable seed-coating agent (FS), water emulsion seed coating agent (EWS), suspended emulsion seed coating agent (SES), microcapsule seed coating agent (CS), dry flowable seed coating agent (DFS), and water-dispersible granule type seed coating agent (WGS) (El-Mohamedy et al., 2008).

13.3.2 IMPROVED DRY PRODUCT TECHNOLOGY

The improved dry product technology in pesticide formulations is briefly explained in the following sections:

13.3.2.1 WATER DISPERSIBLE GRANULE (WG)

Water-soluble granules are also known as dry flowables. Their production uses the newest technology which has a new type of formulation and is being developed as a safer technology, which is more or less commercially attractive alternative to wettable powder and SC (Kim et al., 2003). The popularity of water-soluble granules has increased due to their convenience in packaging and use; usually, they are non-dusty and are generally free-flowing granules which disperse quickly when added to water in a spray tank

(Marcroft et al., 2008). Earlier use of wettable powder and other liquids has been replaced by this technology. Today, it is the most preferred process used by the agrochemical industry as the water-dispersible granules are extruded, this process is quite safe, versatile, and economical. In this formulation, the dispersion time in water is a very important property as it secures that no problem should occur in the spray tank which is necessary for all the granules to disperse completely within 2 minutes according to different degrees of water temperature and hardness. Example: Thiamethoxam 25 WG, Deltamethrin 25 WG.

13.3.2.2 DISPERSION CONCENTRATES (DC)

In this formulation, their active ingredient can be dissolved in a water-miscible and polar solvent together with a dispersing or emulsifying agent. The DC are specially developed to dilute in water giving stable, fine particle size dispersion (Nishiyama et al., 2004). They are easy to use and the process equipment can be cleaned easily and is also simple. They have good characteristics, such as solution-type formulation, stability, and good bioefficacy. The earlier formulations, such as SL, emulsifier concentrates (EC), and ME have been replaced by dispersion concentrated (DC) formulations because of their active ingredient's physical, chemical, and biological properties which can stop the use of conventional formulations.

13.3.3 CONTROLLED RELEASE TECHNOLOGY

Before the controlled release formulation was manufactured, the Controlled Release Society was formed in the USA and it was its first Symposium regarding Controlled Release Technology in 1974. After that several types of demonstrations are prepared using a variety of matrices like synthetic and natural polymers. The Joint FAO/IAEA Division of Nuclear Techniques in Food and Agriculture, in 1983, started an Internationally Coordinated Research Program (CRP) to design controlled release using nuclear techniques. The development and registration of controlled release is a costly process. Over the past 15 years, the United States Department of Agriculture, Agriculture Research Service (USDA-ARS) has been working on starch-encapsulated controlled release formulation of pesticides.

The microencapsulation or capsule suspension (CS) techniques are under the controlled release formulation (Beestman, 2003). They are worked under the principle of interfacial polymerization. Their popularity has increased over the past few years as their active ingredients can be easily controlled by adjusting the droplet size, the thickness of the polymer membrane, and also the degree of cross-linking or the porosity of the polymers. The controlled release formulation is specially designed by trapping in a matrix and it is released at a predetermined rate. Example: Lamda Cyhalothrin 10 CS.

Earlier, the conventional formulation technology was applied at a higher rate in order to have good activity for a reasonable period of time, but controlled release formulation does not require a higher dose of pesticides, and the pesticides trapped in the matrix are easily protected from physical, chemical, and microbial processes. Because of this reason, they are effective for a long period of time and also prolonged the effective duration of non-persistent pesticides. They are eco-friendly requiring less dose and reducing the load of contamination particularly in the nontargeted areas such as surface water and groundwater. The various environmental factors, such as evaporation, photolysis, leaching, and degradation of the soil by the chemicals can be reduced by the controlled release formulations. They are found to be eco-friendly to wildlife, beneficial insects, aquatic invertebrates, plants, animals, and birds. They have improved the efficacy of pesticides as they have a good target and in addition to that they are safer to the applicator when coming in contact with controlled release pesticide formulations. Generally, natural and synthetic polymers are used in the controlled release formulations. The synthetic polymers were used for preparing microencapsulated formulations and the natural polymers were derived from starch, cellulose, lignin, alginates, and other polysaccharide products. For developing countries, controlled release is a new boon as it does not require any precautions during handling and protective clothing. One of the advantages of using controlled release formulation is that the acute toxicity of the pesticide is usually reduced and it also has the potential to reduce the phytotoxicity of rice and Azolla. The newer controlled release systems are controlled release from a crystalline matrix and controlled release from microporous cellulose fibers.

13.3.3.1 CONTROLLED RELEASE FROM A CRYSTALLINE MATRIX

They are commercially available heterocyclic substances sold under the name of melamine. The formulation is prepared from the material which is a low-cost urea and simply heated to cause trimerization by the elimination of ammonia and then to recycle it back into the synthesis of urea. Melamine is mainly used after reacting with formaldehyde and function as a controlled release nitrogen fertilizer (Allan et al., 1985). Melamine can easily break down in the soil which is good for crops having a long growing season at higher temperatures. Melamine is found to be good fertilizer for rice. As it is white crystalline in nature, it can be converted into fertilizer-type granules using rotatory pan granulation equipment. Because of its crystalline nature, it can be dissolved easily in soil moisture and has the capability of moisture control, hence, it can be a good controlled release delivery system. In the future, there will be continued use of this formulation as it requires a minute amount of chemicals, and there is a remarkable increase in crop yield on rice (Anon, 1986).

13.3.3.2 *CONTROLLED RELEASE FROM MICROPOROUS CELLULOSE FIBER*

It was first started from the angiosperm woody plant. The cellulose fibers can be separated from the woody plant by boiling with a 1:1 mixture of ethanol and water in the presence of an acid catalyst (Sarkanen, 1990). The ethanol is prepared by fermentation. The final product of cellulose fibers obtained is the same as those produced by old sulfite pulping processes. The wood is pulped with the removal of lignin and hemicellulose, and it creates a void which is filled with water. If the liquid within the pores contains an agrochemical, then the active ingredient will be trapped between the layers of cellulose within the cell wall. They are also known as never dried fibers which act as a well controlled release delivery system, also biodegradable matrix available at low cost.

13.3.4 COMBINED/MIXED FORMULATION TECHNOLOGY

This is the innovation developed with a combination of capsule suspension (CS) and emulsion concentrate (EW) which is safe for the environment and also safe for wildlife and other forms of life. The mixed formulations usually consist of a combination of capsule suspension of lambda-cyhalothrin and concentrated emulsion of chlorpyriphos insecticides. They are specially designed formulations that contain two different active ingredients in which the active ingredient, that is, chlorpyriphos is quickly effective just after application on target pests giving a quick knockdown effect and the lambda-cyhalothrin will slowly control in an effective manner for long-term target pest management (Hazra et al., 2013). The two pesticides are microencapsulated in a polymer membrane and therefore the applicator can easily apply two insecticides simultaneously in a single application. Usually, the product is water-based and does not cause any risk of flammability or irritation of the skin like the other emulsifier concentrates (EC) formulation which contains a large quantity of organic solvents.

As lambda-cyhalothrin causes skin irritation to the farmers at the time of application, microencapsulation gets rid of skin irritation and itching. Similarly, chlorpyriphos also causes skin irritation and is available in the market only as an emulsifiable concentrate (EC). Hence its water-based emulsion concentrates (EW) show good bioefficacy and eliminate the skin problem. The combination of two insecticides will have broad-spectrum insecticidal activities and will have the potential for controlling insects on a large number of crops. Therefore, the formulation has been transferred to pesticide industries for commercialization. Example: lambda-cyhalothrin 25.0 CS + Chloropyriphos10.0 EW.

13.3.5 NANOTECHNOLOGY-BASED PESTICIDE FORMULATION

The development of nanotechnology is being applied in scientific disciplines by using nanoscale products which are useful for pest control and to reduce the risk to plants, animals, and boost plant growth and exposure to beneficial insects (Kuzma et al., 2006). Different engineered nanoparticles are used which give a positive response to the plants. The engineered nanoparticles (ENP) can increase the biomass, chlorophyll content, and photosynthetic reaction. The introduction of engineered nanoparticles to the plants helps in the expression of different defense enzymes peroxidase (PO), polyphenol oxidase (PPO), β 1-3 glucan and PAL (phenylalamine ammonia lyase), antibiotic enzymes, such as superoxide dismutase (SOD), CAT, glutathione, carotenoids, tocopherols, and phenolics. The enzymes PO and PPO are involved in lignin biosynthesis which enhances protection against different pathogens by strengthening the plant's cell wall barrier. The positive response includes photosterilization, secondary metabolites, specific enzyme conjugate formation, scavenging free radicals, improve N fixation and assimilation, and hormonal balance. The increase in enzyme activity, such as α, β-amylase, ascorbate, peroxidase, and catalase increase malondialdehyde content which gives a protective mechanism under the application of nanomaterial. When engineered nanomaterials are introduced to plants, they stimulate the antioxidant system which improves the plant's resistance and produces defense compounds, such as phenols, flavanoids, terpenoids, and alkaloids. A phenolic structure acts as a metal chelator and participates in reactive oxygen species (ROS) scavenging and helps in disease resistance. This process enhances lignification in stresses and acts as a barrier for pathogens to enter the tissues. Khodakovskaya et al. (2009) reported multiwalled carbon nanotubes to govern tomato seed germination and seed growth by balancing stress-related gene expression. Kumar et al. (2009) studied the chromotoxic effects of Ag-engineered nanomaterial on the mitotic cell division in root tip cells of *Allium cepa*. Also, Ag-engineered nanomaterial when associated with membrane proteins, activate signaling pathways which show the inhibition of cell proliferation (Roh et al., 2012; Gopinath et al., 2010). The new advancement of nanoparticles, such as carbon nanotubes (CNTs) and metal oxide nanoparticles are good for seed germination and plant growth (Nair et al., 2010). The nanogel is a nano-sized hydrogel system and is highly a cross-linked system in nature having a polymer system which is either copolymerized or monomeric (Phillips et al., 2010). The nanogel has the potential to deliver active ingredients in a controlled manner with a sustainable and effective targetable.

13.4 CONCLUSION AND FUTURE PROSPECTS

A knowledge on proper selection of pesticide formulation for a particular crop and particular pest is very much important and a critical step for a pest management program. The properties of various forms of pesticide formulations should also be understood. A concern for environmental quality revealed in an accountable submission needs an understanding of the characteristics of a given formulation and its further possible impact on the surroundings. A proper understanding of the pesticide formulations and their recommended dose and timing of application will reduce the pesticide load in the environment. The development of nano-based pesticide formulation leads to very minimum amounts of active ingredients in reacting to environmental triggers and biological demands through controlled release mechanisms.

KEYWORDS

- **pesticide**
- **formulation**
- **contaminants**
- **residue**
- **environment**

REFERENCES

Akron, O. H. Proceedings 1st. Controlled Release Pesticide Symposium, The University of Akron, Akron, OH 44325, U.S.A., Sept 16–18, 1974.

Allan, G. G.; Freepons, D. E.; Crews, G. M. Fertilizer Compositions, Processes of Making Them and Processes of Using Them. U.S. Patent 4,560,400, December 24, 1985.

Anon. California Rice Growers - A Commercial Fertilizer Program for Increased Yields, Technical Brochure, Melamine Chemicals Inc., P.O. Box 748, Donaldsonville, LA 70346, 1986.

Beestman, G. B. Controlled Release in Crop Protection: Past Experience and Future Potential. In *Chemistry of Crop Protection, Progress and Prospects in Science and Regulation*; Voss, G., Ramos, G., Eds.; Wiley-VCH Verlag GmbH & Co: Weinheim, 2003; pp 272–279.

Rachel, C. *Silent Spring*; Hamish Hamilton: London, 1963.

Castro, B. A.; Riley, T. J.; Leonard, B. R. *Evaluation of Gaucho® Seed Treatment and Soil Insecticides for Management of the Red Imported Fire Ant on Seedling Grain Sorghum During 1994–1996*, Agricultural Center, Louisiana Agricultural Experiment Station Research Report, 101, 1–4, 1998.

Copping, L. G. *The Manual of Biocontrol Agents*, 3rd ed; British Crop Production Council (BCPC): Farnham, Survey UK, 2004.

Dipak K. H. Recent Advancement in Pesticide Formulations for User and Environment Friendly Pest Management. *Int. J. Res. Rev.* **2015,** *2*(2), 35–40.

El-Mohamedy, R. S. R.; Abd El-Baky, M. M. H. Evaluation of Different Types of Seed Treatment on Control of Root Rot Disease, Improvement Growth and Yield Quality of Pea Plant in Nobaria Province. *Res. J. Agric. & Biol. Sci.* **2008,** *4*, 611–622.

Gilden, R. C, Huffling, K. and Sattler, B. Pesticides and Health Risks. *J. Obstet. Gynecol. Neonatal. Nurs.* **2010,** *39*(1), 103–110.

Gopinath, P.; Gogoi, S. K.; Sanpui, P.; Paul, A.; Chattopadhyay, A.; Ghosh, S. S. Signaling Gene Cascade in Silver Nanoparticle Induced Apoptosis. *Colloids Surf. B* **2010,** *77*, 240–245.

Gupta, P. K. Pesticide Exposure-Indian Scene. *Toxicology* **2004,** *198*, 83–90.

Hazra, D. K.; Pant Megha Raza, S. K; Patanjali, P. K. Formulation Technology: Key Parameters for Food Safety with Respect to Agrochemicals Use in Crop Protection. *J. Pl. Protec. Sci.* **2013,** *5*(2), 1–19.

Hiromoto, B. Pesticide Microemulsions and Dispersant/Penetrant Formulations. United States Patent, Patent No: US7297351, 2007.

Jeyasankar, A.; Jesudasan, R. W. A. Insecticidal Properties of Novel Botanicals Against a Few Lepidopteran Pests. *Pestology* **2005,** *29*, 42–44, ISSN: 0970-3012.

Khodakovskaya, M.; Dervishi, E.; Mahmood, M.; Xu, Y.; Li, Z.; Watanabe, F.; Biris, A. S. Carbon Nanotubes are Able to Penetrate Plant Seed Coat and Dramatically Affect Seed Germination and Plant Growth. *ACS Nano*. **2009,** *3*, 3221–3227.

Kim, D. S.; Koo, S. J.; Lee, J. N.; Hwang, K. H.; Kim, T. Y.; Kang, K. G.; Hwang, K. S.; Joe, G. H.; Cho, J. H. In *Flucetosulfuron: A New Sulfonylurea Herbicide*, Proceeding of International Congress, Crop Science and Technology, BCPC, Farnham, Surrey, UK, pp. 87–92, 2003.

Knowles, A. Recent Developments of Safer Formulations of Agrochemicals. *Environmentalist* **2008,** *28*(1), 35–44.Kuzma, J.; VerHage, P. Nanotechnology in Agriculture and Food Production: Anticipated Applications. 2010. Retrieved from www.document] URLhttp://www.nanotechproject.org/ file_download/files

Kumar, M.; Mukherjee, A.; Chandrasekaran, N. Genotoxicity of Silver Nanoparticles in *Allium cepa. Sci. Total Environ.* **2009,** *407*, 5243–5246.

Marcroft, S. J.; Potter, T. D. The Fungicide Fluquinconazole Applied as a Seed Dressing to Canola Reduces *Leptosphaeria maculans* (Blackleg) Severity in South-Eastern Australia. *Austral. Pl. Patho*. **2008,** *37*, 396–401.

Mulqueen, P. Recent Advances in Agrochemical Formulation. *Adv. Colloid Interface Sci.* **2003,** *106*, 83–107.

Nair, R.; Varghese, S. H.; Nair, B. G.; Maekawa, T.; Yoshida, Y.; Kumar, D. S. Nano Particulate Material Delivery to Plants. *Pl. Sci.* **2010,** *179*, 154–163.

Nishiyama, N.; Yamamoto, A; Takei, T. In *Ecological Risk Assessment of Surfactants*, The 38th Annual Meeting of Japan Society on Water Environment (Sapporo, Japan), 2004.

Phillips, M. A.; Gran, M. L.; Peppas, N. A. Targeted Nano-delivery of Drugs and Diagnostics. *Nano Today* **2010,** *5*, 143–59.

Rahman, M. M.; Islam, K. S.; Jahan, M; Uddin, M. A. Efficacy of Some Botanicals in Controlling Brinjal Shoot and FRUIT borer, *Leucinodes orbonalis*. *J. Progressive Agri.* **2009,** *20*(1&2), 35–42.

Roh, J. Y.; Eom, H. J.; Choi, J. Involvement of *Caenorhabditis elegans* MAPK Signaling Pathways in Oxidative Stress Response Induced by Silver Nanoparticles Exposure. *Toxicol. Res*. **2012,** *28*, 19–24.

Sarkanen, K.V. Chemistry of Solvent Pulping. *Tappi J.* **1990,** *73*, 210–215.

Tadros, T. F. *Surfactants in Agrochemicals*; Marcel Dekker: New York, USA, 1995.

Tripura, A.; Chatterjee, M. L.; Panda, R.; Patra, S. Biorational Management of Brinjal Shoot and Fruit Borer, *Leucinodes orbonalis* in Mid Hills of Meghalaya. *J. Entomol. Zool. Stud.* **2017,** *5*, 41–45.

Zabkiewicz, J. A. Adjuvants and Herbicidal Efficacy–Present Status and Future Prospects. *Weed Res.* **2000,** *40*, 139–149, DOI: 10.1046/j.1365-3180.2000.00172.x.

Index

A

B

F

G

H

I

L

O

P

T

U

V

W

Z

For Product Safety Concerns and Information please contact our
EU representative GPSR@taylorandfrancis.com Taylor & Francis
Verlag GmbH, Kaufingerstraße 24, 80331 München, Germany